Philosophy and Medicine

Volume 125

More information about this series at http://www.springer.com/series/6414

Lisa M. Rasmussen • Ana Smith Iltis
Mark J. Cherry
Editors

At the Foundations of Bioethics and Biopolitics: Critical Essays on the Thought of H. Tristram Engelhardt, Jr.

Springer

Editors
Lisa M. Rasmussen
Department of Philosophy
University of North Carolina
Charlotte, NC, USA

Ana Smith Iltis
Department of Philosophy
Center for Bioethics, Health and Society
Wake Forest University
Winston-Salem, NC, USA

Mark J. Cherry
Department of Philosophy
St. Edward's University
Austin, TX, USA

ISSN 0376-7418 ISSN 2215-0080 (electronic)
Philosophy and Medicine
ISBN 978-3-319-18964-2 ISBN 978-3-319-18965-9 (eBook)
DOI 10.1007/978-3-319-18965-9

Library of Congress Control Number: 2015952778

Springer Cham Heidelberg New York Dordrecht London

Printed on acid-free paper

Springer International Publishing AG Switzerland is part of Springer Science+Business Media (www.springer.com)

Mine is to chew on the appropriate texts and make them delectable.
– St. Gregory of Nyssa (335–394 or 395)

To our mentor,
Professor Doctor H. Tristram Engelhardt, Jr., Ph.D., M.D., for training in the chewing of texts and ideas

Dr. H. Tristram Engelhardt, Jr. paying tribute to the Republic of Texas near a plaque commemorating its former embassy building in London, England. Photo courtesy of Jeremy Garrett.

Contents

Contributors

Corresponding Editor
Lisa M. Rasmussen, Ph.D. Department of Philosophy, University of North Carolina, Charlotte, NC, USA

Co-Editors
Ana Smith Iltis, Ph.D. Department of Philosophy, Center for Bioethics, Health and Society, Wake Forest University, Winston-Salem, NC, USA

Mark J. Cherry, Ph.D. Department of Philosophy, St. Edward's University, Austin, TX, USA

Authors
George J. Agich, Ph.D. Philosophy Department, Bowling Green State University, Austin, TX, USA

Jennifer A. Bulcock Department of Philosophy, Rice University, Houston, TX, USA

Mary Ann Gardell Cutter, Ph.D. Department of Philosophy, University of Colorado, Colorado Springs, CO, USA

Ruiping Fan, Ph.D. Department of Public Policy, City University of Hong Kong, Kowloon, Hong Kong

Jeremy R. Garrett, Ph.D. Children's Mercy Bioethics Center, Children's Mercy Hospital, Kansas City, MO, USA

Department of Pediatrics & Department of Philosophy, University of Missouri, Kansas City, MO, USA

Fabrice Jotterand, Ph.D., M.A. Department of Health Care Ethics, Rueckert-Hartman College of Health Professions, Regis University, Denver, CO, USA

Institute for Biomedical Ethics, University of Basel, Basel, Switzerland

Eric Juengst, Ph.D. Center for Bioethics, School of Medicine, University of North Carolina, Chapel Hill, NC, USA

George Khushf, Ph.D. Department of Philosophy, University of South Carolina, Columbia, SC, USA

Laurence B. McCullough, Ph.D. Center for Medical Ethics and Health Policy, Baylor College of Medicine, Houston, TX, USA

John C. Moskop, Ph.D. Department of Internal Medicine, Wake Forest School of Medicine, Winston-Salem, NC, USA

Clinical Ethics Committee, Wake Forest Baptist Medical Center, Medical Center Boulevard, Winston-Salem, NC, USA

D. Christopher Ralston, Ph.D. Christian Institute on Disability, Joni and Friends International Disability Center, Agoura Hills, CA, USA

The Center for Bioethics and Human Dignity, Trinity International University, Deerfield, IL, USA

Stephen Wear, Ph.D. Departments of Medicine, Gynecology-Obstetrics and Philosophy, The University at Buffalo, Buffalo, NY, USA

Center for Clinical Ethics, VA Medical Center, Buffalo, NY, USA

Kevin Wm. Wildes SJ, Ph.D. Loyola University New Orleans, New Orleans, LA, USA

Author Bios

George J. Agich co-founded the International Conference on Clinical Ethics Consultation and recently retired as a Professor from the Department of Philosophy at Bowling Green State University. He received his Ph.D. in Philosophy from the University of Texas at Austin in 1976.

Jennifer A. Bulcock is an Assistant Professor in the Department of Philosophy and Liberal Studies at Cabrini College. She earned her Ph.D. from Rice University in 2015.

Mark J. Cherry is the Dr. Patricia A. Hayes Professor in Applied Ethics and Professor and Chair in the Department of Philosophy at St. Edward's University. He received his Ph.D. in Philosophy from Rice University in 1999.

Mary Ann Gardell Cutter is a Professor of Philosophy at the University of Colorado, Colorado Springs. She received her Ph.D. in Philosophy from Georgetown University in 1989.

Ruiping Fan is a Professor of Philosophy in the Department of Public Policy at the City University of Hong Kong. He received his Ph.D. in Philosophy from Rice University in 1999.

Jeremy R. Garrett is a Research Associate at Children's Mercy Bioethics Center, and Assistant Professor of Pediatrics and Adjunct Assistant Professor of Philosophy at the University of Missouri Kansas City. He received his Ph.D. in Philosophy from Rice University in 2008.

Ana Smith Iltis is an Associate Professor in the Department of Philosophy and the Director of the Center for Bioethics, Health and Society at Wake Forest University. She received her Ph.D. in Philosophy from Rice University in 2003.

Fabrice Jotterand is an Associate Professor in the Department of Health Care Ethics at Regis University and Senior Researcher, Institute for Biomedical Ethics, University of Basel, Switzerland. He received his Ph.D. in Religious Studies from Rice University in 2006.

Eric Juengst is the Director of the Center for Bioethics, and a Professor in the Department of Social Medicine at the University of North Carolina, Chapel Hill. He received his Ph.D. in Philosophy from Georgetown University in 1985.

George Khushf is a Professor in the Department of Philosophy at the University of South Carolina. He received his Ph.D. in Religious Studies/Philosophy from Rice University in 1993.

Laurence B. McCullough is the Dalton Tomlin Chair in Medical Ethics and Health Policy and Professor of Medicine and Medical Ethics at the Center for Medical Ethics and Health Policy, Baylor College of Medicine. He received his Ph.D. in Philosophy from the University of Texas at Austin in 1975.

John C. Moskop is the Wallace and Mona Wu Chair in Biomedical Ethics and Professor of Internal Medicine at the Wake Forest School of Medicine. He received his Ph.D. in Philosophy from the University of Texas at Austin in 1979.

D. Christopher Ralston is the Assistant Manager of Curriculum Development at the Joni and Friends International Disability Center and a Fellow of the Center for Bioethics and Human Dignity at Trinity International University. He received his Ph.D. in Philosophy from Rice University in 2011.

Lisa M. Rasmussen is an Associate Professor in the Department of Philosophy at the University of North Carolina, Charlotte. She received her Ph.D. in Philosophy from Rice University in 2003.

Stephen Wear is an Associate Professor in the Department of Medicine at the State University of New York, Buffalo. He received his Ph.D. in Philosophy from the University of Texas at Austin in 1979.

Kevin Wm. Wildes, SJ is the President of Loyola University New Orleans. He received his Ph.D. in Philosophy from Rice University in 1993.

Introduction

Introduction

> Alternative moralities compete without an apparent principled basis for definitively choosing one as canonical. If truth cannot break through to us and personally direct us, will we not always be lost in a plurality of diverse moral and religious views in the sense of not knowing which norms should govern? Will morality then be more a matter of taste, cultural inclination, and communal preference? … Who but a true post-modern could without regret live in such a context? (Engelhardt 2000, xi).

Professor H. Tristram Engelhardt, Jr., is a philosopher of difference. The impressive range and depth of his work in philosophy, medicine, bioethics, and theology illustrate his profound appreciation that careful and critical analytical work is central to reigning in the untutored desire to claim established canons of political correctness as truth. Professor Engelhardt's work in the philosophy of medicine, for example, demonstrates the ways in which we see the world through taken-for-granted background social, scientific, and value expectations. What is often thought of as the common scientific world view is set within a web of complex, etiologic forces. Which goals will be pursued, which eschewed? Whose standards of evidence and inference will be adopted? Which will ground knowledge claims? Whose account of the normal versus the abnormal will guide research? And, which group of scientists will function as exemplar knowers? This does not mean that investigators do not attempt to understand reality as it is in itself, unconstrained by cultural, social, and historical forces. Indeed, as Engelhardt points out, the goal of undistorted knowledge functions as a heuristic to direct one towards the truth (1996, 191). However, "[o]ur world," as Engelhardt argues, "is structured by a special set of assumptions about the rule-governed character of our experience. These scientific and metaphysical presuppositions fashion for us our everyday expectations. They give shape to our lifeworld" (1996, 190). As a result, even those who seek to see truly must be initiated into a particular way of relating to and experiencing reality. One must already have an idea of what truth is and how to find it, as well as how to interpret what one has found. Medical students, for example, have to be taught how to "read" the slides they examine under microscopes; researchers must be initiated into the assumptions and practices that guide the experimental methodology of their particular fields. As Engelhardt concludes, the world is simply not furnished with uninterpreted facts.

The circumstance that all knowledge claims are culturally and historically conditioned similarly influences our understandings of morality and bioethics. In his best-known work, *The Foundations of Bioethics* (1986, followed by a second edition in 1996), Engelhardt laid bare the empirical reality of the deep and intractable moral pluralism of the modern world. In his later theological counterpart, *The Foundations of Christian Bioethics* (2000), he provided readers with the flip side of this coin: Engelhardt's detailed and serious account of Orthodox Christian bioethics. In each case, Engelhardt concluded that general secular reason is unable to secure a uniquely true account of the right, the good, the virtuous, or the just. It cannot even establish a definitive account of the reasonable or the reliable. As a result, general secular reason is powerless to provide definitive foundations for a content-full secular bioethics, much less settle the deep moral controversies of medicine and health care policy. "In attempts to settle differences by sound rational argument, each side presupposes different fundamental moral premises as well as rules of evidence and inference. Each speaks past the other without a final, rational principled resolution. The result is that bioethical controversies fuel engaging but insoluble debates" (Engelhardt 2000, xi). Consider the seriousness of the challenges at stake: Which account of human nature, and whose explanation of human flourishing and basic goods, should be appreciated as normative? There are even significantly diverse theories for rationally debating the merits of such divergent understandings of human nature. Which consequences should be avoided, which virtues taught, and values embraced, and at what costs to self or others? Whose deep moral intuitions should guide public policy? All attempts to engage particular moral content to guide biomedical decision-making assume what must be proven, argue in a circle, engage in an infinite regress, or otherwise beg the question.[1] All protestations to the contrary, and in the face of numerous (incommensurable) fashionable claims to bioethical truth, a significant diversity of moral visions, accounts of human flourishing, and accounts of social justice compete to guide moral choice in medicine. Bioethics is, as Engelhardt rightly argues, a plural noun (1996, 3).

Without the ability to know the nature of reality undistorted by historical, cultural, and social context – reality as God would know it to be – the challenges for establishing a uniquely true bioethics to guide public policy are insurmountable. All one can know with any certainty is one's own experiential phenomenological world; but this is split into numerous competing accounts of phenomenological reality,

[1]As Engelhardt is well aware the problems that beset secular philosophical analysis have been known for millennia. Consider, for example, Sextus Empiricus (A.D. 160–210) on the *Outlines of Pyrronism*: "The more recent Skeptics hand down the following five modes of *epoché*: the first is the mode based on disagreement; the second is that based on infinite regress; the third, that based on relativity; the fourth, on hypothesis; and the fifth is the circularity mode" (Sextus Empiricus 1996, I.164).

each of which is itself socially, historically, and culturally conditioned. Given such abiding and significant moral pluralism, Engehardt argues that secular moral authority must be created through and, therefore, is limited to, the actual agreements among actual persons. General secular morality, and thus secular bioethics, is libertarian by default – not because of any particular celebration of liberty as a value, nor due to any fantastic assumptions regarding the basic human rights of persons, but as a default moral and political reality. The norms that we ourselves create with free and consenting others, and strictly limited to the extent of such agreement, provide for the possibility of binding together moral strangers into morally authoritative common projects. General secular morality is created, rather than discovered.

In many circles, Engelhardt is perhaps best known for this unflinching libertarianism. Indeed, it is widely assumed that Engelhardt affirms not only the libertarian political framework of his conclusions, but also its frequently libertine personal consequences. As Ana Iltis and Mark Cherry remarked on another occasion: "Many (perhaps most) readers have not taken seriously Engelhardt's own announcements found throughout the two editions of *The Foundations of Bioethics* that general secular morality permits and justifies many activities that he, himself, knows to be deeply sinful (e.g., abortion on demand, human embryonic stem cell research, euthanasia, same gender marriage, and so forth) as well as imprudent (utilizing a chiropractor or doctor of naturopathy for treatment of heart disease)" (Iltis and Cherry 2010, 2). The core challenge, as Engelhardt has argued in great depth, in numerous countries, on many continents, in a wide variety of formats, including lectures, commentaries, articles, book chapters, book-length monographs, and other publications too numerous accurately to count (over years too lengthy for it to be polite to mention), is that there simply does not exist secular moral authority permissibly to prohibit such actions among consenting persons.

At the Foundations of Bioethics and Biopolitics: Critical Essays on the Thought of H. Tristram Engelhardt, Jr. engages Engelhardt's diagnosis of the limits of secular reason and moral political authority, as well as his search for ultimate foundations. This is a book by Engelhardt's students. Over the courses of our various and diverse careers, Professor Engelhardt has nurtured each of us into scholars in our own right. None of the chapters is merely laudatory. Each carefully addresses, explores, and critically assesses key elements of Professor Engelhardt's philosophical labors. Many, if not all, raise objections, make careful distinctions, or argue with core aspects of the Engelhardtian project. Some engage his arguments in the bioethics and philosophy of medicine, others in clinical ethics, still others in the history of ideas, social and political philosophy, or international bioethics. One finds essays across the moral and political spectrum. A few of his students have been willing to bite the bitter pill of post-modernity and accept as its consequence libertarianism by default; a few, indeed, have been willing to convert to Orthodox Christianity. But, such diversity is as it should be for a professor who has had such an immense and international impact on the fields of philosophy, medicine, bioethics, and theology. Our hope in presenting this volume to an international audience is to celebrate a singularly distinguished career marked by an extraordinary depth and breadth of scholarship, together with inspiring and engaging teaching, as well as exceptional generosity and love to his students.

Critical Essays

In "A Critical Appraisal of Engelhardt on the 'Enlightenment Project'," *Laurence B. McCullough* challenges the history Engelhardt customarily offers of a single enlightenment and argues that there were actually multiple "enlightenment" projects. In the English and Scottish versions of the project, McCullough argues, the aim was not a quest for certainty grounded in mere reason, but rather "a quest for reliability." In their "moral science," English and Scottish enlightenment thinkers focus variously on sympathy and intuition rather than certainty or the primacy of reason. As a result of his focus on one strand of Enlightenment thought, McCullough argues, Engelhardt ignores another option for bioethics' search for moral foundations, that of a "reliable" morality that does not require certainty. It is just such a "reliable bioethics" that McCullough recommends.

In "Diversity in Clinical Ethics," *George J. Agich* argues for the centrality of diversity (understood multi-dimensionally, including value and practice diversity) to the field, not just in terms of having helped drive its creation, but also as an internal feature that must be incorporated into the practice. This diversity exists in conceptual and theoretical considerations of the practice as well as in the very structures and processes of the practice. Agich expresses a worry that diversity is sometimes taken to be a mark of a field's immaturity, and, in clinical ethics consultation in particular, this worry contributes significantly to the drive to professionalize the field. Instead, Agich argues, diversity is valuable to the field as a basic characteristic of the practice and as a resource for improvement in the practice. Moreover, and crucially, the deep value pluralism within which clinical ethics consultation is practiced (and which is a deep well of Engelhardt's thought) is too casually treated as a marginal problem "rather than as an essential feature of the field that needs to be understood" (Chap. 2, this volume, p. 15). The field seems engaged in a blind and undefended search for consistency, Agich points out, when instead it should understand, embrace, and incorporate diversity in various ways.

John C. Moskop, in his chapter, "Recognizing the Difference that Faith Makes: H. Tristram Engelhardt, Jr., on Life-Ending Medical Interventions," examines some of Engelhardt's contributions to discussions of end of life care, especially debates concerning the permissibility of employing life-ending medical interventions. Moskop notes that in the secular literature, Engelhardt has defended the right of physicians and patients to employ life-ending medical interventions. Yet Engelhardt rejects the permissibility of life-ending medical interventions in his work in Christian bioethics. Moskop examines Engelhardt's scholarship on life-ending medical interventions from a secular and Christian perspective, comparing the two positions and considering the extent to which they are consistent. He argues that there are ways in which Engelhardt's positions may be understood as consistent with one another. Nevertheless, "the shift in emphasis in his writing from endorsement of individual rights to employ these interventions, to religiously inspired condemnation of any form of medical killing, is a remarkable development in his thought."

Stephen Wear, in "The Foundations of Secular Bioethics," assesses Engelhardt's arguments regarding the failure of the Enlightenment and the implications of this failure. The Enlightenment failed to rationally secure a content-filled, lexically ordered secular morality. As a result, Engelhardt argues that we are left with a minimalist account of ethics focused on the ethics of permission between moral strangers. Wear challenges Engelhardt's view that secular bioethics can give us nothing more than the ethics of permission. With a focus on clinical ethics, Wear argues that "a substantial, secular ethic can be (and has been) legitimately fashioned that provides respectable, coherent guidance for moral strangers." Wear argues that there is room between Engelhardt's libertarian ethics of permission and its opposite, the liberal cosmopolitan view (which both Wear and Engelhardt reject), for a more satisfying account of ethics. This account, Wear demonstrates, is substantive and has been developed through "decades of substantial argument, debate, and experience." It is not impoverished, he says, in the ways Engelhardt would have us believe.

Mary Ann Gardell Cutter reviews Engelhardt's significant contributions to the philosophy of medicine in "Disease, Bioethics, and Philosophy of Medicine: The Contributions of H. Tristram Engelhardt, Jr.," arguing that one of the most notable features of his work in this area is the demonstration of the interreliance of bioethics and philosophy of medicine. In demonstrating, via Engelhardt's work, that the bioethical implications of concepts like "disease" are framed in terms of particular epistemological and ontological commitments articulated in the philosophy of medicine, Gardell Cutter shows that philosophy of medicine is, and should be, always present in the study of bioethics.

Much of Engelhardt's work has addressed the differences between Christian and secular bioethics as well as differences among religious approaches to bioethics. The contribution from *Kevin Wildes, SJ*, "Sanctity of Life: A Study in Ambiguity and Confusion" (reprinted from an earlier volume), demonstrates the significance of such differences with respect to the term "sanctity of life." The term, frequently deployed by secular bioethicists, was first used in religious contexts and, Wildes argues, cannot so easily be repurposed as these secular uses suggest. Using the term "sanctity of life" as a heuristic, Wildes demonstrates this failure by comparing just two religious interpretations of it, from Christian and Buddhist traditions. He shows that not only do these religions conceive of the term differently from each other, there are also diverse possible interpretations within each tradition. Each can lead to different conclusions about what kinds of lives have this "sanctity," and with what other values this sanctity can be balanced. And so, he argues, "when such terms are taken from their basic role they yield very little" (Wildes, Chap. 6, this volume, 84).

George Khushf takes on a significant portion of Engelhardt's corpus in order to articulate the many and complex lines of argument woven throughout. In "A Transcendental Argument for Agreement as the Sole Sufficient Basis of a Philosophical Ethic," Khushf explores one of the most central and famous arguments in Engelhardt's work, the "Principle of Permission," which Engelhardt argues provides both a minimal and sufficient condition for the possibility of secular morality. Khushf's tour de force necessarily explores what he calls "argumentative strands" in Engelhardt's work in order to reconstruct the transcendental argument presented in the *Foundations of Bioethics* (2nd ed.). Along the way, he also investigates the meaning of terms such as "agreement," "conflict," "basis," etc., which factor centrally in the transcendental argument. This demanding essay pays rich dividends for those who want to understand or challenge Engelhardt's work, for it is the only work the editors know of which attempts to bring so many of Engelhardt's claims and argumentative pathways together in order better to understand the whole.

Ruiping Fan focuses on Engelhardt's arguments regarding the requirement for a fair equality of opportunity in "Equality Is Problematic: Engelhardt on Fair Equality of Opportunity, Health Care, and the Family." These arguments, Fan notes, demonstrate that contestable values underlie every conception of fairness or equality upon which a social democracy might base a health care system. As a result, the moral legitimacy of a single-payer health care system is in question. Worse, Fan argues, the reliance on the Rawlsian principle of fair equality of opportunity may undermine the integrity of the traditional family. Because the family is a fundamental source of "nonegalitarian intergenerational obligations," it may come under siege from egalitarian movements. With Engelhardt, Fan argues that the establishment of the modern social welfare state has already contributed to an erosion of the family.

In "Bioethics After the Death of God – Reflections on an Engelhardtian Theme," *Mark J. Cherry* explores ontological issues in bioethics that result from the lack of canonical moral foundations Engelhardt articulates. Epistemically, Cherry argues, canonicity requires a God's-eye perspective, but modern secular morality cannot achieve such a perspective. Lacking that perspective, he argues, any claims to moral values without God or a God's-eye perspective will simply reveal prevailing tastes rather than timeless truth. So, to the extent that bioethicists claim authority to guide public policymaking and clinical decisions, what they in fact provide is "their own idiosyncratic criteria for veracity, rationality, and reliability, as well as rationalizations for ideologically driven political advocacy" (Cherry, Chap. 9, this volume, 159–175).

In her chapter, "The Ethical Conduct of Research: The Legacy of the Three Principles," *Ana Smith Iltis* explores Engelhardt's early contributions to discussions regarding the ethical conduct of research. Engelhardt was among the scholars who wrote background papers for the National Commission for the Protection of Human Subjects of Biomedical and Behavioral Research. In his contribution, Engelhardt argued that human research must respect three principles. The National Commission ultimately adopted three principles, the first two of which bear a resemblance to Engelhardt's first two principles. In articulating three principles, Engelhardt addressed the relationship among them. He argued that the first principle, which required respect for "human subjects as free agents," was prior to the other two principles. His second and third principles involved obligations to "foster the best interest of the individual human subjects" and to "maximize the benefits accruable to society from research" (Engelhardt 1978, pp. 5–6). Engelhardt argued that the principles were not all equal and unranked but rather that the first took precedence over the other two. The National Commission was not as explicit as Engelhardt was in addressing the relationship among the principles it articulated. The subject of whether the three principles stipulated in the Belmont Report, respect for persons, beneficence, and justice, are unranked, equally important principles or whether there is some other relationship among them continues to be debated. Iltis explores this debate and the significance it has for contemporary discussions in research ethics.

Lisa M. Rasmussen focuses on Engelhardt's critiques of clinical ethics consultation in "Non-Certain Foundations: A Clinical Ethics Consultation for the Rest of Us." She summarizes arguments.found in a number of Engelhardt's works, particularly the history he offers to explain the rise of the field. Even accepting that history, Rasmussen shows, does not require drawing the same conclusions Engelhardt does regarding the legitimacy of the field. But further, she argues, the conclusions he draws are themselves reliant upon metaethical premises which can be countered with other possible premises. In particular, she focuses on Engelhardt's assumption that certainty provides the only possible justification for moral claims. Similar to McCullough's observations (Chaps. 1 and 14, this volume), she articulates what "non-certain" justification might look like in clinical ethics consultation, but also acknowledges that many of Engelhardt's recommendations for changes in the practice of clinical ethics consultation are correct and must be addressed.

Fabrice Jotterand explores a question that has consumed much of Engelhardt's scholarly work: in the ruins of the Enlightenment and in the face of irreducible moral pluralism, how may one "rebuild and provide moral guidance to social practices such as medicine, science, and technology despite moral pluralism and the incommensurability of current bioethical debates." As Jotterand notes, Engelhardt diagnosed a fatal problem with modernity and deconstructed its philosophical agenda. The chapter reflects on the implications of this diagnosis as Engelhardt characterizes them, raising critical points concerning the notion of moral strangers and procedural ethics Engelhardt describes.

Jeremy R. Garrett explores the tension between the critical and constructive elements of Engelhardt's work on the foundations of secular bioethics in "Cracks in the Foundations of Engelhardt's Bioethics." He argues that the nature, scope, and depth of Engelhardt's critiques of the "prevailing methods and values of (secular) bioethical discourse" threaten to undermine even his own positive project of anchoring bioethics to the (purportedly) value-free, content-less, default-justified principle of permission. This tension, Garrett argues, reveals a number of significant "cracks" in the foundations of Engelhardt's bioethics, "each tracing back to fundamental underlying assumptions embedded in his view." These cracks already surface when Engelhardt's framework treats paradigm cases in bioethics (i.e., interactions between fully mature, independent, and autonomous individuals), but expand dramatically when this same framework is applied to non-paradigm cases, including standard cases in pediatric bioethics. Garrett concludes that Engelhardt's framework cannot withstand the internal tension between its critical and constructive elements, but instead is pushed either toward a nihilistic rejection of all (secular) morality or toward a thick content-full bioethics.

Finally, the volume concludes with a special section of personal essays. In our experience, there seems to be a reflected branding involved in being one of

Englhardt's students. Having misunderstood the man himself, people can sometimes misunderstand what it what it was like to be one of his students, or what it continues to mean to have been one of his students. It is a pleasure, with this final set of essays, to offer a fuller picture of studying with our mentor, Professor Dr. H. Tristram Engelhardt, Jr.

Conclusion: The Plantation

> Because of our inability rationally to defend a canonical correct, concrete moral order or in general secular terms to establish the moral authority coercively to impose a particular concrete moral vision, and because of the ever-available moral standpoint of consensual association based on permission, we have a morality that allows many moralities to be and have their place. In the ruins, even with moral strangers, we can meet and collaborate with moral authority (Engelhardt 1996, 422).

Over the decades, Professor Engelhardt developed a peculiar and very Texian language for describing the relationship among himself and his students: The Plantation. Even though his earliest students will insist that in their time he had not yet fully developed such a robust description, it remains a plausibly descriptive term for all of us who have so labored. There has been much work to be done: two international academic journals: *The Journal of Medicine and Philosophy* (Oxford University Press) and *Christian Bioethics* (Oxford University Press); two major book series: *Philosophy and Medicine* (Springer) and *Philosophical Studies in Contemporary Culture* (Springer); classes to teach on philosophy, health care policy, medical ethics, and ICU ethics rounds; articles to review; books to write; lectures to present; national and international conferences to organize; travel to manage throughout the United States and to many cities abroad; research to engage; good scotch, fine wine, and the occasional James Bond martini to be drunk; and even sometimes a dog that needed to be fed. Excellence would be insisted upon. No excuses. Sleep, a luxury to be frowned upon. In exchange, Professor Engelhardt's obligation was to work tirelessly for his students, not simply through the drafting of a dissertation and successful defense of their PhDs, but until each had obtained the rank of Full Professor with tenure, and even thereafter to continue to support their careers.

Many of Professor Engelhardt's students were and are moral strangers to him and with each other:

> Moral strangers are persons who do not share sufficient moral premises or rules of evidence and inference to resolve moral controversies by sound rational argument, or who do not have a common commitment to individuals or institutions in authority to resolve moral controversies. A content-full morality provides substantive guidance regarding what is right or wrong, good or bad, beyond the very sparse requirement that one may not use persons without their authorization (1996, 7).

This fact has never meant that they could not be affective friends. It has never implied that they could not meet and collaborate with moral authority. As Professor Engelhardt's students, we share a bond that is deep and significant, but impossible adequately to convey to others. We have shared friendships, common experiences, and, perhaps, a bit of post-traumatic stress disorder. Not once in the many years that we have known him, however, has Professor Engelhardt ever strayed from or ceased aggressively to pursue his duties to his students. The depth of his professional energies has always been the greatest when he has sought to help his students and their projects. Here one might recall G.W.F. Hegel's argument that philosophy justifies itself within a circle of thought: the more encompassing the circle, the more powerful and commanding the explanation (Hegel 1975 [1830], sec 17). The Plantation is a powerful circle of thought, indeed.

Charlotte, NC, USA — Lisa M. Rasmussen
Winston-Salem, NC, USA — Ana Smith Iltis
Austin, TX, USA — Mark J. Cherry

References

Engelhardt Jr., H.T. 1986. *The foundations of bioethics.* New York: Oxford University Press.

Engelhardt Jr., H.T. 1996. *The foundations of bioethics*, 2nd ed. New York: Oxford Unviersity Press.

Engelhardt Jr., H.T. 2000. *The foundations of Christian bioethics*. Lisse: Swets & Zeitlinger.

Engelhardt Jr, Tristram H. 1978. Basic ethical principles in the conduct of biomedical and behavioral research involving human subjects. In *National commission for the protection of human subjects of biomedical and behavioral research*, The Belmont Report, Appendix II, 8.1–8.45.

Hegel, G.W.F. 1975 [1830]. *The encyclopedia of the philosophical sciences*, 3rd ed. Trans. William Wallace, forward by John Findlay. New York: Oxford University Press.

Iltis, A, and M.J. Cherry. 2010. Introduction: At the foundations of Christian bioethics; or, why H. Tristram Engelhardt Jr.'s Christian bioethics *is* so very counter-cultural. In *At the roots of Christian bioethics*: *Critical essays on the thought of H. Tristram Engelhardt, Jr.* ed. A.S. Iltis and M.J. Cherry. Boston: Scrivener Publishers.

Sextus Empiricus. 1996. *Outlines of Pyrronism.* Trans. with Introduction and Commentary, Benson Mates. New York: Oxford University Press, 1996.

References

Part I
Critical Essays

Chapter 1
A Critical Appraisal of Engelhardt on the "Enlightenment Project"

Laurence B. McCullough

The Foundations of Bioethics, in both of its editions, has had a considerable influence on global bioethics, perhaps more than any other book in bioethics. In both editions, what H. Tristram Engelhardt, Jr., calls "the Enlightenment" becomes the foil for his explorations of the foundations of bioethics. More precisely, something that Engelhardt calls "the Enlightenment project" and its failure serves as the foil for his reported discovery of the foundations of bioethics: "The term *Enlightenment project* (and related terms) is used to identify the endeavor to establish a canonical, content-full morality in secular terms justifiable to persons generally" (Engelhardt 1996, 23). Understanding what Engelhardt takes "the Enlightenment project" to have been therefore becomes essential to understanding the project that Engelhardt undertakes in the *Foundations*.

This chapter provides a critical appraisal of Engelhardt's conception of "the Enlightenment project" as a quest for certainty in bioethics. The chapter begins with an elaboration of the elements of this conception and its purported sole alternative in morality, nihilism. The faculty of Reason (the capital 'R' is important, as we shall see) plays the dominant role. The critical appraisal begins with an account of a competitor Enlightenment project. There was no single Enlightenment, as Engelhardt apparently assumes, but multiple national Enlightenments. This is both historically and philosophically important; the Scottish and English Enlightenments, it turns out, provide an alternative Enlightenment project, in which reason (the lower case 'r' is also important, as we shall see) does not play the dominant role. The result is, not a quest for certainty in morality, but a quest for reliability. There is, historically and philosophically, a third alternative to the quest for certainty in bioethics on the basis of Reason and nihilism, the quest for reliability in bioethics. The chapter then

L.B. McCullough, Ph.D. (✉)
Center for Medical Ethics and Health Policy, Baylor College of Medicine,
One Baylor Plaza MS 420, Houston, TX 77030-3411, USA
e-mail: Laurence.McCullough@bcm.edu

L.M. Rasmussen et al. (eds.), *At the Foundations of Bioethics and Biopolitics: Critical Essays on the Thought of H. Tristram Engelhardt, Jr.*,
Philosophy and Medicine 125, DOI 10.1007/978-3-319-18965-9_1

shows how, in a very important sense, Engelhardt's bioethics can be read as itself very much an Enlightenment project. The *Foundations*, it turns out, are deeply in philosophical debt, methodologically, to the German Enlightenment and its conception of Reason. The chapter closes with an account of the attraction of a bioethics of reliability for the ethics of the health care professions and for health policy.

1.1 Engelhardt on the "Enlightenment Project"

The first edition begins by placing "[b]ioethical questions … against the backdrop of a moral crisis," provoked by the Protestant Reformation (Engelhardt 1986, 3): "the crumbling of the presumed possibility of a uniformity of moral viewpoint. One could no longer hope to live in a society that could aspire to a single moral viewpoint governed by a single supreme moral authority" (Engelhardt 1986, 3). Put aside the repressive nature of such a single moral viewpoint, epitomized by the statues on the west facades of medieval cathedrals of a woman, to the left (*sinister*) of Christ in Judgment, blind-folded and her spear broken – the synagogue defeated. And put aside the warnings of Isaiah Berlin (1909–1997) against any morality that claims to know for all of us what our good is, especially perfectionist moralities (Berlin 1958). Such a single morality beckoned, even as "the religious synthesis weakened" (Engelhardt 1986, 3). And this beckoning reached its highest expression philosophically during the Enlightenment: "the Enlightenment hope arose that reason alone (through philosophy) could disclose the character of the good life and the general canons of moral probity" (Engelhardt 1986, 3). Citing the magisterial work of Alasdair MacIntyre, Engelhardt declares that "this hope has proven vain" (Engelhardt 1986, 4). This hope founders on the reality that there are competing, sometimes bitterly competing, moral viewpoints in "secular, pluralistic societies" (Engelhardt 1986, 4). In the intellectual and social context of this failed hope of the Enlightenment, Engelhardt formulates the problematic for bioethics: "The problem is how to fashion an ethic for biomedical problems that can speak with rational authority across the great diversity of moral viewpoints" (Engelhardt 1986, 4).

The second edition begins with the frank recognition of moral pluralism, "bioethics in the face of moral pluralism" (Engelhardt 1996, 3): "Moral diversity is real. It is real in fact and in principle. Bioethics and health care policy have yet to take this diversity seriously" (Engelhardt 1996, 3). No "canonical account" or morality – whether of justice in health care policy or of the "proper physician-patient relationship" – is possible. There is "no concrete morality available to all through rational reflection" (Engelhardt 1996, 3). With the weakening of the "Western Christian religious synthesis," "Enlightenment and progressivist hopes grew that reason (through philosophy and rational reflection generally) could disclose the character of the good life and the general canons of moral probity outside any particular moral narrative" (Engelhardt 1996, 5). The "enormity of the failure of the Enlightenment project of discovering a content-full morality … to ground the objectivity of morality" (Engelhardt 1996, 65) brings us inevitably to the "brink of nihilism" (Engelhardt 1996, 65).

Considered together, the two editions of *Foundations* embrace the following account of the "Enlightenment project." This account relies on a very old philosophical distinction between the logic of discovery, which when applied without mistake results in error-free functioning of Reason, and the logic of invention, which is error-prone and therefore of no value in articulating an objective – not culture- or time-bound – morality (or metaphysics).

1. The Enlightenment project is based on Reason, which is understood to be the pre-eminent faculty of mind that, because it is not culture- or time-bound, is the same in all human beings. Reason is the essence of human nature.
2. There is a logic of discovery for the proper functioning of Reason: fixed, unchanging and therefore discoverable rules (*regulae*) of Reason.
3. When functioning in accordance with the logic of discovery, the processes of Reason are objective, transcending cultures, nations, religions, and other differences that divide human beings and societies.
4. When functioning in accordance with the logic of discovery, the products of reason – propositional claims – are objective.
5. When functioning in accordance with the logic of discovery, the products of Reason are indubitably true. Properly functioning Reason becomes the indispensable tool in the quest for certainty.
6. When functioning in accordance with the logic of discovery, Reason can produce a content-full, certain morality that applies to all human beings and therefore can and therefore should be used in any culture, including a pluralistic culture, to guide moral reflection and behavior based on such reflection.
7. Reason's mortal enemy is the passions or instinct, which therefore must be regulated by Reason. The purest expression of this idea is Spinoza's, who explains that Reason can transform any passion or emotion into an idea, thus depriving it altogether of its motive force in generating impulsive behavior.[1]
8. Anyone who denies these claims suffers from disordered thinking, defined as the displacement of Reason by its mortal enemy, the unregulated passions. Unregulated passions pave the way to human conflict and war. The Enlightenment project, born of the weariness of centuries of religious warfare to resolve differences, paves the way to peace. The Enlightenment project can be understood to be a rational longing for peace on the basis of a morality of certainty (hence Engelhardt's use of 'canonical') issued by Reason in its flawless operations.

Diego Gracia describes the historical roots of this *etica ordine geometrico demonstrata*: "the history of Western ethics has been characterized by a continuous, yet unsuccessful attempt at finding a way of creating a definite, universal, and immutable system of morality" (Gracia 2010, 56). One emendation is needed: substitute 'discovering' for 'creating'. Gracia highlights the main points of this quest for certainty in morality and moral reasoning in Western philosophy, starting with Plato and Stoicism and culminating in Kant (Gracia 2010, 56–58).

[1] In this author's view, this is one of the most beautiful ideas in all of the history of Western philosophy, thus making its claim to be true very attractive.

The epistemic product of the Enlightenment project is predicational knowledge produced by the faculty of Reason following the logic of discovery. If Reason were, instead, following the logic of invention, the upper case 'Reason' would become lower case 'reason' and its products would lack certainty. When Reason is operating under the logic of discovery, its product is knowledge with certainty.

In moral reasoning thus understood, knowledge takes the form of an S is P predication of judgment. A predication of judgment seeks to sort or classify persons, events, circumstances, etc., under conceptual categories, much in the manner of "said of" predications in Aristotle's *Categories*. In such predications of judgment, 'S' names a moral circumstance in which human beings act and 'P' names the moral classification of S: S is either obligatory, permissible, or impermissible. This is a predication of judgment, in which S is classified with certainty under P in one of its only three meanings. These are the only moral categories of judgment and all circumstances and all actions that produce those circumstances are sortable without error into one of the three categories. The complete set of such predications of judgment constitutes morality. Because for every S there is a classification, the resulting morality is content-full (a phrase the meaning of which remains elusive in both editions of *Foundations*, in my judgment). Because for every S there is a classification that is made with certainty, without error, the resulting morality is canonical (a phrase also of elusive meaning in both editions of *Foundations*, in my judgment).

1.2 A Competitor Enlightenment Project

Gracia also describes a competitor tradition in Western philosophy, what he calls the tradition of deliberation (Gracia 2010, 58), which does aim to be content-full but does not aim for certainty. This tradition, Gracia claims, originates in Aristotle and continues through Hume to the American pragmatists (Gracia 2010, 58–61). The product is what Gracia, acknowledging a debt to Max Weber (1864–1920), called "responsibility ethics" (Gracia 2010, 61): "the sense of responsibility obliges one to take into account not only the values at stake but also the circumstances and consequences of the decision to be made" (Gracia 2010, 61).

This tradition conceives of moral reasoning as predications of judgment, aiming, not for certainty, but for reliability. It does aim, like all moralities, for comprehensiveness: every circumstance should be reliably classifiable. However, this tradition adds to the moral categories of obligatory, permissible, or impermissible a fourth category that is not contemplated in the quest for certainty: morally uncertain. This tradition aims to be content-full: all actions and the circumstances that they produce can be sorted into one of these four categories.

It is worth noting that the aim for comprehensive classification, for a content-full morality, is ingredient in the nature of morality per se. Being content-full, or comprehensive, is therefore not unique to Enlightenment morality. Medical ethics, for example, aims to be comprehensive in its account of the professional responsibilities of physicians in patient care, research, education, and health policy.

This tradition of responsibility ethics does not aim for certainty and so is not canonical in nature. Nor need it be. The test is not certainty but success in guiding morally serious judgment and behavior, about which more shortly. The great insight of this tradition is that morality per se need not be canonical. Gracia is correct to characterize the quest for certainty in morality as having failed. It failed because the quest for certainty failed; it did not fail because it was a content-full morality. A content-full morality does not require certainty; it need not be canonical in order to be content-full.

A content-full morality also does not need to be based on Reason. That there were multiple national Enlightenments becomes both historically and philosophically important at this point. On the European continent, in France and especially in Germany, the Enlightenment project was based on Reason. In the United Kingdom, the primacy of Reason was rejected and Reason demoted to reason.

David Hume (1711–1776), the leading figure of the Scottish Enlightenment, provides a vivid illustration of the Scottish Enlightenment (Hume 2000 [1739–1740]). For Hume, as for all of his contemporaries known as "moral sense" philosophers, human nature comprises two components: instinct and reason. 'Instinct' names the natural affinity that we have for each other; Hume elaborates on instinct in his *Treatise on Human Nature*. This was not a text in philosophy as some philosophers now understand it to be, i.e., non-empirical in its methods (Degrazia and Beauchamp 2010). Rather, Hume undertakes inquiry into the science of man, moral science, using the "new method" of discovery, i.e., Bacon's "experience"-based method. Baconian experience-based moral science bases all of morality on the results of natural and controlled experiments.

Hume reports the results of his scientific investigation of instinct in the third book of the *Treatise*, on sympathy. He describes the moral physiology of sympathy as the "double relation of impressions and ideas" and does so in the same style that his contemporaries used to report the results of other scientific investigations (McCullough 1998). Sympathy, properly regulated by its virtues, generates our moral obligations to care for each other, by "relieving man's estate," a phrase used by the Scottish moral scientists to mean making the human condition incrementally better than it was in a poor, often famine-stricken nation. One relieves man's estate by reducing the mortality and morbidity of conditions such as pregnancy, as well as diseases, disabilities, and injuries (especially the injuries incurred in combat). One relieves man's estate by reducing the frequency and intensity of human pain, distress, and suffering. Failure to relieve man's estate is attributed to a cold heart, which results from putting oneself first and others second, by becoming morally bent.

Sympathy properly functioning commits us to the all-encompassing task of relieving man's estate, making us "morally serious" to use contemporary nomenclature. This is precisely the meaning that Tom Beauchamp (2010) has recently attached to 'morally serious,' no surprise for someone whose first calling is as a Hume scholar. Beauchamp (2010) argues that we become morally serious when we commit ourselves to the improvement of the human condition, which is the objective of morality. This commitment generates the ethical obligation to "lessen human misery and preventable death" (Beauchamp 2010, 43) and thus "promote human flourishing" (Beauchamp 2010, 43). "Centuries of human experience" (Beauchamp 2010, 43) teach us the value of this commitment. There is nothing canonical in this view, nor need there be. Moreover, morality does not require foundations; morality requires commitments. In the language of American pragmatism, we make morality true, we do not discover it to be true.

Hume uses 'reason' to name a faculty of human nature: reason is a mere calculating machine. As such, reason's work is nothing more than that of matching means to ends, a not particularly complicated task. Reason is thus a weak faculty. Reason is also an inert faculty; it cannot motivate into action to relieve man's estate. This is because reason cannot generate our ends – that is the work of properly functioning sympathy. Reason cannot motivate us to action; this, too, is the work of properly functioning sympathy. Hume's famous dictum that "is does not imply ought" is an observation that he offers in the course of an examination of the severe limitations of reason. Reason-based propositions, because they are detached from experience-based impressions and ideas, become speculative; they have no foundation in experience. Reason-based speculation is therefore altogether worthless; the is-ought distinction is just one arena in which reason fails to provide a reliable guide to morality. Sympathy does and therefore one can reliably reason from is to ought in sympathy-based moral judgments, i.e., morality properly understood.

The English Enlightenment's pre-eminent moral scientist was Richard Price (1723–1791) (1974 [1769]). Price embraced a metaphysics in which entities in the world do indeed have moral properties. These constitutive properties of things, however, are not part of the essence or nature of a thing. Moral properties are thus supervenient, with (contrary to the usually vague concept of 'supervenience' in much of contemporary moral realism) a precise meaning: moral properties are real, constitutive properties of things that supervene on the essence or nature of things. Reason, the weak, inert faculty, cannot discern these moral realities. Intuition (another term used without a precise meaning in contemporary moral theory) is the other faculty of human nature. 'Instinct' names the capacity of human beings to immediately and directly take up realities in cognition. Because the process is non-discursive, it not as prone to error as discursive reasoning and is therefore highly reliable.

The moral sciences of Hume and Price demonstrate that the Scottish and English Enlightenments will be completely mischaracterized as Reason-based moral philosophy. Their moral science is instinct-based, explicated as sympathy by Hume and intuition by Price. Both would be hostile to claims of the primacy of Reason and would, correctly, resist having their method of moral science classified as "rational reflection" or the disclosure of the good life through the exercise of the faculty of Reason. Both intended their ethics to be transnational, transcultural, and transreligious, without being certain. One important lesson of the Scottish and English Enlightenments is that philosophically transcending one's specific national, cultural, or religious origins does not require certainty. Doing so does require high reliability; Baconian method, followed rigorously, fits the bill.

Engelhardt is therefore mistaken to claim that there was such a thing as *the* Enlightenment project. His use of 'the Enlightenment project,' with its debt to MacIntyre, reflects an older historical scholarship that held that there was indeed a cultural phenomenon called *the* Enlightenment that was essentially the same everywhere in the West (Western Europe, North America, and South America) where it occurred. For the past three decades at least, historians have rejected this view and documented in considerable detail that there were multiple national enlightenments (Porter and Teich 1981).

The error that Engelhardt makes is not merely historical, although the historical error is very serious indeed. The error is to treat *the* Enlightenment project of Reason-based, canonical, content-full morality – i.e., the Enlightenment project in France and, especially, in Germany – as the only philosophically substantive Enlightenment account of morality. Hume and Price powerfully demonstrate that this is not the case. One can produce an instinct-based, highly reliable, content-full (in the meaning explained above) philosophically substantive account of morality without foundations in Reason and its deliverances in the logic of discovery.

The choice for bioethics is therefore not between the quest for certainty and nihilism; the choice is among (a) the quest for certainty, a *canonical* morality, (b) nihilism, and (c) a *reliable* morality. Morality itself does not require that we choose only the quest for certainty. Neither does being morally serious in the Humean sense require the quest for certainty.

1.3 Engelhardt's Enlightenment Project

Readers of the *Foundations* might have been left with the impression that, having shown to his satisfaction that the Enlightenment project has failed, Engelhardt is not an Enlightenment thinker. But he is, through and through, in his transcendental method of philosophical reasoning. This method is implicitly deployed whenever moral inquiry is framed as asking the question, "What is the necessary condition for the possibility of X?," where 'X' names some circumstance of obvious moral value. Engelhardt presumes, plausibly at the end of the bloodiest century in the history of our species, that achieving a peaceable community in which we manage our differences through reasoned discourse rather than violence is of obvious moral value.

Transcendental method originates in the German Enlightenment philosophy of Immanuel Kant (1724–1804). The method was discovered by Kant, who accepted that there were indeed synthetic *a priori* propositions. These are S is P predications of constitutive properties of things, in which the property named by 'P' is not in the concept of the subject named by 'S' but adds to that concept (the "synthetic" in 'synthetic *a priori*') and in which the prediction is necessarily true (the "*a priori*" in 'synthetic *a priori*'). Kant asks, what is the necessary condition for the possibility of synthetic *a priori* propositions? The question is important because there are such propositions, in mathematics and the sciences, making them of obvious value.

The answer, as readers of the *Critique of Pure Reason* know very well, is not simple. The *Critique* is written in the logic of discovery but not discovery of things in themselves. Rather, transcendental method leads to the discovery, with certainty, of the transcendental apparatus of pure reason, the regulatory principles (*regulae*) of pure reason. P.F. Strawson (1919–2006) very nicely summarized the conceptual apparatus thus discovered: "the massive core of human thinking which has no history or none recorded in the histories of thought; there are categories and concepts which, in their most fundamental character, change not at all" (Strawson 1963, xiv). Following the logic of discovery of the regulatory principles of pure reason allows reason to generate synthetic *a priori* propositions.

The *Foundations* should be read as invoking the transcendental apparatus in the quest for certainty regarding morality in pluralistic, secular societies. The result, as readers of *Foundations* know well, is not a content-full morality, but a morality of constraints on those who assert a canonical content-full morality. Like Isaiah Berlin (1958), Engelhardt understands a canonical content-full morality as a threat to peace and freedom. Given the catastrophic history of perfectionist ideologies over the past century or more into our own time, Engehardt's – and Berlin's – resistance to canonical, content-full morality is altogether warranted. The result that animates the pages of *Foundations* is spare, austere, and intellectually very beautiful – as beautiful as Spinoza's account of the transformation of roiling passion into docile ideas by the force of Reason alone. The beauty of the *Foundations* invites us to consider seriously that its account is also true, provided that we restrict the scope of the claim to canonical morality. Restricted canonical morality results from the German Enlightenment requirement that morality be transcendental, if it aims to achieve certainty. German Enlightenment, restricted, canonical morality saves us from nihilism, despair, and the descent into the staggering violence of the recently concluded century of violence too vast and cruel for words. With apologies to Gracia, the *Foundations* can be read as *etica ordine transcendentale demonstrata*. The *Foundations*, however, do not succeed in showing the quest for certainty is our only alternative to nihilism, any more than the German Enlightenment did.

1.4 The Attraction of Bioethics Aiming for Reliability

Despite its great beauty and the attraction that such beauty generates (as Plato described long ago in the allegory of the cave in *Republic*), a bioethics committed to the quest for certainty makes for a poor fit with the health care professions. The ethical concept of the health care professions, the invention of two remarkable physician-ethicists, John Gregory (1724–1773) of the Scottish Enlightenment, and Thomas Percival (1740–1804) of the English Enlightenment, has three components: the health care professional commits to scientific and clinical excellence in clinical knowledge and skills; the health care professional commits to using his or her clinical knowledge and skills primarily to protect and promote the health-related interests of the patient, keeping self-interest systematically secondary; and the health care professional commits to sustaining the health care professions as public trusts for the health of society rather than merchant guilds protecting the economic, social, and political power of their members (McCullough 2006). Making these commitments creates the health care professions. The health care professions are therefore not some timeless entity that sustains itself and is discovered as the deliverance of Reason in its transcendental mode. The health care professions are human practices sustained by these three synergistic commitments.

The first of these three commitments is the commitment to professional integrity, which requires the deliberative practice of medicine, research, education, and health policy. 'Deliberative' means that clinical reasoning is evidence-based, rigorous, transparent, and accountable, a concept with deep roots in Bacon's experience-based philosophy of medicine. Thus understood, deliberative clinical reasoning aims for the highest reliability and, in its absence, the responsible management of uncertainty. The goal in doing so is to improve the quality of medical care: responsibly reduce the variation in the processes of patient care so that they closely track the biopsychosocial variability of conditions, diseases, disabilities, and injuries and the responses of conditions, diseases, disabilities, and injuries to clinical management of them. The goal of deliberative clinical judgment is to reliably classify the patient's condition, disease, disability, or injury and manage it according to that category. Genomics will allow increased precision in doing so. Similarly, the goal of deliberative clinical ethical judgment is to reliably classify a clinical case, research project, educational activity, or health policy proposal and manage it according to the categories of morally obligatory, permissible, impermissible, and morally uncertain. This classification is comprehensive, resulting in a content-full medical morality. Not being a quest for certainty, this classification is not canonical. Nor need it be, as the Scottish and English Enlightenments in medicine and medical ethics teach us. Indeed, to insist that it be canonical would be to graft onto medical ethics – and the ethics of all of the health care professions – a requirement that is utterly alien to them. The attraction of a bioethics aiming for reliability is that it is very well suited to the culture of the health care professions.

Such a bioethics is also very well suited to the making of health care policy. In a country like the United States, with its deep-rooted, sustained suspicion of concentrations of power – in both the public and private sectors – health care policy will be like public policy generally: fragmented, incomplete, contradictory, and temporary. It will therefore not be content-full, much less canonical. Its power over us, the self-governing, will be limited. To hold health care policy to the tests of the canonical, restricted morality of the *Foundations* is to bring the wrong intellectual tools to the task at hand: responsibly managing our differences through the limited institutions of self-government that, because in the American system they fragment power, protect us from the tyrannies of canonical moralities. The framers of our Constitution were steeped in the Scottish and English Enlightenments, not the German Enlightenment.

1.5 Conclusion

Engelhardt is mistaken in his reading of intellectual history and the history of Western philosophy. There was no single Enlightenment but, instead, there were multiple national Enlightenments. There was, therefore, no single Enlightenment project. The national Enlightenments did not agree about the nature, scope, and power of the faculty of Reason or reason. These Enlightenments did aim for a content-full morality, but only because that is the proper aim of any comprehensive morality. The Reason-based Enlightenments did aim to be canonical, with the understanding that 'canonical,' or certainty, has very different scope in non-transcendental and transcendental methods.

The result was competing Enlightenment projects. This point is not simply historical, although the historical point should be underscored. The choices before bioethics as a field are not the quest for certainty vs. a nihilism impotent in the face of violence. The choices are the quest for certainty, the quest for deliberative reliability, and nihilism. The quest for reliability in bioethics has much to recommend it, as I hope to have persuaded the reader in this short essay.

References

Beauchamp, T.L. 2010. *Standing on principles: Collected essays*. New York: Oxford University Press.

Berlin, I. 1958. *Two concepts of liberty: An inaugural lecture delivered before the University of Oxford 31 October 1958*. Oxford: Clarendon.

Degrazia, D., and T.L. Beauchamp. 2010. Philosophy: Ethical principles of common morality. In *Methods in medical ethics*, 2nd ed, ed. J. Sugarman and D.P. Sulmasy, 37–53. Washington, DC: Georgetown University Press.

Engelhardt Jr., H.T. 1986. *The foundations of bioethics*. New York: Oxford University Press.

Engelhardt Jr., H.T. 1996. *The foundations of bioethics*, 2nd ed. New York: Oxford University Press.

Gracia, D. 2010. Philosophy: Ancient and contemporary approaches. In *Methods in medical ethics*, 2nd ed, ed. J. Sugarman and D.P. Sulmasy, 55–71. Washington, DC: Georgetown University Press.

Hume, D. 2000 [1739–1740]. *A treatise of human nature,* ed. D.F. Norton and M.J. Norton. New York: Oxford University Press.

McCullough, L.B. 1998. *John Gregory and the invention of professional medical ethics and the profession of medicine*. Dordrecht: Kluwer Academic Publishers.

McCullough, L.B. 2006. The ethical concept of medicine as a profession: Its origins in modern medical ethics and implications for physicians. In *Lost virtue: Professional character development in medical education*, ed. N. Kenny and W. Shelton, 17–27. New York: Elsevier.

Porter, R., and M. Teich (eds.). 1981. *The enlightenment in national context*. Cambridge: Cambridge University Press.

Price, R. 1974 [1769]. *A review of the principal questions in morals*. Oxford: Clarendon Press.

Strawson, P.F. 1963. *Individuals: An essay in descriptive metaphysics*. Garden City: Anchor Books.

Chapter 2
Diversity in Clinical Ethics

George J. Agich

2.1 Introduction

In this paper, I address the question: "What implications does diversity have for the field of clinical ethics?" It is commonly recognized that development of clinical ethics and consultation was stimulated by the "the complex value-laden nature of clinical decision-making, the pluralistic context of contemporary society that is reflected to various degrees in clinical settings the world over, and, perhaps most importantly, a growing recognition of the rights of individuals and their implications for patient care" (Aulisio and Arnold 2008, 418). It is less commonly considered, however, how and why pluralism or *diversity* of the field itself might be an essential feature of clinical ethics as a practice. In this paper, I argue that diversity is more than an important external factor that gave impetus to the field or that provides the field much of its everyday challenges; in addition, it is an internal feature of the practice itself that needs to be understood in its own terms. The theme of diversity in clinical ethics is, of course, not new, but consideration of "internal" diversity is frequently subsumed in other concerns and not assessed fully as such.[1]

I argue that diversity in clinical ethics presents itself in two related, but distinguishable manifestations; first, in the largely conceptual or theoretical reflections on

[1] To be sure, sensitivity to diversity is now a common concern in health care institutions and training in diversity is commonplace. I shall not address or assess these efforts to raise the sensitivity of health care providers to the practical problems posed by caring for patients and their families with different cultural, ethnic, or religious backgrounds, beliefs, and commitments, because they lie outside the focus of this paper.

G.J. Agich, Ph.D. (✉)
Philosophy Department, Bowling Green State University,
6805 Via Correto Dr, Austin, TX 748749, USA
e-mail: agichg@bgsu.edu

L.M. Rasmussen et al. (eds.), *At the Foundations of Bioethics and Biopolitics: Critical Essays on the Thought of H. Tristram Engelhardt, Jr.*,
Philosophy and Medicine 125, DOI 10.1007/978-3-319-18965-9_2

clinical ethics and ethics consultation and, second, in the practices, structures, and processes of clinical ethics and ethics consultation services. These aspects of diversity in the field are located across three distinguishable domains, which I characterize as the educational, process, and philosophical. Although these domains are interrelated, it is useful to distinguish them for the sake of analysis since the issues posed within these domains have different implications for the field. In discussions about clinical ethics and ethics consultation, these domains are often merged, which contributes to some of the confusions characterizing the field. For analytical and discussion purposes, I discuss these domains as *aspects* of the way that diversity is at issue and not because I believe that the domains are in any substantive sense separate. I claim that diversity within the field of clinical ethics is often taken as a problem to be corrected or as a mark of the field's immature status; hence, diversity in clinical ethics is disvalued. Diversity is seen neither as a characteristic of an innovative spirit in the field nor as a resource for improvement of practices. Finally, the fundamental and irreconcilable value commitments that are the deep challenge that diversity represents are treated, if at all, as marginal concerns.

Historically, attention to diversity in the emerging field of clinical ethics was framed first in terms of questions that have come to dominate the field of clinical ethics and consultation ever since. Should ethics committees, teams, or individual consultants provide ethics consultation services?[2] What kind of professional qualifications and education should the ethics consultant possess?[3] Should consultants be credentialed, formally certified, or licensed?[4] Does clinical ethics and consultation alter the distribution of power among families, physicians, patients, and nurses in ways that compromise patient care and subvert health professional values?[5] Should ethics consultants or advisors ever be tolerated in liberal, democratic societies since such advisors occupy positions of authority that threaten, rather than enhance, personal autonomy?[6] These questions are critical for the field given that ethics consultation services and hospital ethics committees are ubiquitous features of contemporary health care not only in North America where they first developed, but also around the world.[7] That said, it is important to note that

[2] See, e.g., Cohen 1992; Gramelspacher 1991; LaPuma and Toulmin 1989; Ross 1990.

[3] See, e.g., Ackerman 1987; Barnard 1992; Beauchamp 1982; Cranford 1989; Grunfeld 1990; Jonsen 1980; LaPuma and Schiedermayer 1990, 1992; Marsh 1992; Morreim 1983; Thomasma 1991; Zaner 1984.

[4] See, e.g., Engelhardt 2009; Fletcher and Hoffmann 1994; LaPuma and Priest 1992; Spike 2009; Tarzian 2009, 2013; Task Force on Standards for Bioethics Consultation 2010.

[5] See, e.g., Agich 1995, 2000.

[6] See, e.g., Agich 1995, 2009c; Agich and Spielman 1997; Casarett et al. 1998; Slowther et al. 2001b; Spielman and Agich 1999; Wikler 1982; Wildes 1997; Yoder 1998.

[7] See, e.g., Aleksandrova 2008; Borovecki et al. 2005, 2006; Forde et al. 2008; Graf and Cole 1995; Guerrier 2006; Hurst et al. 2007a, b; Lebeer 2005; Melley 2001; Meulenbergs et al. 2005; Mino 2000; Newson et al. 2009; Newson 2009; Parker 2002; Reiter-Theil 2001a, b; Richter 2001; Robles 1999; Schlaudraff 1992; Slowther and Underwood 1998; Slowther et al. 2001a, b; Sorta-Bilajac et al. 2008; Steinkamp 2003; Tan 2002; Wray 2002.

the thrust of much of the literature addressing these questions is toward reducing or, even, to eliminating diversity.

In 1983 only 1 % of US hospitals had ethics committees, but, by 1989, the number had grown to more than 60 %, and to more than 93 % by 1999 (Fox et al. 2007; Youngner et al. 1983). A 2001 study found that all US hospitals with more than 400 beds, federal hospitals, and those that are members of the Council of Teaching Hospitals have some form of ethics consultation available (McGee et al. 2001). Hospital ethics committees are at least as prevalent in Canada as they are in the United States, and their presence is growing elsewhere around the world, sometimes in cultural, legal, and religious settings that are very different from where it originated. It is important to recognize that the expansion of the field of clinical ethics is due more to external factors than factors internal to the field itself. Instead of growing out of the emergence of cadres of specifically trained specialists in clinical ethics and ethics consultation, the field was populated from a wide variety of academic and professional sources: ethics and philosophy, religion, law, medicine, nursing, as well as allied health as health care institutions sought to address external pressures from accreditation, health care professional, legal, and social influences. This has resulted in a field that from its inception has been exceedingly diverse in its professional makeup and orientation as individual health care institutions addressed the growing concerns with ethical issues arising in the direct care of patients.

Clinical ethics has been seen as a response to increasing ethical problems in health care arising from the pace of development in medical sciences and technologies and to changed health care institutional structures. But it is also a response to ethical problems associated with an increased diversity of social views on the purposes of health care services and the erosion of traditional structures of authority in medicine, primarily the authority of the physician. The dynamics of change in the cultural, ethnic, religious, and social values and beliefs impacting health care further complicates developments within medicine and creates a fertile ground for ethical questions. Although the actual responses of health care institutions to these developments are complex and variegated, they have come to be seen within the frame of clinical ethics. The claim I will explore in this paper is that the inherent complexity of the set of activities and practices that has actually developed under the rubric of clinical ethics within healthcare institutions has not been sufficiently reflected upon in its own terms. When it comes up, the complexity and diversity *within* the field of clinical ethics is dealt with either obliquely or as a problem rather than as an essential feature of the field that needs to be understood.

Awareness of diversity in the field is, of course, hardly new. It was noted quite early as a set of concerns about the way that ethics and law "intruded" into the physician-patient relationship to such an extent that "strangers" gained access to the bedside or clinical space and ultimately to some of the "control" of this space (see, e.g., Blake 1992; Fleetwood et al. 1989; LaPuma and Schiedermayer 1990, 1992; Lo 1987; Rothman 2003; Siegler 1986; Siegler and Singer 1988). These discussions or debates were sometimes acrimonious as the following titles indicate: "Reflections of a Reluctant Clinical Ethicist" (Barnard 1992); "What Philosophers Can Offer" (Beauchamp 1982); "Why Physicians Should Not Do Ethics Consultations" (Marsh

1992); and, "Why Philosophers Should Offer Ethics Consultations" (Thomasma 1991). These early discussions have mutated to seemingly less pointed discussions about the qualifications of consultants, the techniques or methods for doing ethics consultation, or about the underlying philosophical basis or meaning and scope of the practice. Here, I argue that at or near the core of these disputes is the phenomenon of diversity *within* the field, which tends, without much supportive argumentation in these discussions, to be regarded as a danger or problem. I challenge this assumption and offer an alternative take on the diversity in clinical ethics.

For convenience, I aggregate these discussions under three points: first, the educational prerequisites or qualifications necessary for doing ethics consultation; second, the ideal technique or method of ethics consultation; and third, the philosophical or theoretical underpinnings of the field of clinical ethics and consultation. The literature addressing these areas, I contend, is implicitly committed to the pursuit of consistency, standardization, or uniformity for the field, features which are valorized without much argument, thereby casting diversity as a deficiency or problem. I will show this in passing since a comprehensive literature review on these points is beyond the scope of the present paper. Instead, I argue that diversity is addressed, if at all, only obliquely through other concerns that seem to be motivated by a deep suspicion. Nowhere is this suspicion adequately defended. I conclude that this concern for and push toward consistency is misplaced and that more attention needs to be paid to the advantages that diversity brings to the field.

2.2 Educational Aspect

There is significant discussion and, even, controversy about what qualifications, specifically the knowledge and skills, individuals who provide ethics consultation services should possess or demonstrate. Correlative concerns include a) what are the best or at least defensible educational or training practices, and b) how should competence in relevant knowledge and skills be demonstrated to assure that ethics consultants are able to provide competent, quality services. Implicit is an advocacy for standardization of training and a tacit assumption that there is a real and, some would say, a pressing need to assure that a set of defensible or accepted standards[8] are met by those offering ethics consultation services. Although not explicitly stated, much less defended, there is a perception or belief that there are deep problems of quality or competence in clinical ethics that underlie the variability of the types of individuals, organizational structures, and processes and procedures used in delivering clinical ethics services. Sometimes, this is articulated as a call to professionalize the field, which implies that a consensus about the qualifications for providing ethics consultations services is either shared or achievable. It is assumed that such qualifications apparently should be universally shared by ethics consultants with

[8] I say "defensible or accepted standards" and note that defensible and accepted are two very different visions of the validity of standards.

little regard for the differences of local culture or institutional setting in which they practice. These discussions, ironically, occur in the face of the historical development of the field, which has experienced not only an incredible expansion of clinical services, including ethics committees and ethics consultation, around the world, but an expansion that involved adaptation to local settings thereby yielding the complexity and differences that worry proponents of standardization.

This much is uncontroversial. First, the field has a set of practitioners—physicians, nurses, and other health care professionals; social workers; chaplains; philosophers; bioethicists; and those who primarily see themselves as specialists or, at least, competent practitioners in the field of clinical ethics and this set of professionals have no common disciplinary or educational background. This fact alone certainly raises a question about the knowledge or capacities of these individuals to provide clinical ethics services, but it is not clear why this question leads to the conclusion that this situation is problematic. Second, although clinical ethics is practiced in the context of patient care, the settings in which patients receive care are incredibly diverse. These settings include healthcare institutions such as general community hospitals and specialty institutions, such as those focused on cardiac or psychiatric care; they include small hospitals with limited services as well as large comprehensive medical centers providing advanced or innovative treatments with research and education of health professionals as important aspects of the institutional mission. Patient care is also provided in nursing homes and outpatient clinics as well as long term care or rehabilitation institutions. In addition, health care institutions exist and operate within widely varying cultural, political, religious, and social environments. The tendency in the literature to focus on specifically *ethical* problems or conflicts, which are seen as somehow different from cultural, ethnic, political, religious, or social commitments has obscured the importance that the actual context has in shaping the problematic that gives rise to the need for clinical ethics. So, why should clinical ethics services be standardized in the face of such a diversity of contexts? Or, not to deny that some features of the competences of ethics consultants might be universally shared across the contexts in which they practice, should context be addressed at all in determining the core competences of ethics consultants and, if so, then isn't the core more complex and variegated than is sometimes assumed?

Beyond the institutional level, there are many different types of units or services in which patients receive care for a wide variety of medical or health problems. In these settings, practitioners function with specialized skills and confront clinical and value problems and concerns that are frequently unique to the particular setting. For example, despite the overlaps and commonalities in end of life decision-making, the clinical, emotional, psychosocial, and, indeed, ethical features of the cases for which ethics consultation services are sought can vary widely from pediatric to neonatal critical care no less than from general pediatric to adult (including varying specialized units). In addition, the underlying "philosophies" and operating procedures of the units themselves, which include the personalities and communicative practices on said units, contribute to, if not shape, how ethical problems arise and are expressed. So, should the core qualifications of one deemed to be qualified to do clinical ethics in a specific patient care setting address these complex factors that contribute to the emergence and articulation of ethical concerns, conflicts, and questions or just some representative range of them? If so, then perhaps the focus on defining qualifications to do *clinical ethics as such* is insufficient and the conversation should further attend to the specific knowledge or skills that are prerequisite for an ethics consultant to function competently *in defined settings*.

Finally, the cultural, religious, social, and legal context in which patient care is provided also varies widely. The diversity of patient populations as well as health care providers themselves represent a wide spectrum of economic, ethnic, political, and religious beliefs and values that further obfuscate a rather murky picture. In light of these points, I conclude that it is not surprising that clinical ethics displays a striking degree of both complexity and diversity since it has emerged as a response to an inordinately complex set of needs and circumstances. These needs and circumstances share a family resemblance, but not necessarily a set of essential features. Clinical ethics and ethics consultation may be a heterogeneous rather than a homogeneous concept. Although commonalities may exist, stressing them to the exclusion of the deep differences that comprise the field of clinical ethics may seriously distort the reality of clinical ethics.

It might be rightly objected that even though diversity undoubtedly is present in clinical ethics, it does not represent an ineliminable problem since there is a core knowledge of ethical concepts, principles, and theories that should be shared by all clinical ethicists. This core knowledge should be the main part of the basic education for being a clinical ethicist and demonstrated proficiency in ethical concepts, principles, and theories should be required of all who operate in clinical ethical roles. This view, which is implicitly held by many commentators, however, overlooks the deep disagreements that characterize the field of bioethics and ethics regarding ethical concepts, principles, and theories. At the level of concepts, theories, and principles, one is hard pressed to point to a demonstrable consensus on these matters. Some might point out that the law or, in its absence, health care ethics policies and/or professional ethical guidelines cuts through this problem by providing socially approved frameworks and substantive guidelines for decision-making. But even if one accepts such a view that social consensus expressed in laws, institutional ethics policies, or professional guidelines provides a reliable and normative guide in clinical ethical matters, we have to recognize that the law, institutional ethics policies, and even professional guidelines are dynamic and are somewhat shaped by clinical ethics itself. These sources of normativity are complex and sometimes inconsistent; furthermore, they are invariably subject to interpretation and application. The appeal to consensus at this level is thus fictive at best, since at the point of interpretation and application complexity and diversity again rears its head. Controversies surrounding the ethical and logical status of the fetus, the questions of access to health care, and end-of-life decision-making such as assistance in dying represent divisions and disagreements that are culturally, ethnically, legally, religiously, and, of course, ethically distinct and seldom reconciled. I do not doubt that there are cases and situations in which resolution is possible since the success of clinical ethics attests to this possibility, but to assume, even tacitly, that this can be brought about simply by a canonical appeal to some normative foundation that underlies the field and comprises the core knowledge base for clinical ethics is preposterous. Controversies about foundational normative commitments are resistant to resolution, because of deep and irreconcilable beliefs and values (Engelhardt 1996, 2006), though they are, of course, subject to discussion and refinement even if not resolution on the theoretical level. Practical or pragmatic accommodation can occur, but such accommodation will not satisfy and will possibly offend those with dogmatic normative value commitments. However, there is a remarkably wide range of interpretations and understandings about what one's beliefs and values involve, especially when one is confronted with a loved one in an experientially novel clinical situation. To achieve situation-specific resolution—and what this means—is a practical problem of the first order that is highly influenced by communication and the empirical facts of the case, but it is ridiculous to assume in advance that any particular theory will be decisive.

The skills for seeking common ground or reasonable accommodation in the face of clinical circumstances can, of course, be achieved, but there is no evidence that there is a best or ideal approach for all of these various contexts and types of situations even at the practical level. Some features of many approaches might be useful and identified for wide adoption by clinical ethicists, but no one has shown that a single best approach exists, though there is enthusiasm for a number of approaches, such as mediation (Dubler and Marcus 1994). There are many reasons, however, to doubt that a standardized approach could work across the wide range of ethical concerns, conflicts, and questions that come to the attention of ethics consultants.

That said, it is important to recognize that clinical ethical concerns, conflicts, and questions may cluster and that specific settings of healthcare institutions or individual patient care units tend to localize these types of clinical ethics problems. Recognizing such clustering, it is important to stress that one can always develop or tailor approaches to address recurring problems. Identifying such "best practices" for clinical ethics and consultation is, of course, desirable. Doing so, which I certainly support as one proper focus in the field, however, does not warrant the conclusion or fortify the hope that an ideal approach will emerge over others which might in some way be standardized. In my view, commonly occurring ethical problems or issues might best be approached not primarily through individual case consultation anyway, but through the development of procedures that are expressed in ethics policies, unit guidelines, staff development, or unit quality improvement projects.

Some would undoubtedly respond that even when irreconcilable differences or conflicts occur, they can be resolved, because the principle of patient autonomy and the concept of patient rights can cut through the morass. Therefore, education of the clinical ethicist in patient advocacy and the promotion of patient rights should be a primary element in the education of clinical ethicists. But surely this does not cut deep enough. For example, a woman who autonomously decides for an abortion or assistance in dying still needs compliant health care professionals to provide such services and also a permissive legal and institutional framework that provides the services. So, again, the commonly assumed normative priority of patient autonomy may in some settings actually cause the clinical ethical issue to arise, but have limited utility in its practical resolution. It is hard to see how, given the diversity of actual clinical settings and social contexts, a reliance on patient autonomy can provide practical traction in non-supportive circumstances.

The ethics consultant is often called to negotiate how the competing rights or, more often, values can be accommodated. Defending a patient's autonomous decision-making is ethically acceptable when deep concerns about the motives of or reasons for a patient's decision are not in question or when health professionals are acting paternalistically. However, when it is reasonably clear that a patient's decision, for example, to forego a clearly beneficial intervention is primarily based on a mistaken assessment of the burdens that impaired survival might pose for loved ones, cutting through the complexity of conflict with the family by appeal to patient autonomy alone would be irresponsible and Procrustean. Similarly, when family or surrogate decision-making is itself compromised by misunderstandings, emotional or other conflicts, but health professionals have reliable knowledge of patient wishes expressed during the course of earlier treatments, then simply accepting the legal surrogate's decision would be as unjustified as would acceding to the preferences of health care professionals based on their "professional" assessment that fails to take into account patient/family values. Much more analysis and discussion is needed for the ethics consultant to ferret out the layers of value conflicts or misunderstandings, a process that surely should accommodate respect for autonomy, but which involves processes of communication and interpretation not derivable from patient autonomy in any standard bioethical understanding.

The assumption or hope that one primary and canonical ethical content can enable the clinical ethics consultant to successfully address the wide range of ethical concerns that arise in the course of patient care is naïve or, at its worst, dangerous. Some conflicts will be irreconcilable just because the beliefs and values are at bottom incompatible and in conflict. As a matter of fact, these kinds of cases are handled in a variety of ways by clinical ethicists and their health care institutions; some of these approaches are arguably based on standard normative or theoretical ethical commitments, but it is another matter to understand how these normative principles or theories actually drive the solutions that are reached. No single principled approach will reliably enable one practically to cut through the empirical morass in all cases and health care settings. Even when there is apparent consensus at the normative level, diversity in approach is not only practically unavoidable, it is necessary. Psychosocial considerations and personal preferences or existing operating procedures will complicate and shape the ways that the cases are handled by ethics consultants. In point of fact, these kinds of restraints cannot be reliably removed or overcome by appeal to any known procedure or theory. Success in clinical ethics as in any practical undertaking is an empirical matter that has to be achieved repeatedly in different circumstances. Success depends both on the actual clinical setting and circumstances of the case and the actual ways that the ethics consultant approaches the case and not on some gold standard approach or method.

For example, a liberal commitment to support autonomous patient choices, defensible though it may be, should not be applied without qualification. Cases in which patient's choices differ from that of the family or health professionals, who find the patient's choices to be unethical and, so, unacceptable because of their understanding of the patient or patient's clinical situation, need an assessment and handling that is more nuanced than simply applying the principle of patient autonomy. For this reason, some have advocated that the education of ethics consultants and clinical ethicists should primarily consist in developing a set of skills that allow the mediation or resolution of value conflicts. This response moves the worries about the disagreements on the normative or substantive value foundations of clinical ethics to one side and instead attempts to address diversity and complexity in clinical ethical situations through a uniform practical approach.

I mention only two problems with any approach that relies on a single technique to the exclusion of others. First, although I have spoken of conflicts and disagreements, which admittedly appear to call forth a set of skills at mediation or resolution, there is no reason to believe that clinical ethics only, or even primarily, deals with conflicts as such. The literature, sparse on this point though it is, does not support this belief nor does informal or personal communication among ethics consultants. If the ethical question is one of the application or interpretation of a law or institutional ethics policy, which the involved patient, family, and health care providers readily agree they will accept and follow, it is hard to see how the skills at mediation or conflict resolution would be helpful. Rather analytical, communicative, interpretive, and educational skills would be necessary and they, of course, would need to be grounded in a firm knowledge of the relevant laws or policies. This is especially important since it is widely recognized that communication problems much more than conflicts in ethical beliefs and values predominately underlie requests for clinical ethics involvement. Clearly, expertise in approaches to resolve conflicts, disagreements, or disputes alone would not be sufficient and, in fact, might be disruptive. Other skills—indeed, a rich and diverse tool bag of skills—are needed to allow the ethics consultant to respond effectively to the wide set of communicative occlusions over patient care decision-making that comprise clinical ethics consultation work. Conceding that such a tool bag would need to include some capacity to arbitrate disputes, mediate conflicts, or settle disputes, however, does not imply that reliance on any of the formal techniques of arbitration, conflict resolution, or mediation would be primary or sufficient. Indeed, such reliance might be seriously disruptive. The important point is that interactional skills that round out knowledge and communication skills are essential in any practical domain, clinical ethics included, but these are not well understood (Collins 2004; Collins and Evans 2002, 2008).

Second, some cases come to the attention of ethics consultants not because patients, their families, or health professionals lack effective communication or understanding and so need help in finding an ethically defensible course of action. Rather, they arise even when there is no conflict. For example, patients, family members, or health care professionals can be confused or uncertain about the ethical justification for certain clinical choices. There is acceptance, sometimes tacit, of the relevant ethical values and/or applicable institutional policy or law, but uncertainty about how to apply them or to draw out the implications that these normative guidelines or principles have in the particular circumstances of the case. They do not seek help to reconcile differences, much less conflicts, but *assurance* about the ethics of a certain course of clinical action. Again, it is hard to imagine that specialized training in mediating disputes or resolving conflicts would be sufficient, much less relevant.

In short, I am both arguing and suggesting that the diverse backgrounds of those providing clinical ethics services, including both their knowledge and their practical skills of analysis, communication, discernment, interpretation, mediation, and a wide array of other specific skills that ethics consultants draw on, might at least be regarded as a strength of the field rather than a weakness or deficiency that needs to be corrected through some process of standardizing the education and training of clinical ethicists. If such diversity is correctly regarded as a strength, which I believe it is, then the project of advancing the field of clinical ethics and consultation is better served by undertaking the difficult task of identifying the *typical* knowledge and skills that effective ethics consultation services in specified or particular settings or contexts rely upon. The complexity and unpredictability of cases coming to the attention of ethics consultation services suggests that the ideal consultation service would include not only a range of expertises—possessed either by expert individuals or a team—but also a commitment to and, perhaps, formal processes for collaboration and consultation among ethics consultants on cases and problems that challenge the responding clinical ethicist. Individual consultants must, too, have flexibility in approaching cases. It is doubtful that any rigidly applied normatively justified approach would work across a wide range of cases.

How can such a capacity be developed? Experience is clearly needed, either direct or indirect. That is why the practical purposes of improving ethics consultation can be partly achieved through the practice of monthly or quarterly review of cases. However, it should be pointed out that such review occurs after the fact and so is removed from the actual dynamic of the clinical circumstances. Such discussion, though useful for learning about ethics consultation, is often regarded, without foundation in my view, as oversight or supervision of the clinical service when done by an ethics committee as a whole. Even so, this practice is a move in the right direction, but it is incomplete if the ethics consultants themselves are not *reflective practitioners* (Schön 1983).

Nothing in what I have said should be taken to imply that I think there is no common knowledge or no basic set of skills that are fundamental to ethics consultation. I am simply questioning the grounding for the claim that there is a single ideal set of such skills. Since the circumstances of clinical ethics practice varies so widely, it may be that a variety of "core" or "basic" skills could—and should—be enumerated and defended as features that a competent clinical ethicist in a particular set of clinical situations should possess. However, this point should not be interpreted to mean that no set of universal "basic" or "core" content for the field exists, but only that we should be open to a range of understandings of what actually constitutes this "core." Rather than defending skepticism about the existence of a basic knowledge and skills for doing clinical ethics, since that is an empirical matter, I am arguing that the field should focus on trying to identify what knowledge, process, and skills are actually essential for competently addressing typical kinds of ethical problems arising in specific types of clinical settings. We should be open to discovering that the "core" is differentiated or variable for various practice settings. We should also recognize the limits of such core knowledge for actually doing ethics consultation since practitioners must learn from experience to mature in competence. Unfortunately, efforts to define an ideal core or common curriculum have overshadowed attention to the varieties of ethics consultation experiences. The American Society for Bioethics and Humanities Clinical Ethics Task Force rightly recommended that the particular education needs for ethics consultants should be framed in terms of a personal assessment of the individual's particular setting and situation (2009), but the significance of this point is seldom appreciated. The relevance of any putative knowledge or skills should be determined at least in part by reference to the practice setting and consultant's existing knowledge and skills. Thus, a solid appreciation of the typical settings and the typical clinical ethics problems faced is needed before we can design effective programs of training ethics or credentialing consultants. We should not be surprised or alarmed if training programs emerge that have different visions of clinical ethics. Proponents of standardization have failed to appreciate that the diversity of ways in which clinical ethics is actually practiced in clinical and healthcare institutional settings may be an asset. A careful and thorough examination might show that differences that first appear questionable or even objectionable may actually be adaptations to specific circumstances. Understanding how ethics consultation is actually practiced would thus provide a more reliable basis for identifying what might be a defensible "core" knowledge base or skill set for clinical ethics. Instead, the tendency in discussions of clinical ethics education is to propose a requisite content without linking that knowledge and skills to the demands of actual practice settings.[9]

Above all, one needs to accept that clinical ethics is a *doing* (Agich 1990, 2005, 2009); it is a practice and the rules that guide its performance are internal to the practice and so need to be internalized by individual clinical ethics consultants and put into action as they engage in doing the various activities that comprise the work of clinical ethics consultation (Agich 1995, 2001, 2009a). Too much attention, in my view, has been devoted to statements of curricular design and content for the education or training of new ethics consultants in abstraction from the performative acts that comprise a successful clinical ethics practice. Furthermore, there is insufficient discussion regarding how to implement ethics consultation services that foster the commitment to continued improvement of the requisite knowledge/skills.

This is evident in discussions around the topic of credentialing or certifying ethics consultants where it is often granted that individuals already doing ethics consultation will or can be "grandfathered" into the field of "qualified" or "certified" practitioners. Although this may be politically expedient, it surely avoids the

[9] Many might point out that many programs accomplish this point by requiring a "clinical" component. But some of these clinical tracks are merely observational experiences in clinical settings led by clinical colleagues. It is hard to understand how an effective clinical ethics training program could succeed in developing competent clinical ethics practitioners without a significant experience in a supervised clinical ethics consultation practice.

important need for ongoing education and improvement and provides no real assurances regarding the competence of those grandfathered into the field. Too much of this discussion is focused on training novices in clinical ethics or setting minimum eligibility requirements for being called a clinical ethics consultant and too little on the need for continued education and improvement. The use of quality improvement (QI) projects in the field is likely a more promising approach for the field to undertake as a measure to improve its stock of competence. Such projects have the virtue of translating ideas into actions that are observable, measurable, and focused on the achievement of specific clinical ethical outcomes or goals in particular patient care settings. In QI the outcomes are achieved through *changes in the practices of patient care*, which is, after all, the central purpose of *clinical* ethics.

Thus, the challenges of education in clinical ethics will not be exhausted even if or when a core curriculum or set of minimal standards for the field is formally established since the singular importance of continuing education and refinement of skills that is the hallmark of competent practitioners in a field as dynamic as clinical ethics would still need sustained attention. In all of this, it is hard to see how a single canonical approach will provide substantive help. To be sure, a common approach would provide a standardized vocabulary and shared framework, but that will not *substantively* resolve diversity. It may cover it up, but that is not a defensible outcome. In the end, responses to the educational needs of clinical ethics will understandably continue to mirror the complexity and diversity of clinical settings and clinical ethical challenges that clinical ethics services are intended to address.

2.3 Process Aspect

This discussion leads me to the second theme that can be discerned in discussions about ethics consultation services, namely how they should be organized and delivered, and which methods, processes, or techniques should be followed in providing and improving clinical ethics services (Agich 2009b, d; Dubler et al. 2009; Swiderski et al. 2010). Clearly, this is correlated with the concerns about the educational requirements or qualifications for doing ethics consultation—the so-called content of the field—but it goes beyond the matter of prerequisites of either knowledge or skill to the procedures or techniques that are actually put into practice in doing ethics consultation. These concerns about the process aspects of ethics consultation are often characterized as methodological concerns, but this is a mistake since the question of method in ethics consultation is a much more complex and involves an elaborate set of issues beyond the practical rules that guide clinical ethicists as they go about their work (Agich 2001). For this reason, I term this aspect the *process* aspect, though many of my comments can be seen as addressing what others would call the methods of doing clinical ethics or ethic consultation.

A common way to respond to questions about which model(s) or approach(es) should ethics consultation services adopt is often expressed in terms of an advocacy for one or another approach to the exclusion of others. There is, of course, little doubt that the skills imparted in various approaches to doing ethics consultation can be helpful, but as the literature on expertise amply shows, skilled or competent consultants draw on a variety of techniques or approaches to a presented problem rather than relying on a single, canonical one (Agich 2009b; Collins 2004; Collins and Evan 2002, 2008; Dreyfus and Dreyfus 1991; Schön 1983). Competent practitioners tailor their actions and responses to the elements of the question or problem at hand and are deployed in an organic rather than mechanical fashion. They do not force problems into preconceived formats. Judgment or discretion is used and often it is exercised unreflectively or even pre-reflectively.

Saying that does not imply that clinical ethics consultants function best when they lack a framework or style of approach, but rather that any style or technique will need to enacted (Agich 2001, 2009a). Furthermore, the approach must be *appropriate or suitable* to the *circumstances of the clinical case*. Consider the following. An ethics consultant finds a health care professional who has moral concern or qualms about a particular clinical decision and decides to address this simply by providing an analytically apt and philosophically sound argument about the ethical justification for the decision. In some situations and for some health care professionals, achieving intellectual clarity on the question would undoubtedly be appropriate, but for others the concern or qualms might require more than cognitive clarification. Instead, assurance that involves a personal or emotional communicative engagement with the distressed individual is needed. So, one should not be surprised if the consultant who merely and routinely offers only analytically sound advice finds that the advice is ignored.

The same situation would be poorly handled by criticizing the health professional for an ill-informed and unjustified belief. Also, simply informing a distressed health care professional that because a clinical decision had already been reached or action already undertaken—say by the attending physician—that there is no conflict of decision-making and therefore no reason for an ethics consultation would be equally questionable. The moral distress of the individual would be thereby devalued and the perception that ethics consultation has nothing to offer in future cases could be reinforced. One could give many other examples of the inappropriateness of a particular approach to a distressed health professional. The lesson here is that application of *any* approach without a careful assessment of its appropriateness to the circumstances is risky. Nonetheless, there is a tendency in the literature to promote preferentially certain approaches in clinical ethics consultation over others.

This tendency to advocate one way over others of doing ethics consultation seems to rest on the belief that a diversity of approaches and ways of handling a problem is a sign of weakness or, worse, incompetence. One reason for this assumption is that many such advocates appear to have vested interested in promoting their way of doing things or simply do not have a broad enough experience in clinical ethics to appreciate that diversity may represent a mark of maturation in a consultant or the adaptation of the field to local circumstances over time. Proficiency in the techniques that they promote is tacitly taken to provide the tools sufficient for competency in clinical ethics as such. They naturally think that diversity in the field represents a failure to adopt and perfect a particular approach rather than being a reflection of the adaptability of the field or of ethics consultants. In addition, it is doubtful that successful consultants who advocate a particular approach over others do in fact *only* use the recommended approach in all cases. That is, of course, an empirical point that is certainly worth investigating, but I will observe that in my experience as a consultant to ethics consultation services and clinical ethics programs success often involves a more complex and wider set of behaviors than the practitioners themselves actually profess.

A more defensible alternative to advocating any one singular approach is to recognize that a tool bag or repertoire of skills and knowledge is a necessary rather than a fashionable accoutrement for the ethics consultant. This implies that the techniques that are best in clinical ethics are those that *work* in the context of the given clinical case situation.[10] It also implies that competent clinical ethicists will be able to recognize not only what is needed, but also what to do if the appropriate approach to the problem is beyond their capacities. In this instance, as I have pointed out, they should have access to and support for collaboration or referral. Deciding what approach will work best cannot be known in advance, which is why flexibility and adaptability is a desirable character trait of competent ethics consultants. Saying this is consistent with trying to identify what are typical clinical situations or problems and what approaches actually tend to work best. That is possible only if there is a detailed body of case reports that document the actual processes and procedures of the ethics consultation. Such a body of case reports can enlarge the experiential basis for ethics consultation. Too often, case reports highlight theoretical ethical points and only summarize or condense the actions of the clinical ethicist, if they document them at all, or the reports follow a format which repeats the mantra of the subscribed approach. For instance, a report that records that a family meeting occurred and that a particular decision was reached, but not a narrative of the actual discussions, is misleading. Even when a narrative is provided, it tends to focus on what others say, is the "content" of the case, but not the actions of the consultant and the reasoning underlying the particular actions. In this regard, the literature on the reflective practitioner might provide valuable insights for how the field of clinical ethics might think about its activities (Schön 1983). Rather than promoting a single best approach to clinical ethics and consultation to the effacement of others, the field needs to create a space for dialogue about common kinds of challenges that clinical ethicists face and the techniques shown to be effective in dealing with typical challenges.

[10] This point, of course, needs considerable discussion. A wide range of questions are immediately raised such as are there normative standards for "what works" or how and by whom is success determined. These are, unfortunately, beyond the scope of this paper to explore.

2.4 Philosophical Aspect

Third, advocacy for a particular methodology and training regimen is paralleled by a concern to establish or clarify the philosophical foundation of the field. A firm theoretical foundation is thought to be essential for establishing the legitimacy of clinical ethics generally and ethics consultation specifically. This concern has a long history, originating with the emergence of bioethics as an interdisciplinary field, and this concern has understandably been ported to clinical ethics and consultation (Arras 2009; Beauchamp and Childress 2001; Gert 2004; Gert et al. 1997; Jonsen 1991; Jonsen and Toulmin 1988; Kuczewski 1998; Toulmin 1982, 1981). I find this concern paradoxical to say the least given the attack on foundationalism by so many contemporary thinkers.

Perhaps more to the point, the field of clinical ethics and consultation has by all measures done exceedingly well without a foundational theory.[11] It has expanded greatly from its origins in the United States and Canada to the rest of the world, a development which would seem improbable at best and impossible at worst if a canonical method or philosophical foundation was essential. Interestingly, none of the proponents of a unified method or philosophical foundation for clinical ethics have taken the trouble to show that a meta-method or meta-account of clinical ethics is possible for a field as diverse as clinical ethics operating across such diverse clinical and cultural settings. Principlism, casuistry, common morality, reflective equilibrium, among others, have their devotees who aim to tame the unruly field of clinical ethics. They are like evangelists proselytizing for converts to their position. But none of these approaches has achieved demonstrated ascendency in bioethical theory (Arras 2009). There is also a more important set of deeper divisions over fundamental beliefs and values within contemporary societies that philosophical theories will not resolve. So it is unreasonable to expect bioethical or clinical ethical theory to make headway in resolving these contradictory commitments at the level of theory. The public space is to an extent, but not as universally as Engelhardt has insisted, a place of moral strangers (Engelhardt 1996, 2006). It includes not only moral strangers, who share no fundamental moral commitments, and moral friends, who do, but also moral acquaintances who share much more than a libertarian theory such as Engelhardt's will allow (Loewy 1993, 1997). It is unsurprising, then, that ethics consultants confront cases in which there are deeply divergent beliefs and values in conflict, yet are able in many cases to find accommodation. In these

[11] Some might object that this claim is untrue since there is a broad set of commitments such as promotion of patient rights that underlies the clinical ethics movement. I agree, but only to a limited extent. This set of commitments are quite general and do not comprise a normative foundation or theory. Talk of the principles of bioethics, for example, does not make for a theory, and principlism in itself is not really a foundational theory. Common morality theory is, of course, another alternative with enthusiastic proponents. Alleged foundational values like patient autonomy, patient rights, or respect for life are too vague and contested to provide the stable base that foundationalists desire, so it is fair to say that very idea of a solid theoretical foundation for bioethics is highly contested (Arras 2009). My point is that there has been no compelling theory of the field of clinical ethics and consultation and I doubt that there could be one.

situations of deep sectarian commitments, it is vain to think that *philosophical or ethical theory* as such will provide the traction needed. Since, even in circumstances of deep conflict, some *ad hoc* practical accommodations seem to be achievable, resolution of conflicts of fundamental value commitments or beliefs might be sought at either of two levels different from theory. As Engelhardt stresses repeatedly in his work, no moral accommodation is possible at the fundamental level when there is incompatible commitments; however, this does not mean that important values such as tolerance for the divergent views of others cannot be ethically invoked in practical terms. The important function for clinical ethics would then be to help devise a plan that minimizes disruptions to the process of patient care in the involved unit and that also respects the patient or family position while it preserves the integrity of the health care providers.

One could, of course, focus on either side of the conflict to the exclusion of the other. In philosophical terms, the dispute is irreconcilable, because it is grounded on moral values or views that conflict in principle; however, in practical terms, the ethical problems posed by the conflict, such as the distress of health care professionals or the resulting disruptions to the care of other patients might still admit temporary or ad hoc solutions. Faced with insurmountable obstacles in the way, one can always alter one's path or direction of travel. Such practical adaptation is, it can be concluded from the success of ethics consultation, possible despite the existence of divergent beliefs. This can occur precisely because no single theory normatively constrains the clinical ethics consultant to achieve workable solutions in all cases and because individuals espousing a commitment to values that are irreconcilable do so more often as a matter of living than in theory. Ethics consultation functions in this existential realm of living with tragedy and choosing in the face of the brutal reality of empirical conditions, and not in the sphere of theory.

It is important to stress that diversity manifests itself not only in situations of irreconcilable moral difference, which is commonly discussed in the literature, but also in the more common situations in which ethical conflicts, problems, and questions arise in patient care among individuals who share a set of beliefs and values. For example, typical ethics consultation cases involving conflicts among family members and health professionals occur not because there are irreconcilable moral values at issue, but because there is uncertainty or disagreement about how the accepted values apply. A son and daughter might each say, "You must respect my father's wishes and values," but they cannot agree about what those wishes and values are. Coming to a resolution of this conflict is a problem of ethical discourse involving an incorporation of the clinical facts within their decision-making as conversations ensue about who the patient was and what were the patient's commitments. This often entails a narrative approach to understanding how values were lived in a parent's life. In such cases, no single ethical theory or method provides a reliable guide. The ethics consultant who effectively assists in this situation uses a customized combination of analytic, communicative, interactional, and interpretive skills along with knowledge of applicable normative guidelines to help the children and the health care team to come to a decision, *a decision that is their own given the circumstances*. It is important to stress that this is not to say that ethical concepts, principles, or theories are irrelevant, but only that none uniquely to the exclusion of others have any reliable practical precedence in the variable circumstances of the case.

This leads one to question why commentators find diversity in the field so problematic. Many reasons will surely occur to any thoughtful reader, but I will mention only one: it seems that commentators abhor the diversity of approaches and find comfort in the idea of a clear theoretical underpinning for the field because they see lack of such a foundation as a weakness or flaw. If diversity is regarded from the start as a problem or flaw, then it is not surprising that diversity *in* clinical ethics becomes a trait that should be erased. Diversity in clinical ethics, however, might not be a blemish, but an asset. When a diversity of values presents itself in clinical cases, it is a problem that can be addressed not at one fell theoretical swoop, but only in a practical way. The problem of diversity as it arises *for* clinical ethics can only be approached by experienced judgment. That is why I suggested earlier that collaboration and the use of quality improvement might help ethics consultation services to develop a shared competence in the absence of a sufficient caseload. As with any practice, experience in doing ethics consultation is required for refining the practical skills that an expert has. The practical problem of a diversity of issues arising in a clinical ethics practice is that it makes acquiring such a broad experience more difficult, yet essential. Diversity in the way clinical ethical problems are handled, however, might be a symptom that best practices are either not identified or widely shared. The differences in approach to addressing ethical problems, however, should not be uncritically assumed to be a weakness, since the differences might be adaptations to particular features of the circumstances of the case in which the problems occur. But this is an empirical matter and, frankly, the field has been remiss in not devoting resources to identifying and eliminating inelegant and inefficient ways of handling common or typical problems.

2.5 Conclusion

I have argued that diversity in clinical ethics can be regarded from the perspective of education, process, or philosophy. Each of these aspects has theoretical as well as practical dimensions, which need to be distinguished. Failure to do so foments confusion about the field of clinical ethics, and these confusions create a fertile ground on which the efforts to consolidate the field through the political mechanism of professionalization seem to germinate (Spielman 2001). In doing so, I am not arguing against professionalization or standardizing credentialing or certification procedures. My concern is simply to bring attention to the ways that diversity in the field is tacitly approached and how these taken-for-granted ways of dealing with diversity support the movement toward standardization. Too frequently, diversity is seen either from a developmental perspective as a symptom of immaturity, where maturity represents a phase characterized by consensus and standardization, or from a normative perspective, where diversity is seen as a fatal flaw in the field because of the belief that without a normative foundation the field lacks legitimacy. I am a skeptic when it comes to the efforts to professionalize the field, because these efforts seem, like all movements toward achieving the status of a formal profession, to harbor the goals of dominance and control over clinical ethics work. Because there is scant sound evidence that clinical ethics and consultation are rife with incompetence and riddled with problems, attaining professional status does little to actually improve the practice. Improvement is nonetheless possible and needed, but not because of demonstrated systemic failure or weakness, which seems to be the common view. Rather improvement is both possible and desirable because clinical ethics is first and foremost a practice (Agich 2001, 2005, 2009a).

References

Ackerman, T. 1987. The role of an ethicist in health care. In *Health care ethics*, ed. G. Anderson and V. Glesnes-Anderson, 308–320. Rockville: Aspen Publishing Company.

Agich, G.J. 1990. Clinical ethics: A role theoretic look. *Social Science and Medicine* 30: 389–399.

Agich, G.J. 1995. Authority in ethics consultation. *Journal of Law, Medicine, and Ethics* 23: 273–283.

Agich, G.J. 2000. Why should anyone listen to ethics consultants? In *The philosophy of medicine*, ed. H.T. Engelhardt, 117–137. Dordrecht: Kluwer Academic Publishers.

Agich, G.J. 2001. The question of method in ethics consultation. *American Journal of Bioethics* 1: 31–41.

Agich, G.J. 2005. What kind of doing is ethics consultation? *Theoretical Medicine and Bioethics* 26(1): 7–24.

Agich, G.J. 2009a. Clinical ethics as practice. *Journal International de Bioéthique* 20(4): 15–24.

Agich, G.J. 2009b. The issue of expertise in clinical ethics consultation. *Diametros* 22(December): 3–20.

Agich, G.J. 2009c. Understanding criticisms of clinical ethics and ethics consultation. *Formosan Journal of Medical Humanities* 10(1–2): 3–15.

Agich, G.J. 2009d. Why quality is so rarely addressed in clinical ethics consultation. *Cambridge Quarterly of Healthcare Ethics* 18(4): 333–337.

Agich, G.J., and B. Spielman. 1997. Ethics expert testimony: Against the skeptics. *Journal of Medicine and Philosophy* 22: 381–403.

Aleksandrova, S. 2008. Survey on the experience in ethical decision-making and attitude of Pleven University hospital physicians towards ethics consultation. *Medicine, Health Care and Philosophy* 11(1): 35–42.

American Society for Bioethics and Humanities Clinical Ethics Task Force. 2009. *Improving competencies in clinical ethics consultation: An education guide*. Glenville: American Society for Bioethics and Humanities.

Arras, J.D. 2009. The hedgehog and the Borg: Common morality in bioethics. *Theoretical Medicine and Bioethics* 30(1): 11–30.

Aulisio, M., and M. Arnold. 2008. Role of the ethics committee: Helping to address value conflicts or uncertainties. *Chest* 134(2): 417–424.

Barnard, D. 1992. Reflections of a reluctant clinical ethicist: Ethics consultation and the collapse of critical distance. *Theoretical Medicine* 13: 15–22.

Beauchamp, T.L. 1982. What philosophers can offer. *Hastings Center Report* 12(3): 13–14.

Beauchamp, T.L., and J.F. Childress. 2001. *Principles of biomedical ethics*. New York: Oxford University Press.

Blake, D.C. 1992. The hospital ethics committee health care's moral conscience or white elephant? *Hastings Center Report* 22(1): 6–11.

Borovecki, A., S. Oreskovic, and H. ten Have. 2005. Ethics and the structures of health care in the European countries in transition: Hospital ethics committees in Croatia. *British Medical Journal* 331(7510): 227–229.

Borovecki, A., S. Oreskovic, and H. ten Have. 2006. Ethics committees in Croatia in the healthcare institutions: The first study about their structure and functions, and some reflections on the major issues and problems. *HEC Forum* 18(1): 49–60.

Casarett, D.J., F. Daskal, and J. Lantos. 1998. The authority of the clinical ethicist. *Hastings Center Report* 28(6): 6–11.

Cohen, C. 1992. Avoiding cloudcuckooland in ethics committee case review: Matching models to issues and concerns. *Law, Medicine & Health Care* 20(4): 294–299.

Collins, H.M. 2004. Interactional expertise as a third kind of knowledge. *Phenomenology and the Cognitive Sciences* 3(2): 125–143.

Collins, H.M., and R. Evans. 2002. The third wave of science studies: Studies of expertise and experience. *Social Studies of Science* 32(2): 235–296.

Collins, H.M., and R. Evans. 2008. *Rethinking expertise*. Chicago: University of Chicago Press.

Cranford, R. 1989. The neurologist as ethics consultant and as a member of the institutional ethics committee. *Neurological Clinics* 7: 697–713.

Dreyfus, H., and S. Dreyfus. 1991. Towards a phenomenology of ethical expertise. *Human Studies* 14: 229–250.

Dubler, N.N., and L. Marcus. 1994. *Mediating bioethics disputes: A practical guide*. New York: United Hospital Fund of New York.

Dubler, N.N., M.P. Webber, D.M. Swiderski, and The Faculty and the National Working Group for the Clinical Ethics Credentialing Project. 2009. Charting the future: Credentialing, privileging, quality, and evaluation in clinical ethics consultation. *Hastings Center Report* 39(6): 23–33.

Engelhardt Jr., H.T. 1996. *The foundations of bioethics*, 2nd ed. New York: Oxford University Press.

Engelhardt Jr., H.T. 2006. *Global bioethics: The collapse of consensus*. Salem: M & M Scrivener Press.

Engelhardt Jr., H.T. 2009. Credentialing strategically ambiguous and heterogeneous social skills: The emperor without clothes. *HEC Forum* 21: 293–306.

Fleetwood, J.E., R.M. Arnold, and R.J. Barrin. 1989. Getting answers or getting questions? The problematic role of institutional ethics committee. *Journal of Medical Ethics* 15: 137–142.

Fletcher, J., and D. Hoffmann. 1994. Ethics committees: Time to experiment with standards. *Annals of Internal Medicine* 120(4): 335–338.

Forde, R., R. Pedersen, and V. Akre. 2008. Clinicians' evaluation of clinical ethics consultations in Norway: A qualitative study. *Medicine, Health Care and Philosophy* 11(1): 17–25.

Fox, E., S. Myers, and R.A. Pearlman. 2007. Ethics consultation in United States hospitals: A national survey. *American Journal of Bioethics* 7(2): 13–25.

Gert, B. 2004. *Common morality: Deciding what to do*. New York: Oxford University Press.

Gert, B., C.M. Culver, and K.D. Clouser. 1997. *Bioethics: A return to fundamentals*. New York: Oxford University Press.

Graf, H., and D. Cole. 1995. Ethics-committee authorization in Germany. *Journal of Medical Ethics* 21(4): 229–233.

Gramelspacher, G. 1991. Institutional ethics committees and case consultation: Is there a role? *Issues in Law and Medicine* 7: 73–82.

Grunfeld, G. 1990. Non-MD ethics consultants? *Journal of Clinical Ethics* 1: 325–326.

Guerrier, M. 2006. Hospital based ethics, current situation in France: Between "Espaces" and committees. *Journal of Medical Ethics* 32(9): 503–506.

Hurst, S., R. Perrier, R. Pegoraro, et al. 2007a. Ethical difficulties in clinical practice: Experiences of European doctors. *Journal of Medical Ethics* 33(1): 51–57.

Hurst, S., S. Reiter-Theil, A. Perrier, et al. 2007b. Physicians' access to ethics support services in four European countries. *Health Care Analysis* 15(4): 321–335.

Jonsen, A.R. 1980. Can an ethicist be a consultant? In *Frontiers in medical ethics*, ed. V. Abernathy, 157–171. Cambridge, MA: Bollinger Publishing Co.

Jonsen, A.R. 1991. Casuistry as methodology in clinical ethics. *Theoretical Medicine* 12(4): 295–307.

Jonsen, A.R., and S.E. Toulmin. 1988. *The abuse of casuistry*. Berkeley: University of California Press.

Kuczewski, M. 1998. Casuistry and principlism: The convergence of method in biomedical ethics. *Theoretical Medicine and Bioethics* 19(6): 509–524.

LaPuma, J., and E. Priest. 1992. Medical staff privileges for ethics consultants: An institutional model. *Quality Review Bulletin* 18: 17–20.

LaPuma, J., and D. Schiedermayer. 1990. Must the ethics consultant see the patient? *Journal of Clinical Ethics* 1(1): 56–59.

LaPuma, J., and D. Schiedermayer. 1992. The clinical ethicist at the bedside. *Theoretical Medicine* 12: 285–292.

LaPuma, J., and S. Toulmin. 1989. Ethics consultants and ethics committees. *Archives of Internal Medicine* 149: 1109–1121.

Lebeer, G. 2005. Clinical ethics support services in Europe. *Medicinska Etika a Bioetika* 11(Suppl): 8–11.

Lo, B. 1987. Behind closed doors: Promises and pitfalls of ethics committees. *New England Journal of Medicine* 317: 46–50.

Loewy, E.H. 1993. *Freedom and community: The ethics of interdependence*. New York: Suny Press.

Loewy, E.H. 1997. *Moral strangers, moral acquaintance, and moral friends: Connectedness and its conditions*. New York: Suny Press.

Marsh, F. 1992. Why physicians should not do ethics consultations. *Theoretical Medicine* 13(3): 285–292.

McGee, G., A. Caplan, J. Spanogle, et al. 2001. A national study of ethics committees. *American Journal of Bioethics* 1: 60–64.

Melley, C. 2001. Clinical ethics consultation in Germany: A philosopher's prognosis. *HEC Forum* 13(3): 306–313.

Meulenbergs, T., J. Vermylen, and P. Schotsmans. 2005. The current state of clinical ethics and healthcare ethics committees in Belgium. *Journal of Medical Ethics* 31(6): 318–321.

Mino, J. 2000. Hospital ethics committees in Paris. *Cambridge Quarterly of Healthcare Ethics* 9(3): 424–428.

Morreim, H. 1983. The philosopher in the clinical setting. *Pharos* 46: 2–6.

Newson, A.J. 2009. The role of patients in clinical ethics support: A snapshot of practices and attitudes in the United Kingdom. *Clinical Ethics* 4: 139–145.

Newson, A.J., G. Neitzke, and S. Reiter-Theil. 2009. The role of patients in European clinical ethics consultation. *Clinical Ethics* 4: 109–110.

Parker, M. 2002. The development of clinical ethics support in the United Kingdom. *Politeia* 8(67): 82–86.

Reiter-Theil, S. 2001a. Ethics consultation in Germany: The present situation. *HEC Forum* 13(3): 265–280.

Reiter-Theil, S. 2001b. The Freiburg approach to ethics consultation: Process, outcome and competencies. *Journal of Medical Ethics* 27(Suppl. 1): i21–i23.

Richter, G. 2001. Ethics consultation at the university medical center – Marburg. *HEC Forum* 13(3): 294–305.

Robles, P. 1999. Evaluation of healthcare ethics committees: The experience of an HEC in Spain. *HEC Forum* 11(3): 263–276.

Ross, J.W. 1990. Case consultation: The committee or the clinical consultant? *HEC Forum* 2: 289–298.

Rothman, D.J. 2003. *Strangers at the bedside: A history of how law and bioethics transformed medical decision making*. New York: Basic Books.

Schlaudraff, U. 1992. Ethics committees in Germany. *Bulletin of Medical Ethics* 80: 40–44.

Schön, D. 1983. *The reflective practitioner: How professionals think in action*. New York: Basic Books.
Siegler, M. 1986. Ethics committees: Decisions by bureaucracy. *Hastings Center Report* 16(3): 22–24.
Siegler, M., and P.A. Singer. 1988. Clinical ethics consultation: Godsend or 'godsquad?'. *American Journal of Medicine* 85: 759–60.
Slowther, A., and M. Underwood. 1998. Is there a demand for a clinical ethics advisory service in the UK? *Journal of Medical Ethics* 24(3): 207.
Slowther, A., C. Bunch, B. Woolnough, and T. Hope. 2001a. Clinical ethics support services in the UK: An investigation of the current provision of ethics support to health professionals in the UK. *Journal of Medical Ethics* 27(Suppl. 1): i2–i8.
Slowther, A., T. Hope, and R. Ashcroft. 2001b. Clinical ethics committees: A worldwide development. *Journal of Medical Ethics* 27(Suppl. 1): i1.
Sorta-Bilajac, I., K. Bazdaric, B. Brozovic, and G.J. Agich. 2008. Croatian physicians' and nurses' experience with ethical issues in clinical practice. *Journal of Medical Ethics* 34(6): 450–455.
Spielman, B. 2001. Has faith in health care ethics consultants gone too far? Risks of an unregulated practice and a model act to contain them. *Marquette Law Review* 85(1): 161–221.
Spielman, B., and G.J. Agich. 1999. The future of bioethics testimony: Guidelines for determining qualifications, reliability, and helpfulness. *San Diego Law Review* 36(4): 1043–1075.
Spike, J. 2009. Resolving the vexing question of credentialing: Finding the Aristotelian mean. *HEC Forum* 21: 263–273.
Steinkamp, N. 2003. European debates on ethical case deliberation. *Medicine, Health Care and Philosophy* 6(3): 225–226.
Swenson, M.D., and R.B. Miller. 1992. Ethics case review in health care institutions: committees, consultants, or teams? *Arch Intern Med* 152: 694–697.
Swiderski, D.M., K.M. Ettinger, M. Webber, and N.N. Dubler. 2010. The clinical ethics credentialing project: Preliminary notes from a pilot project to establish quality measures for ethics consultation. *HEC Forum* 22: 65–72.
Tan, S.Y. 2002. Hospital ethics committees: Will America's model work in Asia? *Annals of the Academy of Medicine, Singapore* 31: 808–812.
Tarzian, A.J. 2009. Credentials for clinical ethics consultation – Are we there yet? *HEC Forum* 21: 241–248.
Tarzian, A.J. 2013. Health care ethics consultation: An update on core competencies and emerging standards from the American Society for Bioethics and Humanities' core competencies update task force. *American Journal of Bioethics* 13: 3–13.
Task Force on Standards for Bioethics Consultation. 2010. *Core competencies for health care ethics consultation*, 2nd ed. Glenville: American Society for Bioethics and Humanities.
Thomasma, D. 1991. Why philosophers should offer ethics consultations. *Theoretical Medicine* 12(2): 129–140.
Toulmin, S.E. 1981. The tyranny of principles. *Hastings Center Report* 11: 31–39.
Toulmin, S.E. 1982. How medicine saved the life of ethics. *Perspectives in Biology and Medicine* 25(4): 736–749.
Wikler, D. 1982. Ethicists, critics, and expertise. *Hastings Center Report* 12(3): 12–13.
Wildes, K. 1997. Healthy skepticism: The emperor has very few clothes. *Journal of Medicine and Philosophy* 22(4): 365–371.
Wray, E. 2002. The Padua bioethics service: A model of excellence in clinical ethics? *Bulletin of Medical Ethics* 183: 13–15.
Yoder, S.D. 1998. The nature of ethical expertise. *Hastings Center Report* 28: 11–19.
Youngner, S.J., D. Jackson, C. Coulton, et al. 1983. A national survey of hospital ethics committees. *Critical Care Medicine* 11: 902–905.
Zaner, R.M. 1984. Is ethicist anything to call a philosopher? *Human Studies* 7: 71–90.

Chapter 3
Recognizing the Difference that Faith Makes: H. Tristram Engelhardt, Jr., on Life-Ending Medical Interventions

John C. Moskop

In his long and prolific career, bioethics pioneer H. Tristram Engelhardt, Jr. has published books, articles, and book chapters on virtually every significant topic in the field of bioethics. Included in this prodigious scholarly output are multiple articles and book chapters examining the justifiability of medical interventions that result in death, most notably abortion, euthanasia, physician-assisted suicide, and the limitation of life-prolonging treatment. A striking feature of Engelhardt's publications on life-ending medical actions is the fact that they articulate two very different positions on the morality of these actions. In a series of articles and chapters published before 1990, Engelhardt defends the rights of physicians and patients to employ these life-ending medical interventions, with limited exceptions (Engelhardt 1973, 1974, 1975a, b, 1976, 1977, 1986b, c, d, 1989; Moskop and Engelhardt 1979). In articles and chapters published after 1990, he condemns the use of these life-ending medical interventions, again with limited exceptions (Engelhardt 1993, 1998, 2000d, e; Engelhardt and Iltis 2005).

In this contribution, I will describe first the former and then the latter position Engelhardt defends on the morality of life-ending medical interventions, and the reasons he marshals in support of those two positions. Then I will compare the two positions and consider the extent to which they are inconsistent. Finally, I will reflect on the significance of the dramatic shift in Engelhardt's writings on this subject and its implications for moral conclusions about these interventions.

J.C. Moskop, Ph.D. (✉)
Department of Internal Medicine, Wake Forest School of Medicine, Winston-Salem, NC, USA

Clinical Ethics Committee, Wake Forest Baptist Medical Center, Medical Center Boulevard, Winston-Salem, NC 27157, USA
e-mail: jmoskop@wakehealth.edu

L.M. Rasmussen et al. (eds.), *At the Foundations of Bioethics and Biopolitics: Critical Essays on the Thought of H. Tristram Engelhardt, Jr.*, Philosophy and Medicine 125, DOI 10.1007/978-3-319-18965-9_3

3.1 The Former Position

In a series of publications that appeared during the first two decades of his career, Engelhardt develops and defends a uniform and linked set of conclusions about multiple life-ending medical interventions. His conclusions in these publications rely heavily on two key premises. The first of the two premises is an appeal to the Kantian doctrine of the unique moral status of persons, understood as self-conscious and rational free agents. After a quotation from Kant's *Foundations of the Metaphysics of Morals* on the distinction between persons and things, Engelhardt (1976, 52) asserts, "I will follow Kant on this point, and will maintain that persons do have rights which are not reducible to values and goods and, moreover, that only rational beings are persons." Foremost among these rights for Kant and Engelhardt is respect for the freedom of persons to make and carry out their own moral choices. This respect, Engelhardt argues, is a necessary condition for the establishment of a peaceable society in which moral agents claim reciprocal rights to respect for their own choices and actions.

Engelhardt's second key premise is a basic distinction between human biological and human personal life. Engelhardt cites the recent (in the 1970s) recognition in medicine and law of neurologic criteria for the determination of death as compelling evidence for this distinction. He points out that persons declared dead based on the irreversible loss of brain function may still retain multiple biological functions, including respiration, circulation, digestion, and even reproductive capacity (Engelhardt 1977, 1986b). As Engelhardt (1977, 18) explains,

> The issue of brain-death thus introduces two levels of life—the vegetative life of brain damaged bodies, and the full personal life of normal adult humans. The first level, the vegetative life of a brain-dead body, appears as the remnant part of a personal human life, but distinguishable and separable from such personal human life.

Neurologic criteria for death identify the death of the person as the irreversible loss of mental function, established by the death of the brain, the essential physical substratum of mental activity. This distinction, Engelhardt maintains, justifies very different treatment by physicians of persons and of biologically alive human bodies after neurologic determination of personal death.

Analogous to the regression from personal to mere biological human life recognized via the loss of brain function, Engelhardt (1977) argues, is a progression from biological to personal life in human ontogeny, the pre- and post-natal development of human beings. Engelhardt points out that human biological development is continuous, from gametes to zygotes, fetuses, infants, children, and adults, but that human personal life, in the Kantian sense of self-conscious rational agency, does not apply to all of these stages of human development. Human gametes, zygotes, and (at least) early fetuses lack consciousness, and fetuses and infants lack self-consciousness, the recognition of self as a continuing subject of experiences. "Vegetative, sentient, and even conscious life do not either mark off humans as a morally distinct class of entities, nor provide the basis for a moral life. Only self-conscious life can be responsible" (Engelhardt 1977, 19). We do not, Engelhardt concludes, owe the same duties to individuals in the biological but non-personal early stages of human life as we do to persons in the morally significant Kantian sense.

Engelhardt recognizes that scholars like Michael Tooley have appealed to this Kantian conception of personhood to argue that human fetuses and infants lack rights to life, and to conclude that abortion and infanticide are therefore morally permissible practices. Engelhardt resists Tooley's conclusion, however. Instead, he introduces a second distinction, between "strict" and "social" senses of personhood. By 'strict personhood,' Engehardt means the status of self-conscious, rational agents who are able to make moral choices, claim moral rights, and assume moral responsibilities. These abilities remain the primary source of moral worth for Kant and for Engelhardt. By 'social personhood,' Engelhardt means the status of some human beings, including post-viability fetuses and infants, who have the ability to enter into social relationships with persons in the strict sense. Engelhardt (1977, 25) explains the moral import of this social sense of personhood as follows:

> I have in mind here the infant's role in the mother-child relation which is a necessarily social, not a biological relationship (e.g., the mother-fetus relationship). In this role the infant is treated as if it were a person in virtue of this human social relation. The status 'person' is imputed to it because of the value of the mother-child relationship, and to protect the integrity of similar social relations and the role 'person'. Here, I believe, lies a conservative argument against positions such as Michael Tooley's defense of infanticide.

From the two central premises described above, Engelhardt draws conclusions about a variety of life-ending medical actions. Regarding abortion, Engelhardt (1974, 233) strongly affirms the rights of women to choose to continue or to terminate a pregnancy:

> The rights of the mother regarding abortion are paramount. After all, she is the only actual person involved. As a fully actual, developed rational human, it is she, surely, who has prima facie rights over her own body. Consequently, it is to her that one owes overriding obligations.

These rights take precedence over claims made on behalf of pre-viable fetuses, since those fetuses have only human biological, not personal life. As noted above, Engelhardt (1974) ascribes an extended "social" sense of personhood to viable fetuses, based on their ability, after birth, to survive, assume the new and valuable role of "child," and enter into social relationships with family members and others. This extended sense of personhood, he argues, can justify the recognition of obligations to viable fetuses and the general proscription of abortion after fetal viability. Because viable fetuses are not yet persons in the full sense, however, Engelhardt (1975a) would permit their abortion in cases of profound fetal disability or of risk to the health of the woman.

Engelhardt (1975b, 1986b) appeals to the same social sense of personhood to justify both the provision of needed medical care for infants and young children and a general prohibition of infanticide. He points out that infants and young children cannot make medical treatment choices for themselves, and so we authorize parents to make those decisions on behalf of their children. He argues that these decisions should be based on assessment of the future quality of life of the child and the costs of the contemplated treatment, and that parents should refuse life-sustaining treatment when continued survival would constitute an injury rather than a benefit to the child. Engelhardt recognizes that relief of severe suffering could in principle also justify a limited practice of infant active euthanasia. He does not endorse this practice, however, on the grounds that it may be susceptible to significant abuse: "it is doubtful that active euthanasia could be established as a practice without eroding and endangering children generally, since … children cannot speak in their own behalf" (Engelhardt 1975b, 188).

The centrality of respect for the free choices and actions of adults as rational moral agents directs Engelhardt's conclusions regarding the rights of competent patients to choose life-ending medical interventions. Engelhardt (1986c) endorses robust rights to free and informed consent and refusal of treatment, including life-sustaining treatment. He argues further that the paternalistic interests of states do not provide sufficient reason to override personal freedom of choice and action by prohibiting competent persons from committing suicide or prohibiting others from assisting a competent person's suicide or administering active euthanasia at a competent person's request. Engelhardt acknowledges that most individuals who attempt suicide do so as a result of serious mental illness, and he asserts that "they should receive psychiatric treatment, not aid in committing suicide." He goes on, however, to observe that

> … there are debilitating diseases and final stages of decrepitude where competent men and women may reasonably decide that enough is enough. One may not wish to live out the last few weeks of a death from cancer complicated by multiple metastases and multiple organ difficulties. One may not wish to be warehoused in a nursing home or other facility where one will have very little, if any, conscious awareness of the circumstances of one's life. (Engelhardt 1986c, 315)

Physician-assisted suicide or euthanasia may be a "reasonable alternative" for some patients, according to Engelhardt (1986d, 108), "because medical science is often able to save life at a quality that the patient may not wish to accept or simply to prolong the process of dying."

3.2 The Latter Position

In multiple articles and book chapters published after 1990, Engelhardt develops and defends a position that condemns all medical interventions whose intention is to cause the death of a human being. Not surprisingly, the arguments Engelhardt employs to defend this latter position are very different from those he offered in defense of the former position. The latter position is firmly rooted in his affirmation of a "traditional Christian" faith. Engelhardt (2000c) describes traditional Christianity as a faith that adheres to the unchanging tradition of belief and practice of the first millennium of Christianity, guided by the Holy Spirit, explicated by the Fathers of the early Church, exemplified by the saints, and interpreted today by the spiritual fathers of the Orthodox Christian Church. He clearly distinguishes traditional Christianity from both "secular morality" and "post-modern" or "post-traditional" Christianity. Secular morality, as described in these publications, is essentially Engelhardt's former position, with its overriding emphasis on respect for the freedom and dignity of self-conscious moral agents. Unlike traditional Christianity, "post-traditional" Christianity, as practiced in both the contemporary Roman Catholic and many Protestant churches, does not view its traditions and doctrines as unchanging, but rather allows or even encourages ecumenism and "doctrinal innovation," based on discursive reasoning or on Biblical exegesis. Engelhardt (2000b, 148) envisions the possibility of a unified, ecumenical future coalition of Christian churches, presided over by the Roman Catholic pope, and offers the following comment:

> Such a post-Enlightenment, post-Christian Christianity offers many the hope, if not for perpetual peace, then at least for the resources for better mutual understanding and peaceable collaboration in a thick community of common moral commitments. The libertarian and the traditional Christian can only regard this moral and political ideal with horror and foreboding.

In stark contrast to both secular morality and post-modern Christianity, Engelhardt (1998, 148) explains, traditional Christianity affirms its own distinctive epistemology, metaphysics, axiology, and account of history. Its overriding moral imperative is the pursuit of holiness as the single path to union with God and eternal salvation. All choices and actions, therefore, must be evaluated based on their contribution to or interference with holiness. To pursue the path to holiness, Christians must turn away from the sins of pride and passion and submit themselves to the will of God, living lives of asceticism, prayer, repentance, almsgiving, and obedience. Faithful practice paves the way for divine grace and illumination, a form of spiritual knowledge that transcends that of secular human discursive reasoning (Engelhardt 1998, 150). Engelhardt describes this spiritual knowledge as "noetic," that is, non-discursive, experiential, mystical, and grounded in a personal relationship with God. He elaborates on this noetic sense of knowledge in the following passage:

> Noetic experience is the sufficient condition required to establish a Christian bioethics, as well as to identify true theologians. Theologians *sensu stricto* are those who know in a direct, non-discursive fashion the content of Christian morality. Only a direct noetic experience of God's transcendent reality can bring human knowledge beyond the sphere of immanence and provide access not only to ultimate personal meaning, but to the experience of the canonical character of a morality. Christianity brings the indwelling of God, so that spiritual comprehension is achieved. (Engelhardt 2000c, 169)

Traditional Christianity's understanding of the moral life, as summarized above, has direct implications for the role of health care in general, and for the use of various life-ending medical interventions. Health care, like any other human activity, must be evaluated according to its effects on the overriding moral Christian imperative, the pursuit of holiness. People should accept health care whenever it contributes to holiness. This can occur in many ways, as, for example, by removing physical or psychological barriers to prayer or to participation in the liturgy and sacraments. People should refuse health care whenever it interferes with the pursuit of holiness.

This can also occur in many ways, as, for example, by refraining from the use of medical aids that enable a person to engage in illicit sexual activity. Engelhardt observes that the use of life-prolonging medical treatments near the end of life can have either effect, depending on the specific situation and on the patient's intention. Prolongation of life by medical means may, for example, give a dying person a valuable opportunity for reflection, confession, and Holy Communion before death. In contrast, the use of burdensome and invasive treatments in order to prolong life at all costs may constitute an idolatrous pursuit of physical life. Engelhardt (2000e, 317) explains that

> … medicine should not be used if it significantly distracts from our life of prayer or brings us to being obsessed with preserving this life… It is not that such treatments *may* be withdrawn or withheld. They *should* be withdrawn or withheld. The postponement of death, the avoidance of suffering, and the correction of disabilities should not become all-consuming projects.

With co-author Ana Iltis, Engelhardt describes the traditional Christian evaluation of life-ending medical actions in the following unambiguous and comprehensive way: "The Christian pursuit of holiness through humble submission to God excludes intentionally bringing about death through either omission or commission" (Engelhardt and Iltis 2005, 1045). Taking control over the time or manner of one's own death or of the death of another is viewed as contrary to the example of Christ's humble acceptance of crucifixion and to the suffering and martyrdom of the saints. For the Christian, in other words, the proper approach to dying is not a matter of self-determination, dignity, and comfort, but rather of submission to the will of God, humility, repentance, prayer, and, if necessary, acceptance of suffering as expiation for sin.

This prohibition of killing in any form and for any reason includes abortion, physician-assisted suicide, euthanasia, and forgoing life-prolonging treatment with the intention of hastening death. Engelhardt (1998, 2000d) cites authoritative testimony condemning these practices from the canons of early Church Fathers. He also elaborates on how this general prohibition applies to a variety of specific situations. Engelhardt asserts that the early Christian tradition does not recognize exceptions to the proscription of abortion, and he follows that example, concluding that abortion cannot be condoned in cases of rape or serious fetal handicap. Even in cases of spontaneous abortion and removal of a cancerous uterus to save the life of a pregnant woman, he concludes that "causal proximity in the death of another must be regarded as an involuntary evil for which spiritual therapy should be sought" (Engelhardt 2000d, 277–278).

Engelhardt (2000e, 327–332) also takes pains to respond to a suggestion by David Thomasma (1998) that assisted suicide and euthanasia may be justifiable in a Christian context on the grounds that they are similar to Christ's acceptance of crucifixion and the acceptance of death by the Christian martyrs. He argues that Thomasma's suggestion represents a fundamental misunderstanding of the Christian tradition. Christ's humble submission to crucifixion and the acceptance of martyrdom by the saints are holy acts of obedience to God's will. Suicide and euthanasia, in contrast, are sinful acts of pride that violate the divine prohibition of homicide and separate the sinner from union with God (Engelhardt 2000e). Engelhardt concludes that, when the suffering of a terminal illness is intractable, rather than resorting to suicide or euthanasia, Christians should accept that suffering as an opportunity for repentance and for emulating the example of Christ.

3.3 Comparing the Positions

On one level, the two positions defended by Engelhardt at different times in his career are clearly inconsistent. They articulate and defend nearly diametrically opposed views on the moral warrant for life-ending medical interventions. As described above, the former position asserts that life-ending medical interventions are reasonable alternatives that should be permitted, and the latter position asserts that they are immoral practices that should be condemned. The two positions differ even about the moral significance of the topic. In 1986, for example, Engelhardt (1986b, 242) asserts: "Despite its capacity to attract major public interest and sustain bitter public debate, abortion is not a serious moral issue." Seven years later, in a brief response to a review of Christian concepts of personhood, Engelhardt (1993, 420) remarks parenthetically that "the author of this response is a committed Christian who recognizes the profound wrongness of abortion."

At a different level, however, the two positions Engelhardt adopts may be understood as consistent with one another. In the final pages of the first edition of *The Foundations of Bioethics* (1986a, 385), Engelhardt summarizes his project in that volume as offering "a vision" of "a secular pluralist ethic." He argues that this secular pluralistic ethic, based on mutual respect among moral agents, is unavoidable, but recognizes that it is severely limited, that it cannot resolve the many moral disagreements that exist between members of different concrete moral communities, be they committed Christians, Jews, Hindus, Buddhists, Communists, or others. He acknowledges further that the fullness of the moral life can only be experienced as a member of one of these concrete moral communities. The second edition of *The Foundations of Bioethics* (1996) retains the arguments for the unavoidability of affirming a secular pluralistic ethic, but emphasizes much more explicitly the relative poverty of that ethic and the significance of membership in a community of substantive shared moral beliefs. This strong emphasis on the poverty of the secular pluralistic ethic is clearly expressed in following passage from the Preface to the second edition:

> If one wants more than secular reason can disclose—and one should want more—then one should join a religion and be careful to choose the right one. Canonical moral content will not be found outside of a particular moral narrative, a view from somewhere. Here the reader deserves to know that I indeed experience and acknowledge the immense cleft between what secular philosophical reasoning can provide and what I know in the fullness of my own narrative, but realize it cannot be given by reason, only by grace. I am, after all, a born-again Texas Orthodox Catholic, a convert by choice and conviction, through grace

> and in repentance for sins innumerable (including a first edition upon which much improvement was needed). My moral perspective does not lack content. I am of the firm conviction that, save for God's mercy, those who willfully engage in much that a peaceable, fully secular state will permit (e.g., euthanasia and direct abortion on demand) stand in danger of hell's eternal fires. (Engelhardt 1996, xi)

In *The Foundations of Christian Bioethics* (2000a), Engelhardt undertakes a different project, that of articulating the ethics and bioethics of the particular concrete moral community to which he belongs, the community of traditional (Orthodox) Christianity. From the perspective of traditional Christianity, Engelhardt offers a highly critical appraisal of the values of secular pluralistic societies, but he does not repudiate his earlier arguments for the moral imperative of respecting the freedom of other moral agents, including "moral strangers" who affirm very different moral beliefs and values. Despite a strong conviction that life-ending moral interventions are morally wrong, therefore, traditional Christians may still acknowledge a duty of forbearance, allowing others who hold different beliefs to employ those interventions. In turn, Christians may claim respect from others for their decisions not to accept or participate in these interventions. Engelhardt's former position can then be understood as the claim that life-ending medical interventions should be permitted for those who choose them and should be tolerated by others. His latter position can be understood as the claim that those who embrace the faith of traditional Christianity must reject life-ending medical interventions as incompatible with the central beliefs and values of their faith. Understood in this way, the two positions are not inconsistent.

3.4 Final Comments

Even if the two views on the morality of life-ending medical interventions described above are consistent parts of Engelhardt's broader moral vision, the shift in emphasis in his writing from endorsement of individual rights to employ these interventions to religiously inspired condemnation of any form of medical killing is a remarkable development in his thought. Engelhardt has been one of the most original, articulate, intellectually rigorous, and prolific contributors to the field of bioethics over the past four decades. Younger scholars may not be as familiar with his publications from the 1970s and 1980s articulating the former position described above as they are with his more recent writings defending the latter position. Those who share with Engelhardt a commitment to the traditional Christian faith he confesses are likely to find his thorough examination of early and contemporary Christian sources and his arguments based on those sources both instructive and persuasive.

It is less clear, however, how those who are not members of the traditional Christian faith community are likely to respond to Engelhardt's latter position. Engelhardt (2000d, 283) recognizes and acknowledges that the traditional Christian bioethical arguments and conclusions he offers will seem strange and "radically out of step" for those who are non-believers or who are post-traditional Christians. He points out, moreover, that traditional Christians reject the ecumenical efforts of post-modern Christian churches to find common ground with those of other faith traditions, based on discursive reasoning or on shared values (Engelhardt 2000c, 202–208). Rather, traditional Christianity challenges all to pursue a life of repentance, asceticism, prayer, and obedience, rejecting human reason and human passions in favor of achieving a spiritual knowledge of God that is personal, experiential, and ineffable. Traditional Christianity asserts that the stakes in this choice are of the highest importance, namely, eternal salvation or eternal damnation. Nevertheless, because the gulf between the epistemological, metaphysical, and moral commitments of traditional Christians and others is very wide, accepting this challenge will require of those others a virtually blind and therefore difficult leap of faith. Even though most bioethics scholars and students will not make this leap, Engelhardt's eloquent expression of traditional Christian bioethics should command both their attention and their respect for his position.

References

Engelhardt Jr., H. Tristram. 1973. Viability, abortion, and the difference between a fetus and an infant. *American Journal of Obstetrics and Gynecology* 116: 429–434.
Engelhardt Jr., H. Tristram. 1974. The ontology of abortion. *Ethics* 84: 217–234.
Engelhardt Jr., H. Tristram. 1975a. Bioethics and the process of embodiment. *Perspectives in Biology and Medicine* 18: 486–500.
Engelhardt Jr., H. Tristram. 1975b. Ethical issues in aiding the death of young children. In *Beneficent euthanasia*, ed. M. Kohl, 180–192. Buffalo: Prometheus Books.
Engelhardt Jr., H. Tristram. 1976. On the bounds of freedom: From the treatment of fetuses to euthanasia. *Connecticut Medicine* 40: 51–55.
Engelhardt Jr., H. Tristram. 1977. Ontology and ontogeny. *The Monist* 60: 16–28.
Engelhardt Jr., H. Tristram. 1986a. *The foundations of bioethics*. New York: Oxford University Press.
Engelhardt Jr., H. Tristram. 1986b. The endings and beginnings of persons: Death, abortion, and infanticide. In *The foundations of bioethics*, 202–249. New York: Oxford University Press.
Engelhardt Jr., H. Tristram. 1986c. Free and informed consent, refusal of treatment, and the health care team: The many faces of freedom. In *The foundations of bioethics*, 250–335. New York: Oxford University Press.
Engelhardt Jr., H. Tristram. 1986d. Suicide and the cancer patient. *CA—A Cancer Journal for Clinicians* 36: 105–109.
Engelhardt Jr., H. Tristram. 1989. Fashioning an ethic for life and death in a post-modern society. *Hastings Center Report* 19(1(suppl)): 7–9.
Engelhardt Jr., H. Tristram. 1993. Personhood, moral strangers, and the evil of abortion: The painful experience of post-modernity. *Journal of Medicine and Philosophy* 18: 419–421.
Engelhardt Jr., H. Tristram. 1996. *The foundations of bioethics*, 2nd ed. New York: Oxford University Press.
Engelhardt Jr., H. Tristram. 1998. Physician-assisted suicide reconsidered: Dying as a Christian in a post-Christian age. *Christian Bioethics* 4: 143–167.
Engelhardt Jr., H. Tristram. 2000a. *The foundations of Christian bioethics*. Lisse: Swets & Zeitlinger.

Engelhardt Jr., H. Tristram. 2000b. Christian bioethics as a human project: Taking immanence seriously. In *The foundations of Christian bioethics*, 127–155. Lisse: Swets & Zeitlinger.
Engelhardt Jr., H. Tristram. 2000c. Bioethics and transcendence: At the heart of the culture wars. In *The foundations of Christian bioethics*, 157–231. Lisse: Swets & Zeitlinger.
Engelhardt Jr., H. Tristram. 2000d. Procreation: Reproduction, cloning, abortion, and birth. In *The foundations of Christian bioethics*, 233–308. Lisse: Swets & Zeitlinger.
Engelhardt Jr., H. Tristram. 2000e. Suffering, disease, dying, and death: The search for meaning. In *The foundations of Christian bioethics*, 309–351. Lisse: Swets & Zeitlinger.
Engelhardt Jr., H. Tristram, and Ana S. Iltis. 2005. End-of-life: The traditional Christian view. *Lancet* 366: 1045–1049.
Moskop, John, and H. Tristram Engelhardt Jr. 1979. The ethics of suicide: A secular view. In *Suicide: Theory and clinical aspects*, ed. L.D. Hankoff and B. Einsidler, 49–57. Littleton, MA: PSG Publishing.
Thomasma, David C. 1998. Assisted death and martyrdom. *Christian Bioethics* 4: 122–142.

Chapter 4
The Foundations of Secular Bioethics

Stephen Wear

As long as we stick to secular bioethics, the basic view of H. Tristram Engelhardt, Jr. is quite clear and simple: given the failure of the Enlightenment to rationally secure a content-filled, lexically ordered secular morality, we are left with a bare-bones sort of ethics of permission between moral strangers (Engelhardt 2000). This is all very unfortunate, as Engelhardt himself repeatedly notes, in that it utterly fails to provide ultimate meaning and specific moral guidance to people within the often bewildering and tragic realm of health care. But whereas many other commentators rail against such a constrained view of bioethics, he commends it as all that is possible from a secular perspective.

Engelhardt has, in many ways, made a career out of retailing this skeptical vision of the limited possibilities of secular bioethics and the underlying critique that supports it. While others run out their content filled visions of bioethical "truth" from whatever perspective, be it pragmatic, communitarian, Georgetown mantra, etc., there is Engelhardt, over and over again, pointing out the unjustified assumptions advanced, the questions begged, infinite regresses triggered, and the overall "news from nowhere" that he sees much of contemporary bioethics as offering. In a nutshell, the Enlightenment project has failed and perhaps nowhere is this more apparent than in contemporary bioethics, where so many issues of profound moment to so many people enjoy no real consensus whatsoever.

In what follows, I will address three interrelated issues: (1) is the secular, libertarian ethics of permission that Engelhardt sees us left with as impoverished as he and others hold it is – is this really all a secular bioethics can say?; (2) does this

S. Wear, Ph.D. (✉)
Departments of Medicine, Gynecology-Obstetrics and Philosophy,
The University at Buffalo, Buffalo, NY USA

Center for Clinical Ethics, VA Medical Center,
3495 Bailey Avenue, Buffalo, NY 14215, USA
e-mail: wear@buffalo.edu

L.M. Rasmussen et al. (eds.), *At the Foundations of Bioethics and Biopolitics: Critical Essays on the Thought of H. Tristram Engelhardt, Jr.*,
Philosophy and Medicine 125, DOI 10.1007/978-3-319-18965-9_4

libertarian view somehow naturally tend to evolve into the liberal cosmopolitan view that he abhors as much as I do?; and (3) is there nothing between the libertarian and liberal cosmopolitan views within which those who lack the content-filled sort of vision that Engelhardt aspires to might rest relatively satisfied? I will proceed to address these issues first by briefly describing my own personal perspective on all this, initially regarding the urge toward transcendence that I believe figures prominently in Engelhardt's thought, an urge which I simply do not share. The overall point here will be that as I can not follow where Engelhardt has led my task will then be to reflect on whether the result, for those of us stuck in the realm of the immanent, is really as impoverished and unsatisfying as Engelhardt believes it to be.

4.1 Immanence Versus Transcendence

For all his advocacy of a secular ethics of permission, one gets the sense that Engelhardt, like many other commentators, does not think people can live with such an ethics, however much he feels they must when faced with moral strangers. Other commentators often want to emphasize that a true appreciation of the realm of health care, with its asymmetrical power relationships, the diminished competence of patients due to illness, the bewildering medical assembly line, etc., calls for a great deal beyond a procedural ethics of permission. Without robust commitments to beneficence and virtue on the part of health care providers, patients are just too vulnerable and not seldom prey to those who provide their health care. Similarly, the libertarian political account that undergirds Engelhardt's bioethics is seen by others as completely inadequate to support and fund the equality of access that they see as absolutely necessary for an ethical health care system. Engelhardt's point, however, is that as long as it is moral strangers who are interacting an ethics of permission is all that can be justified, and the political claims of equality and justice on moral strangers are similarly quite minimal. An impoverished setup perhaps, but anything more flies in the face of the deep seated diversity of moral views that exist in post-modern society.

But there remains a sense of the transcendent in Engelhardt, as one reads him closely over the years, that appears to lurk in the immediate background of all he says. That is: Engelhardt repeatedly talks of the deeper hunger for more than the minimal secular ethic that he otherwise supplies (Engelhardt 2000, 209); put simply "the immanent cannot still the hunger for the transcendent" (Engelhardt 2000, 313). He clearly believes that without a belief in God there is no ultimate personal meaning to life, no adequate account of suffering, etc. Not that he straightforwardly asserts that there is some clear logical or existential necessity operating here; at one juncture, he asserts that the inadequacy of secular bioethics is only truly appreciated from a religious (read: Christian) perspective (Engelhardt 2000, 73). But however much his minimal ethics may work as a way to deal with moral strangers, he simply does not see it as being adequate for one's personal ethics; it is not a place where one can live, at least comfortably. The point is that however much secular ethics must be seen as radically limited, ways of life that people can rest in are not, can not be. There seems to be a psychological "necessity" here for Engelhardt, however much other people may more or less still or ignore such deep urges.

By way of initial response: as Engelhardt often appeals to personal knowledge and experience, I will do the same. That is: whereas Engelhardt at various junctures lets us know that he is a born again Texican, I am a lapsed Unitarian from New Hampshire. Unitarian by raising because of a childhood experience of a rather locally grown "Christianity" with not a little deference to our neighbors Emerson and Thoreau down the road. Lapsed because when I finally got the nerve to ask the minister what God was, his response of "whatever you think he is" was so unsatisfying that I quickly concluded that Unitarianism merited neither further study nor allegiance.

So, as far as the hunger for the transcendent goes for me, that was that, then and now. Whatever ethics I have had or have seems to have welled up out of the Yankee farmer background I was nurtured in, with a lot of emphasis on self-reliance and personal responsibility, not much regarding one's obligations to strangers of a positive variety, and as far as the urge to or need for community goes, agreement with another neighbor, viz. the Yankee poet Robert Frost, that "stone walls good neighbors make".

I have certainly attempted to entertain something beyond all this, from graduate study that focused heavily on medieval philosophy, to various excursions into spiritual realms that might be expected of a college student of the 1960s. Particularly to the point here, perhaps, I once even joined Professor Engelhardt in attending an Orthodox Christian service, and was certainly impressed. Unlike what I have seen when I have had occasion to attend various Christian services over the years, this was clearly the real thing. Rather than the usual "Mary is his mother/vote democrat always and often" fare, the incense was smoking, religious images and icons were everywhere, and the priest was deadly serious. Like I said, serious stuff....the real thing.

But a little too real for a lapsed Unitarian. The images and icons served instead to reawaken my iconoclast heritage, and that heritage is equally insistent about not allowing priests, or presbyters, or elders between the individual believer and God. Nor did this experience rekindle any religious urgings in me. It instead instructed me that if I was ever to undergo some sort of religious transcendence it would not be in such a place. Where that place is, I have no idea; nor am I currently seeking it.

So that leaves me, and many other people I suspect, stuck in the realm of immanence, with only our brains, and experience of life, to guide us. In the rest of this essay, I will attempt to respond to Engelhardt's portrayal of its impoverished character, among other things, by arguing that it is not necessarily as impoverished and unsatisfying as he seems to think it is.

4.2 Reflections on the Enlightenment Project

Such a background has always made me a rather mixed bag in relation to Engelhardt, who has been variously my mentor, colleague, and good friend over the years. We have always disagreed sharply in a sectarian way, viz. about the significance of the War of Southern Insurrection (mislabeled the American Civil War by some, the War of Northern Aggression by others), Engelhardt believing the southern cause a righteous one, whereas I think it is most unfortunate that Sherman was not allowed to make a few more passes through the south to make them all howl that much louder. I suspect when he thinks of Picket's charge at Gettysburg, he sees glory in the Southern ranks; I see it in the rows of cannon with which my ancestors shredded those rebel ranks with grapeshot.

But, beyond this, whereas Engelhardt has always seemed to see his ethics of permission as somehow unfortunate and not a place one can live, I always thought he had it pretty much accurate as to where I and many others do live. That is: my own upbringing, and experience of life, has instructed me that I should assume other people are in fact moral strangers to me, whatever cultural community I may happen to seem to share with any individual. That we may happen to share moral views thus needs confirmation to the extent we interact and this explicitly. In other words: my experience is that knowing another's religion, cultural background, etc. is unlikely to indicate what that person's moral views are, within health care or outside it. And this is the case with other lapsed Unitarians, or even fellow Yankees. There is just no telling......

This appears to leave me, and many others, in the position of being "libertarian cosmopolitans," as Engelhardt labels the syndrome. That is: a content-filled, lexically ordered vision of the right and the good is not available to us. The Enlightenment having failed to provide this via reason unaided, and the necessary relief of religion being rejected, then our ethics can be procedural only, an ethics of permission which wholly lacks specific guidance within health care as well as without. And, according to Engelhardt, this is not a good result:

> What if the chaos of the moral life is such that many people possess no coherent understanding of the right, the good, and virtue? Thoroughly post-modern persons that not only have no moral narrative to share with others but also no coherent moral account of their own lives are exactly such individuals. Life happens to them, including their passions. They are persons without a moral plot for their own bibliographies. They have desires, impulses, urges, needs, wants, and concerns, but no moral projects that shape and unite their lives as a whole. In particular they have no coherent sense of good and evil to structure their life projects. This does not mean that such persons lack coherence to the point of suffering from a moral thought disorder disabling them from acting as moral agents. They can quite coherently and accountably seek satisfaction, fulfillment, and happiness. They simply lack a coherent substantive personal moral narrative. Instead their life is a sequence of happenings. (Engelhardt 2000, 137)

Whoa! There appears to be a great deal packed into the notion of coherence here. And what does it mean to say, in the same breath, that such people "have no coherent sense of good and evil to structure their life projects" but concurrently allow that "they can quite coherently and accountably seek satisfaction, fulfillment, and happiness." How should we interpret the preceding?

First, Engelhardt clearly holds that only via a "noetic" experience of the foundations of Christian bioethics can the "difficulties besetting secular morality" be escaped. As he indicates, this is so because these foundations are: "(1) anchored in an experience of God, (2) apperceived as true, and (3) sustained in a community maintained in this experience over the centuries" (Engelhardt 2000, 168). In sum, without this noetic experience, any secular morality must be either a minimal ethics of permission, or if it presumes to more content, ends up offering the incoherent, question begging, news from nowhere that Engelhardt sees much of contemporary bioethics as offering.

Now, as already indicated, I and others do not have this noetic experience to rely on, nor are we willing to take Engelhardt's word for it. So if this is the real ground of the critique of secular ethics that Engelhardt offers, we might just as well shrug our shoulders and return to our own moral reflections, such as they are. At least we can "quite coherently and accountably seek satisfaction, fulfillment, and happiness."

But there is another barrel in Engelhardt's shotgun, viz. his portrayal of the failure of the Enlightenment project. And my main point about this particular salvo is as follows: Engelhardt's assertion of the failure of the Enlightenment project to secure a content-filled, lexically ordered vision of the right and the good from reason unaided by tradition or revelation takes on an altered significance once one appreciates that in the end he holds that only a noetic religious experience can succeed in such a project. That is: his view of the Enlightenment thus succeeds only by begging the question of what counts as success in ethics once it is severed from any appeal to revelation and tradition. If we instead, divorced from that tradition and revelation as some of us are, tired of priests and presbyters telling us how to act, etc., adopt the project of seeking to ascertain what a reason-based ethics might provide, however minimal, then success may well turn out to lie in a quite marginal, barebones sort of ethic, perhaps Engelhardt's ethic of permission. This would not then be a failure for those of us not vouchsafed Engelhardt's noetic experience. Nor would any such result be a failure because the expectations of the early Enlightenment thinkers were much too high. That reason unaided may well supply much less than religion does is not only not failure; it should have been expected.

My point, in the end, is that Engelhardt may well have let the cat out of the bag here, i.e., if Engelhardt's critique of the Enlightenment project ultimate rests on the fact that it could not provide the same sort of content-filled morality, on the same terms and principles, as the noetic Christian experience, then his critique is itself circular. Pursuit of the Enlightenment project might still legitimately continue, however humbled. It might just have to recognize that a more basic and humbler sense of what constitutes ethical justification will be part of the result. And, if we keep in mind Aristotle's dictum that one should not expect more of a type of inquiry than it is capable of providing, then this will not be failure either.

If the above rings true, and for those of us stuck in the realm of immanence without any such noetic instruction I believe it must, then I submit that in the end we must conclude that Engelhardt is not even playing the same game as the rest of us. If, for example, he wants to insist that any reason-based ethics must provide the same level and depth of satisfaction, logical and existential, that religion does then we simply need to part company with him, indicating that we are willing, and bound, to pursue whatever result we can manage. Failure is simply not shown, for us, if the result does not measure up to what he has found, which we have not. In the remainder of this essay, I will presume to summarize what I believe some of this legitimate result amounts to, first within the bioethical realm where moral strangers interact as patients and providers.

4.3 Justification and Belief in Clinical Bioethics

Stuck in immanence as some of us are, the primary issue thus comes to regard what a secular ethics can provide, with an expectation that the result will probably not approximate the sort of ordered content that Engelhardt has obtained for himself. And this question will, in part, turn on what constitutes legitimate justification and belief in bioethics. Here a further conjecture regarding Engelhardt may be helpful. That is: if we attempt to place Engelhardt as a philosopher, we might initially tend to think of him as some sort of chastened Hegelian. Aside from his well known Germanophile tendencies, his dissertation was on Hegel and he presents, in many ways, the aspect of the dedicated rationalist, however much the end result of that rationalism is severely limited. I believe we should, however, look further north from Germany for a philosophical soul mate for Engelhardt, viz. to Kierkegaard. In sum, where Kierkegaard spends most of his writing illustrating how no secular "way of life" can rest easy in itself, that all involve contradiction in some way or another, I submit that Engelhardt may be seen as engaged in exactly the same project regarding all forms of secular ethical argument. For both Engelhardt and Kierkegaard, the end result is that one is left to somehow jump to what one might rest easy in or, of course, remain in the realm of the incoherent and ultimately unsatisfying.

But the point of my argument thus far is, rather, that if we cannot jump with either of these thinkers, then we are stuck with coming up with the best account that we can (some accounts may be more adequate than others). And the search for such an account may well also involve rethinking what constitutes legitimate justification and belief in ethics, with the expectation that it may end up much less austere than what Engelhardt insists upon.

More specifically: take Engelhardt's reflection on Tom Beauchamp's attempt to hold that "there is a common morality shared by all." Engelhardt notes that Beauchamp "tries to make his case by listing fourteen rules he holds to be universal," such as "(1) Tell the truth. (2) Respect the privacy of others … (6) Do not kill" (Engelhardt 2000, 31). So far, so good, one would think, but not for Engelhardt. Aside from the obvious fact that no such principle enjoys unanimity across cultures, Engelhardt observes that Beauchamp is explicitly offering an "open-ended list of moral considerations that different persons from different communities may rank differently." And this will not do for Engelhardt; being open-ended, Beauchamp's principles "cannot provide any moral guidance" (Engelhardt 2000, 31).

What should we think of this? On the one hand, I believe Engelhardt is correct in making the important point that Beauchamp's system fails in an important sense in that many of the specific dilemmas that called for the growth of the field of bioethics are simply not going to be solved by such an approach. This is so because they are dilemmas precisely in that they involve head-on conflicts between the very principles that Beauchamp advocates. And thus, as Engelhardt endlessly points out, as Beauchamp and others cannot provide a lexical ordering of such principles whereby we might decide between them when they conflict, then his principalism does not provide the guidance sought. At most Beauchamp's principalism helps us better understand the nature of the dilemmas at hand, but cannot resolve them without begging the question as to the relative ranking of any of these principles, in general or in any particular case.

Fair enough; I believe Engelhardt has Beauchamp dead to rights in this regard. What Engelhardt fails to allow is that Beauchamp may still be meaningfully and helpfully offering a great deal of guidance about how to proceed in the usual case where we are not faced with intractable dilemmas, but are merely trying to fathom how we should usually act. And Beauchamp's acceptance of the open-ended character of all this does not constitute abject failure of his system; it is just a prudent recognition of its limits. Beauchamp may thus succeed in an importance sense for all that Engelhardt says.

Let us consider the possibilities here more specifically. Consider the notion of truth-telling. Now it is clear that truth-telling does not enjoy universal acceptance across cultures, so descriptive ethics will not help us. Further, à la Engelhardt, when truth-telling conflicts with some other basic principle, e.g. beneficence (as when we hesitate to tell a patient who is currently suffering a myocardial infarction that this is so for fear we will worsen the effect somehow), it may well be that reflection on the case at hand will leave us with a "six of one, half dozen of the other" result regarding the weight of the conflicting principles. But that hardly indicates that the principle of truth telling is not ethically mandatory in the usual case. That ethical guidance has limits does not mean that it does not provide guidance.

Now this is hardly the place to attempt to argue that truth-telling is a universal ethical maxim. Nor is this needed, as I believe sufficient argument has already been supplied. To my mind, Sissela Bok, in her classic piece on "Lies to the Sick and Dying," has done a quite adequate job of marshalling a "preponderance of the evidence" in favor of embracing truth-telling as an appropriate ethical rule, as what we should do in the usual case (Bok 1978). Admittedly it is primarily a consequentialist argument, and would surely remain open-ended, as Beauchamp's principles are. But still it does provide guidance in the usual case, and its consequentialist character seems hardly objectionable for those of us stuck in immanence and thus intent only to "coherently and accountably seek satisfaction, fulfillment, and happiness," as Engelhardt puts it.

More generally, and toward the issue of how impoverished any legitimate secular ethics must be, I submit that such consequentialist argumentation can and, in fact, has produced a full blown "system" of clinical ethics that is not impoverished, contains much guidance in the usual case, and legitimately guides the behavior of many in the clinical realm. As I have suggested elsewhere:

> Whether it be in the "Patients Bill of Rights" (American Hospital Association 2001) which is hung on the walls all over the institution, or in well known legal or reviewing body (i.e. JCAHO) statements that many staff can parrot, or in the policies and procedures of the institution, many bioethical issues appear in the form of established truths. The right of competent patients to informed consent and confidentiality, to be told the truth, to refuse any and all treatment, and so forth, are seen as guiding principles for everyone. They are seen as no more up for grabs, intellectually or morally, than the clinical guidelines for managing diabetes, or the proper methods for assessing and responding to multi-infarct dementia.
>
> Teaching within such a framework of accepted moral truths goes into considerable clinical and ethical detail before any true controversy arises. Explaining how competence, i.e. decision making capacity, should be evaluated, the rank-order of surrogates for incompetent patients, what sort of interventions require informed consent and what the elements of any such disclosures should be, and so on, are clearly delineated for staff who, in the main, want to know how to proceed in the usual case. One often spends time "talking tactics," whether this regards how (not whether) to tell bad news to patients and families, when and how to encourage patients to designate a surrogate or generate advance statements regarding extraordinary scenarios, or how to document what one has discussed or determined so it will be available and useful during subsequent care.
>
> The further point here is that, however much actual bioethics teaching at (or near) the bedside may incidentally key to, or be triggered by, actual controversy or disputes, a very broad and complex background of what is accepted or assumed guides most such discussion. Often these moral truths completely control what is then done, as when a patient's specific prior statements are held, per hospital policy, to overrule contrary wishes or views of family members, or when "self-destructive" patients who are deemed competent are allowed to continue those behaviors, however much staff and family would like somehow to prevent them from occurring. Even when policy does not clearly stipulate the proper ethical course, tactics aimed at dispute mediation – attempting to restore staff-patient/family interaction – or simply talking it out in the hopes that consensus might be reached, are utilized to resolve problems as often as anything one would recognize as ethical reasoning. That such latter tactics as converting dispute into consensus often work suggests that the accepted truths mentioned above are not only those of the institution and its staff, but are often shared equally by patients and their families. The diversity of ethical beliefs and values itself appears to be more a creature of the lecture hall, than a presence at the bedside (Wear 2002).

My basic point here was and is that there actually exists a canon of clinical ethics that is widely shared (at least in the West), provides specific and comprehensive guidance in the usual case, seems to satisfy most patients and providers, and enjoys substantial support from the last few decades of argument, debate and, experience in Western clinical circles.

I will further presume to suggest that this canon might well amount to a universal clinical ethics, for all that Engelhardt says. It is surely the case that certain other cultures do not subscribe to this canon, e.g., if Engelhardt is correct, that of Orthodox Christianity. I can not see, however, that this is a telling criticism. For once one marshals one's arguments, e.g. as Bok does in favor of truth-telling, one might legitimately conclude that different practices are, in fact, unethical, however much they enjoy the sanction of some particular culture. Not that we are obliged to be overly imperialistic about this; in the end, all we need seek is a system of ethical norms that can guide us in our interactions with moral strangers. They can then still inform us if they see things differently, e.g. that they would not want the sort of aggressive truth-telling that this Western canon advocates. But we still have guidance for ourselves and, via patient's bill of rights statements and otherwise, can advise moral strangers of how we intend to conduct ourselves absent contrary instructions from them.

Nor do I think such an approach is as impoverished and merely procedural as Engelhardt portrays it. We can surely presume to emphasize the need for beneficence and virtue in health care providers given the vulnerability and diminished competence seen in patients. More substantially, there are many situations where such providers might legitimately go beyond the merely procedural as when they perceive that patients are making uninformed, foolish, or needlessly tragic choices; as when a patient with moderate emphysema indicates he or she never wants to be on a "breathing machine," and the provider appropriately advises the patient that such an exclusion may well not be wise as they may later present with an eminently treatable acute infiltrate for which a short term trial intubation might successfully return them to base line. More substantially, I submit, there is a whole constellation of "standard practices" in medicine that are no less so because the occasional patient rejects them, and it seems fair to say that the debate of medical futility has clearly succeeded in identifying scenarios where medical aggressiveness is just not appropriate, however much certain patients or families may have different views, and however one wants to deal with the discrepancy. This canon, then, is neither impoverished, nor merely procedural, for all that Engelhardt says. Nor is it simply arbitrary "news from nowhere" however much unanimity about it does not exist, and it does not fully satisfy the hopes and expectations of the early Enlightenment.

4.4 Avoiding Liberal Cosmopolitanism

So much for the supposedly impoverished nature of secular bioethics. Insufficient for "transcendentlists" like Engelhardt, fine and dandy for us lapsed Unitarians and others of our ilk. It is what we expected and works just fine. But to proceed: the other fly in the ointment here regards Engelhardt's claim that secular bioethics must also embrace a libertarian sort of political philosophy; anything beyond this is seen as unjustified and more "news from nowhere." I propose to address this issue by referring to Engelhardt's further argument that such a libertarian political view naturally tends to evolve into some sort of much more insidious liberal cosmopolitanism, a result we both abhor. As Engelhardt says:

> This libertarian cosmopolitanism is libertarian in drawing authority from the permission of those who collaborate, and not from any particular valued state of affairs, much less from a lexical priority or value given to freedom or liberty. It is cosmopolitan in the sense of providing a framework that can be invoked outside of any particular socio-historical context, tradition, or moral community by drawing simply on the consent of those willing to be involved. Such a sparse moral foundation may be endurable only if the individuals who collaborate primarily place their own lives within functional moral communities where they confront others as moral friends, persons with whom they share a content-filled moral vision… (Engelhardt 2000, 43).

Engelhardt proceeds to observe, however, that:

> Because increasingly people do not find themselves in such communities, and because they often find themselves hungering for community, value and meaning, the default position becomes, not as a matter of strict necessity, but as a matter of moral desire, a liberal cosmopolitanism (Engelhardt 2000, 43).

What we thus find, according to Engelhardt, is another form of the urge to transcendence, as previously discussed:

> This shift from a libertarian cosmopolitanism to a liberal cosmopolitanism involves a radical change of moral and metaphysical perspective. It establishes a fundamentally different context for a bioethics. A libertarian cosmopolitanism advances no criticism of particular moral communities, as long as those who participate can from the outside be seen as giving their permission…it constitutes the moral point of view of moral strangers…involves no particular ranking of values…eschews moral imperialism. This is in contrast to a liberal cosmopolitanism, which assigns a cardinal value to a particular understanding of autonomous choice and holds that all persons should likewise. The liberal cosmopolitan ethos requires that people decide to be autonomous, self-determining individuals. The failure to pursue this ethos of autonomy becomes an indication of false consciousness…Liberal cosmopolitanism locates self-determination centrally in its account of human flourishing. In the absence of a transcendent moral truth, the focal point of the moral life becomes autonomous self-determination. The good life is not found in submitting to and being determined by the good and the true. Autonomy instead becomes integral to the good (Engelhardt 2000, 43).

There is much here that I would second; in fact, it signals a parallel thread of argument in Engelhardt that can be profitably read even by those who have no particular Christian commitments but are more simply unable to follow along with the liberal cosmopolitanism that Engelhardt is at such pains to explicate. It, in effect, traces the movement from a minimal ethics of permission, which relies on freedom as a side constraint, to a full-blown liberalism that makes freedom the primary value and focus of the moral life. This in turn, as Engelhardt observes, moves us from a tolerant minimal secular ethic to an intolerant, imperialistic ethic that demands that all support a society, and health care system, that keys to a much richer notion of human flourishing as its touchstone. And the arguments for the welfare state, socialized medicine, confiscatory taxation, etc., are not far behind.

But within the confines of this essay, we should now pause to reflect on what we have to say about this shift from libertarian to liberal cosmopolitanism. Is it driven by some clear logical or existential necessity? Engelhardt does not think so, although his own tendency toward transcendence is appearing here in another form, however misguided, by his lights. But is this liberal cosmopolitanism still the "news from nowhere" that he has always claimed it is, especially given the arguments we have previously considered?

I do not think so. Once we have placed ourselves wholly in the realm of the immanent, with our ethics charged with ascertaining how we might best "coherently and accountably seek satisfaction, fulfillment, and happiness," then it would seem that restricting ethics to considering freedom as a side constraint is no longer mandatory, and a reflection on whether and how a given society might consider supporting the liberal view of human flourishing becomes as legitimate as any ethical reflection. In effect, if we reject Engelhardt's austere employment of the Enlightenment project for a much more garden variety ethics that humbly pursues the issue of how best to "coherently and accountably seek satisfaction, fulfillment, and happiness," then the liberal cosmopolitan view merits as much of a hearing as anything else. It is NOT a reduction ad absurdum as Engelhardt seems to imply; it may well be very wrong but, if so, it will be for other reasons that a "secular immanentist" can reflect upon.

Lapsed Unitarian Yankee that I am, I am as committed as Engelhardt to opposing this shift to a liberal cosmopolitan view. But not for the philosophical reasons that he gives. As previously argued, I submit that his critique of secular ethics itself fails once we recognize that its only real satisfaction, and wellspring, comes from a view of the Enlightenment project that can be satisfied only be returning to some sort of religious vision. Given this, and thus proceeding with our more garden variety ethical reflections, I would feel obliged to admit that a society might legitimately and coherently opt for liberal cosmopolitanism. The basic argument for this could appeal to Engelhardt's own notion of permission but expand its basis, viz. that permission may be sufficiently secured from a democratically based, majority rules sort of permission. As I have argued extensively elsewhere (Wear 2005), the "representatives of the people assembled" might legitimately choose to support such a system, including forcing it on others who do not agree with it. In effect, I believe permission is necessary, but see no reason why this must be limited to only the explicit permission of specific individuals.

I would then, for my part, presume to argue that such a liberal cosmopolitanism is unwise and imprudent for many of the sort of reasons that contemporary American conservatism offers. A good start for such an argument might be obtained by reviewing F. A. Hayek's book *Road to Serfdom* (1944). I will not supply such argument here, but be content to merely signal it towards my rejection of the idea that such a much more circumscribed political view need not, for good and sufficient reasons, shift to that of liberal cosmopolitanism, as Engelhardt seems to believe. It would, most likely however, once the "representatives of the people assembled" have reflected on the issues and arguments, result in a polis, and a health care system, that goes way beyond the libertarianism that Engelhardt erroneously holds we are limited to in the secular realm.

4.5 Concluding Remarks

My basic argument here is that for those of us who, unlike Engelhardt, remain stuck in the secular world of immanence, a substantial, secular ethic can be (and has been) legitimately fashioned that provides respectable, coherent guidance for moral strangers. His critique, in effect, succeeds in claiming that the Enlightenment has failed only by appealing to a transcendental result that many of us do not accept. Ethics then can legitimately and helpfully proceed in a more garden variety fashion whereby we seek, as best we can, to "quite coherently and accountably seek satisfaction, fulfillment, and happiness," as Engelhardt himself expresses it. Nor need the result of such an inquiry be objectionably impoverished; in fact, we have about three decades of substantial argument, debate, and experience that has resulted in a detailed canon of clinical ethics that seems to work for many, and may well constitute a universal ethic for all that Engelhardt says. Similarly, I would submit that legitimate political permission can be gained from the "representatives of the people assembled" to a polity that goes way beyond the limited libertarian polis that Engelhardt advocates for. This in turn might be kept from shifting to an objectionable liberal (read: imperialistic socialist) polity by the usual conservative prudential arguments.

References

Bok, S. 1978. *Lying: Moral choice in public and private life*, 232–255. New York: Vintage Books. See especially chapter XV: Lies to the sick and dying.

Engelhardt Jr., H.T. 2000. *The foundations of Christian bioethics*. Lisse: Swets & Zeitlinger Publishers.

Hayek, F.A. 1944. *The road to serfdom*. Chicago: University of Chicago Press.

Wear, S.E. 2002. Teaching bioethics at (or near) the bedside. *Journal of Medicine and Philosophy* 27(4): 433–445.

Wear, S.E. 2005. Ethical expertise in the clinical setting. In *Ethics consultation*, ed. Lisa Rasmussen, 243–258. Dordrecht: Springer.

Chapter 5
Disease, Bioethics, and Philosophy of Medicine: The Contributions of H. Tristram Engelhardt, Jr.

Mary Ann Gardell Cutter

5.1 Introduction

When I was asked to contribute an essay in this volume in honor of physician-philosopher H. Tristram Engelhardt, Jr. and his contributions to bioethics and the philosophy of medicine, I asked myself, what could a former student of Dr. Engelhardt's say about her esteemed teacher? What could someone like me who spends most of her time teaching undergraduates biomedical ethics say about the immense influence of Engelhardt's expansive work in bioethics and philosophy of medicine and for the practice of medicine? Ah, herein lay the start of an essay. In this essay, and in honor of Engelhardt's contributions to bioethics and philosophy of medicine, I argue that what makes Engelhardt's body of work notable in the history of philosophy is his recognition that understanding the medical concept of disease illustrates the dual dependence of bioethics and philosophy of medicine on each other. In what follows, I review Engelhardt's analysis of disease in terms of its descriptive, explanatory, evaluative, and social dimensions. I show how the concept of disease carries bioethical implications, and such bioethical implications are framed in terms of ontological and epistemological considerations made explicit in the philosophy of medicine, thus highlighting the reliance of bioethics on philosophy of medicine.

M.A.G. Cutter, Ph.D. (✉)
Department of Philosophy, University of Colorado, 1420 Austin Bluffs Parkway, Colorado Springs, CO 80918, USA
e-mail: mcutter@uccs.edu

L.M. Rasmussen et al. (eds.), *At the Foundations of Bioethics and Biopolitics: Critical Essays on the Thought of H. Tristram Engelhardt, Jr.*, Philosophy and Medicine 125, DOI 10.1007/978-3-319-18965-9_5

5.2 Engelhardt's Philosophy of Disease

5.2.1 Why Study Disease?

A study of *disease* is important because disease is a centralizing notion in health care, one that has direct and important consequences for daily life. How medicine classifies, describes, explains, and diagnoses disease guides treatment (Engelhardt 1981, 1984, 1996, Chap. 5; Engelhardt and Wildes 2003). It guides what actions are advised to be taken and which ones are not, as well as who is charged with what tasks over what recommended period of time. If one is diagnosed with breast cancer, for instance, there will be certain avenues of treatment (e.g., mastectomy, chemotherapy, radiation) that will be recommended. If one is diagnosed with congestive heart failure, one will not receive treatment recommended for breast cancer but a different intervention (e.g., minimally invasive heart surgery).

Further, how medicine classifies, describes, explains, and treats disease evokes the allocation of vast amounts of societal and personal resources. In 2009, the United States spent 17.6 % of its Gross Domestic Product (GDP) on health care, a figure that translates into 2.5 trillion dollars. In the United States, Medicare spent 20 % of the total of national health expenditure, resulting in $502.8 billion dollars and Medicaid spent 15 % of the total of national health expenditure, resulting in $373.9 billion dollars. Private spending constituted 32 % of the total national health expenditure, resulting in $801.2 billion dollars. The federal government's share of health care spending increased just over 3 % in 2009 to 27 %, while the share of spending by households (28 %), and state and local governments (16 %) fell by 1 % each (Department of Health and Human Services 2011).

In addition, a study of disease connects us to our historical past and provides insight into directions for future development. How we understand disease is a reflection of past ways of knowing. Eighteenth century Swedish clinician and botanist Carolus Linnaeus (1707–1778) developed a taxonomy of disease that included eleven classes of disease based on clinical signs and symptoms. These included *exanthematici* (e.g., smallpox), *critici* (i.e., critical fevers), *phlogistici* (i.e., inflammations), *dolorosi* (i.e., painful diseases), *mentales* (i.e., mental disturbances), *quietales* (i.e., impairment of voluntary actions), *motorii* (i.e., convulsive diseases), *suppressorii* (i.e., suppression of bodily fluids), *evacuatorii* (i.e., discharge of fluid), *deformes* (i.e., physical wasting), and *vitia* (i.e., skin diseases) (Linnaeus, translated by Bowman 1976, 9). These eleven classes of disease reflect how clinicians used to classify the clinical complaints patients brought in to the clinic. Not knowing about viruses, Linnaeus classified rabies as a mental disturbance, and not knowing about vitamin deficiency, he classified rickets as physical wasting (Bowman 1976, 10). Early taxonomies reflect a way to organize patient signs and symptoms into categories that were useful for diagnosis and treatment.

In contrast to Linnaeus, and in the nineteenth century, German physician and cellular pathologist Rudolf Virchow (1821–1902) began to classify disease in terms of anatomic and physiological underpinnings (Engelhardt 1996, 214). The early

Virchow understood diseases not as "self-subsistent, self-contained entities" (1981 [1858], 188) or invading organisms, but as states indicative of underlying pathophysiological processes. Disease represented "only the course of corporeal appearances under changed conditions" (1981 [1858], 188). For the early Virchow, disease should not be confused with its symptoms or its causes: "[s]cientific medicine has as its object the discovery of changed conditions, characterizing the sick body or the individual suffering organ. Its object is also the delineation of deviations experienced by the phenomena of life under certain conditions…" (1981 [1858], 188). On this view, disease is understood in terms of its pathophysiological processes as deviation from normal bodily functions, as opposed to signs and symptoms as reported by patients.

The transformation of clinical classifications from an enterprise that catalogues patient signs and symptoms to one that organizes anatomic and physiological observations and measurements allowed a reorganization of clinical classifications and the diagnoses and treatments that followed from such classifications (Engelhardt 1982). Conditions such as fever and pain were no longer considered diseases in their own right, but symptoms associated with underlying physiological processes that were in turn given a name (e.g., tuberculosis, malaria). Previously discriminated problems (e.g., coughing up blood, shortness of breath) could now be brought together under the same rubric (e.g., tuberculosis). Clinical complaints that had not been discriminated (e.g., pain) could now be distinguished in terms of their anatomical (e.g., hernia), physiological (e.g., arthritis), and molecular (e.g., genetic) factors.

Today, anatomical, physiological, and genetic specialties have grown and the number of taxonomies employed by clinicians far exceeds the eleven offered by Linnaeus. The International Classification of Disease–10th edition (ICD-10) used today contains approximately 12,000 categories of disease and serves numerous health care purposes. They are based on:

1. *symptomatology – manifestations*: known pattern of signs, symptoms, and related findings
2. *etiology*: an underlying explanatory mechanism
3. *course and outcome*: a distinct pattern of development over time
4. *treatment response*: a known pattern of response to interventions
5. *linkage to genetic factors*: e.g., genotypes, patterns of gene expression
6. *linkage to interacting environmental factors* (Production of ICD-11: The Revision Process 2007)

Disease nosologies vary in part because they are used for a range of purposes, including the analysis of the general health situation of population groups and monitoring of the incidence and prevalence of diseases and other health problems in relation to other variables, such as the characteristics and circumstances of the individuals affected, reimbursement, resource allocation, and guidelines. They are used internationally to classify diseases and other health problems recorded on many types of health and vital records, including death certificates and health records. In addition to enabling the storage and retrieval of diagnostic information for clinical, epidemiological, and quality purposes, they provide the basis for the compilation of national mortality and morbidity statistics by World Health Organization (WHO) Member States (World Health Organization 2013). The next version of the ICD (version 11) will be released in 2016.

Beyond the clinical, a study of disease tells us something about what it means to be human. While it is possible that a human might not have been born, once born, at least at present, every human being experiences disease or illness. Because the disease experience is shared, a study of disease tells us something about how we understand human biological and psychological abilities and functions. Beyond this, and because of the inevitability of experiencing disease, a study of disease is an opportunity for self-exploration and reflection upon one's own life, how one knows and judges it, and how one thinks one ought to act in the face of such knowledge and judgments. As such, thinking about disease is as much a personal as it is an academic exercise of a shared human experience.

In addition, disease tells us something about how we differ. Even though we all experience disease, humans do not share the same view and experience of disease. As previously indicated, there are different views of disease in the history of medicine. These include symptomatic, etiological, prognostic, therapeutic, genetic, and environmental, to name a few major approaches to disease. Within these groups, there are different views of disease held by different cultures. Anorexia nervosa is considered a disease in a nutrition-rich culture but not in a nutrition-depleted one; dyslexia is a disease in a literate society, but not one in a pre-literate society (Reznek 1987, 169). Further, there are different views of disease among individuals. An individual may interpret his or her disease as somatic dysfunction while another may interpret it as a psychological condition, a violation against Divine Law, or a family curse. Finally, there are different experiences of the so-called same disease. Some may experience great pain and disability with a certain disease while others do not, as is evident in any conversation about symptoms between patients.

In short, as Engelhardt teaches us, there are many reasons to study disease. Disease is a centralizing notion in medicine, commands significant amounts of resources in a culture, connects us to our historical past, tells us something about directions for future developments, and informs us about what it means to be human.

5.2.2 *The Descriptive, Explanatory, Evaluative, and Social Dimensions of Disease*

The ways in which we speak of, react to, and experience medical reality such as disease are shaped and directed by a number of interests. According to Engelhardt, these interests include descriptive, explanatory, evaluative, and social dimensions. These interests reflect "four conceptual dimensions" or "modes of medicalization" (1996, 195). They constitute the "language of medicine" (1996, 195) in that they provide the "grammar" and "rules," so to speak, for constructing meaning about and practical guidelines for addressing clinical problems.

5.2.2.1 Descriptive Dimension

More specifically, disease is "seen through a set of descriptive assumptions" (Engelhardt 1996, 207). In medicine, description takes place by providing "facts." The term "fact" is derived from the Latin "factum" (*The Complete Oxford* 1994, 560) and refers to "a thing done" or a "reality of existence," that is, to something that has really occurred or is actually the case. Such a view assumes that there is a reality "out there" to be discovered, a position called a *realist* view in philosophy. In medicine, a typical test for a fact is verifiability, which seeks to confirm whether the facts correspond to experience. Such a view assumes that matter is the basis of reality, a position called a *materialist* view in philosophy. In clinical medicine, and over time, a materialist view of disease has proceeded from the organ systems (e.g., reproductive) to individual organs (e.g., breast), tissues (e.g., breast tissue), cells (e.g., breast cells), and the molecular level (e.g., biomarkers and genes) to describe a clinical condition. The position in which objects are reduced to their component parts is known in philosophy as *reductionism*. Here the properties of the whole are the addition or summation of the properties of the individual parts. In the case of breast cancer, the National Cancer Institute describes breast cancer as the abnormal growth and build-up of extra cells in the breast lobes, ducts, tissues, lymph vessels, and lymph nodes that form a mass called a lump or tumor (National Cancer Institute, *What* 2014b, 2). A tumor is then submitted to pathological tests to determine its size, shape, and, if available, biomarkers and/or genetic characteristics. The description of breast cancer assumes that breast cancer is a reality out there to be discovered and composed of empirical or physical matter. Such matter can be reduced from the whole to its parts and can be studied.

Engelhardt (1996, 208) reminds us that the so-called "facts" in medicine are not neutral. They are seen through theoretical frameworks (also see Fleck 1979 [1935]). "Descriptions require standardization of terms" (Engelhardt 1996, 208), and as such are framed by prior discussions, presumptions, and claims within particular frameworks. For instance, surgeons describe clinical problems in terms of surgical features, geneticists describe clinical problems in terms of genetic factors, and pathologists describe clinical problems in terms of pathological criteria. Such descriptions can and do change. One might think of the change that the American Joint Committee on Cancer (AJCC) Breast Cancer Task Force made from the fifth to the sixth edition in 2003 in recommending that the N (node) category of the TMN (tumor, metastases, node) cancer staging system be changed from one to three categories based on the number of axillary (i.e., under the arm) lymph nodes that are present (American Joint 2010, 423). This change came about in part because of a theoretical shift in understanding the role of lymph nodes in determining the extensiveness of breast cancer and the need for more specific diagnoses of breast cancer so that treatments for breast cancer can better be tailored.

Given that the so-called "facts" of medicine are not neutral, it may be misleading to say that disease is "out there" to be discovered as reducible matter. Rather, disease in part reflects the "lenses" the clinical knower brings to this so-called reality. Disease is in part the experience of a disability, dysfunction, and/or suffering reported by a conscious agent that hinders the achievement of certain goals. On this *idealist* view of reality, disease may not be fully reducible to matter that can be studied using laboratory tests. It is not a thing but an idea of a *holistic* event in the life of a patient. In the case of breast cancer, breast cancer reflects not simply a collection of mutated cells that affects biological function in certain ways, it constitutes an experience reported by a patient, even when the patient may be asymptomatic.

5.2.2.2 Explanatory Dimension

Second, the "facts" of disease are structured around explanatory claims and assumptions. In this way, disease is an explanatory concept, and as such "brings coherence…to the multiplicity of events we encounter in medicine" (Engelhardt 1996, 209). It brings coherence to the signs and symptoms that bring patients into the clinic, and the pathoanatomical and pathophysiological data that are generated by laboratory findings by gathering and interpreting empirical data within the framework of interpretations that have been handed down. This approach is known in philosophy as *empiricism*. Here clinical facts are verified by repeatable experiments and data and they maintain an accepted status until they are shown to be non-verifiable. In this approach, disease relates "two worlds of observations" (Engelhardt 1996, 209), the world of the clinic and the world of the laboratory. "The findings of the clinician are related to the observations of the pathoanatomists and pathophysiologists and take on a new significance through these anatomical and pathological observations" (Engelhardt 1996, 209–210). For instance, in the nineteenth century, and with the shift to pathophysiological explanations of disease, consumption, King's evil, and Pott's disease were gathered together under the rubric of tuberculosis. Alternatively, in the twenty-first century, molecular science shows us that breast cancer is no longer a single disease. With these shifts, what results is "an expansion…of the explanatory powers of medicine" (Engelhardt 1996, 210) and the ability to diagnose and treat clinical problems with greater accuracy.

In contemporary medicine, disease is often explained in terms of an etiological account. In other words, medicine explains disease in terms of a causal relation between that which brings the disease about and the disease itself, the result of which is used to predict the onset and severity of disease. But one might note, following Engelhardt, that the notion of cause in medicine is far from simple and involves appeal to *empirical* as well as *rational* or logical criteria. "The term *cause* can be used to identify conditions that are sufficient to produce effects, necessary to produce effects, or that contribute to the likelihood of an effect's occurring" (Engelhardt 1996, 223). Here, and first, "[f]actor X is a *sufficient* determinant, if X always leads to Y," which "does not imply that Y is always preceded by X" (Wulff 1981, 57). As an example, mutations in what is called the "BRCA1" and "BRCA2" genes are typically (but not always) sufficient to bring about breast cancer, but breast cancer occurs by other causal factors as well.

Second, "[f]actor X is a *necessary* determinant, if X always precedes Y, which of course does not imply that Y always succeeds X" (Wulff 1981, 56) Here, in some sense, breast tissue is a necessary condition for breast cancer, but as any clinician or patient knows, this account of the necessary condition of breast cancer fails to provide a basis for specific tests for the diagnosis of breast cancer. While there is emphasis on the role of particular genetic and lifestyle factors that link to breast cancer, breast cancer specialists continue to search for necessary conditions for breast cancer in order to provide more specific accounts of the causal relation between that which brings breast cancer about and breast cancer.

Third, "[s]ome determinants of disease are neither necessary nor sufficient, but only *contributory*. This term is used if some factor X leads to an increased probability of Y, though X does not lead to Y and Y is not always preceded by X" (Wulff 1981, 57). Recent research on breast cancer shows that there are statistical links between ER+ (estrogen receptor positive) and HER2+ (human epithelial growth factor receptor-2 positive) biomarkers (or cellular proteins) and breast cancer. Beyond this, research continues to establish such links and to provide more specific causal determinants of breast cancer. Although medicine may aspire to discover sufficient and necessary causes of disease, "[i]n medicine, where the data are often statistical [or contributory], causal factors are frequently identified in the last sense" (Engelhardt 1996, 223).

Given that medicine usually offers contributory accounts of disease causation, a tension arises. A statistical causal account of disease provides less certainty than, say, a necessary causal account. This recognition of the "limitations of human reason" causes "tension," as Engelhardt puts it, "between the universal aspirations of knowers and the particular context in which real individuals actually know and frame explanations" (Engelhardt 1996, 218). Stating what is not known in an academic discussion may not give us pause; stating what is not known in medicine and in the context of treating patients is notably difficult for many clinicians and patients. Often, clinicians are not comfortable with stating what they do not know, and patients are not comfortable with accepting medical advice and treatment given uncertain etiologies for their conditions. Informed consent documents are filled with claims about what cannot be guaranteed in medical procedures, yet patients and clinicians alike still hope for greater certainty in medical diagnosis and treatment. Such is the condition of knowing enough about what we do not know about and working with an empirical methodology that cannot guarantee 100 % certainty.

5.2.2.3 Evaluative Dimension

Third, disease is an evaluative concept. "To see a phenomenon as a disease, illness, or disability is to see something wrong with it" (Engelhardt 1996, 197). Disease is experienced as a failure "to achieve an expected state, a state held to be proper to the person afflicted" (Engelhardt 1996, 197). This may be a failure to achieve an expected freedom from pain or suffering, an expected level of function or ability, a realization of human form or grace, and/or an expected span of life. This may also be a failure to achieve a state requested by a patient, determined to be beneficial to a patient, and/or in keeping with the standards of moral integrity and the virtues of the health care profession. In other words, the "facts" of disease are inextricably tied to the "value" of disease and its treatment, where value is understood as an important and enduring sign of significance or worth, belief, or ideal.

For Engelhardt, disease is an evaluative concept because disease is not simply reducible to physical dysfunction. Consider the case of osteoporosis.

> The species-typical character of calcium metabolism for post-menopausal women is one of negative calcium balance. More calcium is absorbed than deposited, leading to the development of osteoporosis and painful debilities such as collapsing vertebrae and greater exposure to risks of fractures. Such phenomena are as species typical as menopause itself. Yet, one would usually want to say that osteoporosis in postmenopausal women is a disease (Engelhardt 1996, 203).

Osteoporosis is a disease not because it is abnormal function but because the pain and disability experienced by individual patients leads them to seek the treatments offered by health care professionals. In seeking clinical help, patients determine that their condition is, all things considered, disvalued and harmful to their life experiences and, as a consequence, changes are in order.

As seen here, the clinical determination of what is considered "normal" and "abnormal" is complex. "Normal" and "abnormal" are recognized as such "within a particular context of expectations" (Engelhardt 1996, 199). Because the measure of an individual organ, tissue, or cell is highly uninformative, a clinician has no idea from a single case what range of data can count as "normal" or "abnormal." Group data must be taken into account. What is needed is a range of measurements for a range of populations that clinicians and researchers can agree upon that constitutes "normal" or "abnormal." Here, "normal" can mean that which is statistical, average or mean, typical or expectable, conducive to the survival of a species, innocuous or harmless, commonly aspired to, or excellent in its class (Murphy 1976, 117–133). Alternatively, "abnormal" can mean what is statistical, average, not typical, not conducive to the survival of a species, harmful, not commonly aspired to, or deficient in its class. The determination of what is "normal" and "abnormal" requires value judgments about thresholds between points of data. In medicine, such judgments typically concern the value of how to maximize patient welfare and minimize patient harm.

A concern here is that if disease is evaluative, then it must be even less certain a concept in medicine than we had ever imagined. As Engelhardt tells us, this does not follow. One will encounter a wide range of clinical problems that lie along a continuum. "On one end of the continuum there are circumstances likely to be disvalued in whatever culture an individual lives, and in terms of whatever goals are possessed by individuals or societies" (Engelhardt 1996, 204). Here one need not assume transcultural values regarding the proper range of human form and function. "One need only recognize that certain circumstances are likely to be impediments to the realization of goals (1) in nearly any foreseeable environment and (2) in terms of any likely cluster of human purposes" (Engelhardt 1996, 204). For instance, unusual breast tenderness and pain, breast lumps or rashes, swelling in the armpit, and death among women who report these clinical events over time are likely to make breast cancer a clinical problem across cultures.

On the other hand, there will be clinical problems that are not disvalued in every culture. Here one might think of "culture-specific diseases" or "culture-bound syndromes," a general term for a combination of psychiatric and somatic, recurrent, locally-specific symptoms (Culture-Specific Diseases 2013). The concept of culture-bound syndrome was recognized in the fourth edition of the American Psychiatric Association's (APA) *Diagnostic and Statistical Manuel of Mental Disorders* (2000). An example of a culture-specific disease includes *amok* or *mata elap* from Malaysia, a dissociative episode characterized by a period of brooding followed by an outburst of violent, aggressive, or homicidal behavior. *Amok* is similar to *cafard* or *cathard* in Polynesia and *mal de pelea* from Puerto Rico. Beyond these types of classification, and as already mentioned, clinical conditions such as anorexia nervosa and dyslexia are recognized as diseases in certain contexts and not in others.

5.2.2.4 Social Dimensions

Fourth, in addition to its descriptive, explanatory, and evaluative dimensions, disease entails a social dimension. A designation of disease takes place within the social practices of developing professional clinical standards, devising educational requirements and licensure agreements, formulating funding options, and instituting health laws and policies. To claim that a patient has a disease "is to cast that individual in social roles where certain societal responses are expected" (Engelhardt 1996, 217). Some of the social responses include assigning individuals a sick role, expecting that such persons seek help from socially recognized therapists, excusing sick persons from responsibilities for certain tasks while recovering from disease conditions, and expecting that care for certain diseases is covered by medical insurance plans and is protected by particular health care policies.

The social interests that frame disease can be formal or informal (Engelhardt 1996, 219). Formal practices, such as meetings of the American Joint Committee on Cancer (AJCC) (2010), organize rules for thc use of descriptive terms in diagnostic categories. In such circumstances, the social reality of disease is settled by informed votes based on clinical evidence within committees. "The decisions in such circumstances are made not simply in terms of the character of reality as it is taken really to be, but also in terms of which modes of classification will be most useful in organizing treatment and care" (Engelhardt 1996, 219). Choices to divide stages and sub-divide cancer stages and grades into a certain number turn on cost-benefit calculations and understandings of prudent actions that have direct implications for the ways patients are treated within social contexts.

Informal practices also frame the web of social interests in understanding disease. Clinical researchers and practitioners make decisions about what they wish to focus on in their practices, what groups of clinical professionals to join, where they wish to locate, what patient populations they wish to see, and how much they need to charge for their services. With the rise of the internet, patients have increased influence in medicine as they blog about their health care providers, clinical conditions, treatments, and suggestions. Advocacy groups also have significant influence on what diseases are given funding and research attention and which ones are not. Clinical researchers and practitioners, and subjects and patients, contribute to the informal social networks that frame medicine and its nosologies and nosographies.

In some sense, then, medical reality is in some sense a democratization. The choice among different understandings of reality within medicine is a matter of communal interest. As a consequence, "[c]ommunities must begin with a recognition of the constructed character of medical reality. This recognition underscores our choices and indicates our responsibilities as individuals who not only know reality but also know it in order to manipulate it" (Engelhardt 1996, 226). In this way, the social dimension of disease is constituted by the descriptive, explanatory, and evaluative dimensions of disease. It is framed by the clinical evidence of the time, how the evidence is explained, and what values are central to clinical medicine as well as the patients who seek medical services. With this comes the responsibility on the part of clinical professionals to provide the best care that is possible within the boundaries of available resources. Patients have responsibilities as well; they are charged with being responsible decision-makers and co-navigators of their path to health.

The democratization of medical reality is not necessarily a bad thing. Note that Engelhardt titles Chapter 5 of *The Foundations of Bioethics* "The Languages of Medicalization." By "medicalization," Engelhardt means a process by which medical and nonmedical problems become defined and treated as medical problems. In this process, medical and non-medical problems are defined in medical terms, described using medical language, understood through the adoption of a medical framework, and treated with a medical intervention (also see Conrad 2007, 5). While "medicalization" usually refers to a disvalue, to one in which non-medical and "normal" events, such as pregnancy and aging, are transformed into a medical problem, Engelhardt shows us that medicalization may not always be negative. It is not negative when a patient's condition is labeled as harmful and medicine can offer the means to alleviate the pain and suffering that accompanies it (Engelhardt 1996, 223). Insofar as this goal remains central in medicine, medicalization and democratization of medical reality can be seen to be a positive human endeavor.

5.2.2.5 Facts, Theories, Values, and Social Contexts

According to Engelhardt, the descriptive, explanatory, evaluative, and social dimensions of disease are not separate and distinct, but rather frame and limit each other (1996, 196). Facts are theory-laden, the fact/theory dyads are evaluative, and the fact/theory/value triads are socially framed. Consider again the case of breast cancer. In classifying breast cancer, a decision has to be made regarding how many cells with deviant changes of a certain kind in the biopsied breast tissue must be present before the cells are labeled as "carcinoma" or "cancer." An explanation is given about the relation between the mutating cells and the result called "carcinoma of the breast." To be too liberal in classifying cells as "carcinoma" will lead to unnecessary treatment, which harms women, costs money, and wastes resources. To be too conservative in the classification will lead women to receive treatment too late, which leads to increased pain and suffering, as well as unnecessary deaths among women. On this view, the lines among "benign," "suspicious," and "carcinoma" are in part discovered and in part created. They involve appeals to fact/theory/value triads within social frames of reference.

5.3 Disease and Bioethics

Given that disease is an evaluative concept, it follows that how we understand disease entails bioethical implications. Bioethics (Gr. *bio*, life + Gr. *ēthikē*, ethical, or study of standards of conduct) is the study of the ethical or moral implications of biomedical discoveries and practices. It gained notoriety at the end of the twentieth century for its incisive analyses and critiques of practices in medicine (Jonsen 1998). The term "bioethics" was coined by Dr. Van Rensselaer Potter, a research oncologist at the University of Wisconsin in the early 1970s (Jonsen 1998, 27). Potter published an article entitled "Bioethics, The Science of Survival" (1970) and, in 1971, followed it with his book *Bioethics: Bridge to the Future*. In it, Potter defined "bioethics" generally as "a new discipline that combines biological knowledge with a knowledge of human value systems" (1971, 2). Bioethics has since become influential in western medicine especially as many have become concerned about the role, power, and limits of medicine in their lives and as bioethicists enter into mainstream medical school teaching and research to offer analyses and critiques of medical practices (Engelhardt 2012).

According to Engelhardt, the success of bioethics at the end of the twentieth century comes from a variety of sources. First, "there was a cultural hunger to locate medicine within larger cultural concerns" (Engelhardt 2000b, 2). Given that in the late twentieth century health care in every developed country was claiming a larger portion of the gross domestic product, nations and states began to grapple with challenges regarding the allocation of funds and resources in medical care and research. Second, "new technologies…pressed for clarity about issues" (2000b, 2). Moral problems raised by new technologies, such as organ transplantation and gene therapy, spawned significant discussions in bioethics. Third, "old" moral problems, such as abortion, "became more acute because the technologies that occasioned them had become safer" (2000b, 2). As a consequence, there arose the need to rethink some formally settled moral matters in medicine. Fourth, "there appeared to be purely philosophical issues such as the nature of disease and illness that were addressed neither by philosophy of medicine nor even the philosophy of biology" (2000b, 2). That, of course, changes with Engelhardt's work, as he shows us why and how a philosophical analysis of disease is critical in bioethics in the twenty-first century.

A philosophical analysis of disease is critical because, given that disease is not a value-neutral concept, it frames as well as is framed by discussions in bioethics. According to Engelhardt, one will not be able simply to discover, by appeal to factual issues alone, what diagnoses and treatments are indicated, what diagnoses and treatments are appropriate, and what treatments are ordinary versus extraordinary. "Integral to such judgments will be appeals to particular hierarchies of values and to peaceable processes for resolving disputes in these matters" (Engelhardt 1996, 221). In the case of breast cancer, one will not be able simply to discover by appeal to factual issues alone which diagnoses are indicated and what ones are minimal or excessive. Determining the difference, for instance, between stage IB and IIA breast cancer in the case of a patient with a two centimeter tumor and one positive axillary node involves more than a factual judgment. It involves appeals to what benefits to the patient ought to be sought, what risks ought to be avoided, what medical resources ought to be expended, and what goals are sought to be achieved. In democratic nations, determinations are made within the context of treating patients who are seen to be ends-in-themselves as decisionmakers and judges of their own best interests.

In other words, according to Engelhardt, how we understand and undertake disease involves a host of bioethical considerations regarding the rights of patients and health care professionals and the boundaries of what constitutes duties not to impose harm and to be beneficent to patients. More specifically, they involve what Engelhardt calls "two major moral principles" (1996, 121; also see Engelhardt 2006, 25) for guiding our reflections and actions in clinical medicine and the treatment of disease. These include "the principle of permission" and "the principle of beneficence." The principle of permission states that:

> Authority for actions involving others in a secular pluralist society is derived from their permissions. As a consequence,
>
> i. Without such permission or consent there is no authority.
> ii. Actions against such authority are blameworthy in the sense
>
> of placing a violator outside the moral community in general, and making licit (but not obligatory) retaliatory, defensive, or punitive force (Engelhardt 1996, 122).

The principle of permission expresses the circumstance that authority for resolving moral differences in a secular, pluralist society can be derived only from the agreement of the participants. Applied to our discussion of understanding disease, how we understand disease is a function of the agreements made among the parties involved. Different ways of understanding and treating disease depend on the choices made about different understandings of the facts, theories, values, and social interests involved. Further, health care professionals cannot force patients to come in to the clinic for care. They cannot force patients to receive medical care or continue with the medical care that they are receiving. Alternatively, patients cannot force health care professionals to practice in ways that go against what they think and believe. Bioethically speaking, the principle of permission frames how medical reality is understood and manipulated in contemporary culture.

Second, the principle of beneficence guides our reflections and actions in clinical medicine and the treatment of disease. It states that:

> The goal of moral action is the achievement of goods and the avoidance of harms. In a secular pluralist society, however, no particular account or ordering of goods and harms can be established as canonical. As a result, within the bounds of respecting autonomy, no particular content-full moral vision can be established over competing senses (at least within a peaceable secular pluralist society). Still, a commitment to beneficence characterizes the undertaking of morality, because without a commitment to beneficence the moral life has no content. As a consequence,
>
> i. On the one hand, there is no general content-full principle of beneficence to which one can appeal.
> ii. On the other hand, actions without regard to concerns of beneficence are blameworthy in the sense of placing violators outside the context of any particular content-full community. Such actions place individuals beyond claims to beneficence. In particular, malevolence is a rejection of the bonds of beneficence. Insofar as one rejects only particular rules of beneficence, grounded in a particular view of the good life, one loses only one's own claims to beneficence within that moral community; in either case, petitions for mercy (charity) can still have standing (Engelhardt 1996, 123).

The principle of beneficence expresses the circumstance that the promotion of patient welfare and the avoidance of harm to a patient are central to the goals of medicine. Applied to our discussion of understanding disease, the principle of beneficence serves as a basis for health care professionals' determinations regarding what diagnoses and treatments are in the patient's best interest. In these determinations, there is a moral mandate to minimize patient harm (through non-malevolent acts) and maximize patient benefits (through beneficent acts), the details of which need to be worked out within the embrace of particular moral communities (see, e.g., Engelhardt 2000a).

One can begin to understand why Engelhardt raises concern about the subordination of the clinic to the laboratory in contemporary medicine. The concern is that changes in clinical classifications lead to the notion that the goals of medicine are found not in the mere reporting of patient signs and symptoms, but rather in comprehending the pathophysiological processes of such reportings (Engelhardt 1982; Foucault 1973 [1963]). As Engelhardt says about the shift from the clinic to the laboratory, "[p]atient problems came to be understood as bona fide problems only if they had a pathoanatomical or pathophysiological truth value. Absent a lesion or a physiological disturbance to account for the complaint, the complaint was likely to be regarded as male fide" (1996, 216). This is no small matter if one understands, as Engelhardt does, medicine's unique role in interpreting patient complaints and treating patients. What distinguishes medicine from, say, biology, is the role it plays in addressing patient signs and symptoms for purposes of alleviating the pain and suffering that accompanies clinical problems (also see Pellegrino 2008). In the move from the clinic to the laboratory, the significance of patient complaints are discounted, the goal of medicine to *care* for patients may be compromised, and laboratory tests dictate what is attended to in the clinic. From the standpoint of the principle of beneficence, this type of reorientation in medicine raises concern. Even though laboratory tests provide medicine more specific diagnostics, tests by themselves cannot guide the *care* of patients.

In short, the bioethical principles of permission and beneficence are not tangentially related to the development and use of disease classifications and descriptions; they are central to their development and use. Disease classification and description come about through careful clinical research and a framework of choices that are made about how medical reality is understood in light of what patient interests are at stake and for what purposes within particular social contexts.

5.4 Disease, Bioethics, and Philosophy of Medicine

Not only is our understanding of disease framed in part by bioethical considerations, a bioethical analysis of disease is framed by considerations drawn from philosophy of medicine. By philosophy of medicine, Engelhardt means a discipline that provides critical reflection on the concepts, presuppositions, and methods unique to medicine. It addresses a wider range of ontological (concerning being), epistemological (concerning knowledge), and value issues in medicine (see Wulff 1986; Marcum 2008). It "brings together disparate philosophical pursuits because 'medicine' has such broad meaning" (Engelhardt and Erde 1980, 366). Medicine includes the basic sciences (e.g., how the breast functions), theoretical endeavors (e.g., the development of explanatory models for breast cancer), and actual practices (e.g., the treatment of breast cancer care). "As a result, one may have philosophical puzzles about theories of function and disease (e.g., as found in physiology and pathology); and about theories of treatment (e.g., as found in pharmacology); and about ways in which health practitioners engage in their preventive or therapeutic activities (e.g., the ways internists make clinical judgments)" (Engelhardt and Erde 1980, 365). In this way, the grouping of issues concerning medical reality, knowledge, and values proves useful in the philosophy of medicine.

But the grouping of issues in philosophy of medicine is more than useful. Philosophy of medicine has a unique role in addressing the character of disease. As Engelhardt says, philosophy of medicine "offers promise of a special set of conceptual issues bearing on the status of concepts such as disease, illness, and health" (1977, 104). Such concepts "have unique explanatory and evaluative uses" (Engelhardt 1977, 104). Given that disease serves to explain dysfunction and disability for purposes of treating it, disease is a concept unlike one found in the hard sciences. The concept of disease "deals with suffering and one becomes interested in the physiological bases of diseases because of concern for the sufferings and debilities associated with illnesses" (Engelhardt 1977, 101).

Philosophy of medicine has a long history tied to the writings of well-respected clinicians who offer varied philosophical accounts of disease. Physician-philosophers from Hippocrates (5th c. B.C.E), Aristotle (384–322 B.C.E), and Galen (approx. 130–210 CE) through Edmund D. Pellegrino (1920–2013) and H. Tristram Engelhardt, Jr. (1941-present) have concerned themselves with philosophical analyses of medicine. The Greek physician Hippocrates (1943) is noted for rejecting supernatural explanations about health and disease and emphasizing the role of observation in medical diagnosis, thus associating him with the Aristotelian school of thought. The Greek physician Aristotle (384–322 B.C.E.), the son of a physician, believed that medicine could aid in philosophic and moral tasks to a large degree (Owens 1977). The Greek physician Galen of Pergamum (129–161) (1968) devoted much time to developing his theory of humors in order to explain health and disease. English physician Thomas Sydenham (1624–1689) (Romanell 1974, 69–91) advocated for a more rigorous empirical methodology in medicine and encouraged clinicians to classify disease in order to develop better diagnostic and therapeutic guidelines in the clinic. German physician Rudolf Virchow (1821–1902) followed in these steps and emphasized the importance of isolating the anatomic and physiological underpinnings of disease. Today, Pellegrino (1976, 1978, 2004, 2008) and Engelhardt, Jr. (1977, 1986, 2000b), as well as Kenneth Schaffner (1993), and William E. Stempsey (2000, 2004), are part of a tradition of physicians in the U.S. trained in philosophy who have been instrumental in reenergizing discussions in philosophy of medicine, and particularly those about the nature of disease.

Today, philosophy of medicine draws from discussions in bioethics in order to attend to the normative issues facing health care practitioners, patients, and society. According to philosopher George Khushf, a former student of Engelhardt's, there is a "constructive dialectic" (Khushf 1997, 147) between bioethics and philosophy of medicine, between the normative and the corresponding meta-normative, epistemological, and ontological issues raised in and by medicine, and in particular our understanding of disease. As Khushf puts it, in thinking about disease, "medicine, bioethics, and the philosophy of medicine are all reconceptualized in such a way that they are no longer externally related to one another" (1997, 147).

Consider, once again, the case of breast cancer. A study of breast cancer involves the "facts" of cellular mutation in the breast tissue, lobes, and ducts, as well as etiological explanations for why and how cellular mutation occurs in the breast tissue. It involves as well evaluative and social interests found in medicine, the lives of patients, and culture at large. Breast cancer serves as a treatment warrant and, as such, frames treatment options geared toward benefiting the patient. It frames the development of particular insurance policies, health policies, and guidelines for sick leave. Given that breast cancer affects approximately 230,000 females per year in the U.S. (National Cancer Institute, "Breast Cancer," 2014a, 1) and 1.5 million females in the global world (American Joint 2010, 425), breast cancer raises normative challenges in medicine, such as how to navigate between over- and under-treatment in breast cancer care and how to provide access to expensive medical care for those who are affected. It reminds us about what continues to be central in medicine, i.e., the treatment of patients in a morally defensible way. Seen here, a study of breast cancer raises ontological, epistemological, and normative issues, thus illustrating the dual reliance of bioethics and philosophy of medicine on each other.

The dual dependence between bioethics and philosophy of medicine come about for a number of reasons. These reasons harken back to those justifying the emergence of bioethics in the late twentieth century. First, bioethics finds itself part of discussions in philosophy of medicine and the much larger concerns about how a culture understands itself. These concerns are not only ethical ones, but ontological and epistemological ones concerning the nature of reality and how we understand it, respectively. Second, bioethics needs philosophy of medicine in order to find clarity on new bioethical issues that challenge our sense of reality and knowledge and require new ways of thinking. Understanding how worldviews and associated knowledge frameworks change and evolve lend insight into what claims and assumptions fuel the bioethical debates and which ones are open to revision and rethinking. Third, bioethics needs philosophy of medicine in order to address "old" moral problems that reemerge in contemporary culture in new ways. "Old" settled moral problems are no longer so settled given shifts in what constitutes clinical reality and how we know it. Exploring these dimensions of the debate provides new insights into old problems and opportunities for conflict resolution. Fourth, bioethics needs philosophy of medicine in order to reorient itself to the central focus of medicine, i.e., treating the clinical conditions that patients bring into the clinic or hospital. This is not simply a moral claim, but one lodged in understanding the nature of medicine and how medicine knows what it claims to know. In short, as Khushf puts it, "bioethics will need the philosophy of medicine because, at a deep level, it must involve the philosophy of medicine" (1997, 147).

5.5 Closing

Through his analysis of disease, Engelhardt shows us the connection between the project of understanding disease, of deciding what ought to be done in medicine, and of submitting our claims about facts, theory, values, and social concepts in medicine to philosophical analysis. Put another way, Engelhardt's account of concepts of disease establishes the dual reliance of bioethics and philosophy of medicine on each other. This, I suggest, is a notable part of Engelhardt's legacy. It is a legacy that appears in my own work (Cutter 2003, 2012) as well as the work of others found in this volume who celebrate Engelhardt's contributions to the study of disease and the fields of bioethics and philosophy of medicine.

References

American Joint Committee on Cancer. 2010. *AJCC cancer staging handbook*. New York: Springer.

American Psychiatric Association. 2000. *Diagnostic and statistical manual of mental disorders*, 4th ed, text revision. Washington, DC: American Psychiatric Association.

Bowman, Inci. 1976. Classification of disease. *The bookman* 3(6): 1–10.

Conrad, Peter. 2007. *The medicalization of society: On the transformation of human conditions into treatable disorders*. Maryland: Johns Hopkins University Press.

Culture-Specific Diseases. 2013. http://anthro.palomar.edu/medical/med_4.htm. Accessed 22 Feb 2013.

Cutter, Mary Ann G. 2003. *Reframing disease contextually*. Dordrecht: Springer.
Cutter, Mary Ann G. 2012. *The ethics of gender-specific disease*. New York: Routledge.
Department of Health and Human Services. 2011. *National health expenditure data*. http://www.cms.hhs.gov/NationalHealthExpendData/25_NHE_Fact_Sheet.asp#TopOfPage. Accessed 31 July 2011.
Engelhardt Jr., H. Tristram. 1977. Is there a philosophy of medicine? PSA 1976(2): 94–108.
Engelhardt Jr., H.Tristram. 1981 [1975]. The concepts of health and disease. In *Concepts of health and disease: Interdisciplinary perspectives*, ed. A.L. Caplan et al., 31–46. Reading: Addison-Wesley Publishing Company.
Engelhardt Jr., H. Tristram. 1982. The subordination of the clinic. In *Value conflicts in health care delivery*, ed. B. Gruzolski and C. Nelson, 41–57. Cambridge, MA: Ballinger.
Engelhardt Jr., H. Tristram. 1984. Clinical problems and the concept of disease. In *Health, disease, and causal explanations in medicine*, ed. L. Nordenfelt and B.I.B. Lindahl. Dordrecht: D. Reidel Publishing Company.
Engelhardt Jr., H. Tristram. 1986. From philosophy and medicine to philosophy of medicine. *Journal of Medicine and Philosophy* 11: 3–8.
Engelhardt Jr., H. Tristram. 1996. *Foundations of bioethics*, 2nd ed. New York: Oxford University Press.
Engelhardt Jr., H. Tristram. 2000a. *The foundations of Christian bioethics*. Salem: M & M Scrivener Press.
Engelhardt Jr., H. Tristram. 2000b. The philosophy of medicine and bioethics: An introduction to the framing of a field. In *The philosophy of medicine: Framing the field*, ed. H.T. Engelhardt Jr., 1–15. Dordrecht: Kluwer Academic Publishers.
Engelhardt Jr., H. Tristram. 2006. The search for a global morality: Bioethics, the culture wars, and moral diversity. In *Global bioethics: The collapse of consensus*, ed. H. Tristram Engelhardt Jr., 18–49. Salem: M & M Scrivener Press.
Engelhardt Jr., H. Tristram. 2012. *Bioethics after four decades: Looking to the future*. Portugal Talk, 16 March 2012. http://www.Apbioetica.org/fotos.gcal/1331984832discurso.pdf.
Engelhardt Jr., H. Tristram, and Edmund L. Erde. 1980. Philosophy of medicine. In *A guide to the culture of science, medicine, and technology*, ed. P.T. Durbin, 364–461. New York: Free Press.
Engelhardt Jr., H. Tristram, and Kevin Wm. Wildes. 2003. Health and disease: Philosophical perspectives. In *Encyclopedia of bioethics*, 3rd ed, ed. S.G. Post, 1101–1106. New York: Macmillan Reference Books.
Fleck, Ludwik. 1979 (1935). *Genesis and Development of a Scientific Fact,* ed. T.J. Trenn and R.K. Merton. Trans. F. Bradley and T.J. Trenn. Chicago: University of Chicago Press.
Foucault, Michel. 1973 (1963). *The Birth of the Clinic: An Archeology of Medical Perception.* Trans. A.M. Sheridan Smith. New York: Pantheon Books.
Galen. 1968. *On the Usefulness of the Parts of the Body.* Trans. M.T. May. New York: Cornell University Press.
Hippocrates. 1943. *Regimen*. Trans. W.H.S. Jones. Cambridge: Harvard University Press.
Jonsen, Albert R. 1998. *The birth of bioethics*. New York: Oxford University Press.
Khushf, George. 1997. Why bioethics needs the philosophy of medicine: Some implications of reflection on concepts of health and disease. *Theoretical Medicine and Bioethics* 18: 145–163.
Marcum, James A. 2008. *An introductory philosophy of medicine: Humanizing modern medicine*. Dordrecht: Springer.
Murphy, Edmond. 1976. *The logic of medicine*. Baltimore: Johns Hopkins University Press.
National Cancer Institute, National Institutes of Health. 2014a. *Breast cancer*. http://www.cancer.gov/cancertopics/types/breast
National Cancer Institute, National Institutes of Health. 2014b. *What you need to know about breast cancer.* Bethesda: National Cancer Institute. http://www.cancer.gov/publications/patient-education/WYNTK_breast.pdf
Owens, Joseph. 1977. Aristotelian ethics, medicine, and the changing nature of man. In *Philosophical medical ethics: Its nature and significance*, ed. S.F. Spicker and H.T. Engelhardt Jr., 127–142. Dordrecht: D. Reidel Publishing Company.

Pellegrino, Edmund D. 1976. Philosophy of medicine: Problematic and potential. *The Journal of Medicine and Philosophy* 1: 5–31.
Pellegrino, Edmund D. 1978. Philosophy of medicine: Towards a definition. *Journal of Medicine and Philosophy* 11: 9–16.
Pellegrino, Edmund D. 2004. Forward: Renewing medicine's basic concept. In *Health, disease, and illness: Concepts in medicine*, ed. A.L. Caplan et al., xi–xiv. Washington, DC: Georgetown University Press.
Pellegrino, Edmund D. 2008. *Philosophy of medicine reborn*, ed. H.T. Engelhardt, Jr. and Fabrice Jotterand. Indiana: University of Notre Dame Press.
Potter, Van Rensselaer. 1970. Bioethics: The science of survival. *Perspectives in Biology and Medicine* 14: 127–153.
Potter, Van Rensselaer. 1971. *Bioethics: Bridge to the future*. Englewood Cliffs: Prentice-Hall.
Production of ICD-11: The revision process. 2007. http://www.who.int/classifications/icd/ICDRevision.pdf. Accessed 4 Oct 2010.
Reznek, Lawrie. 1987. *The nature of disease*. London: Routledge and Kegan Paul.
Romanell, Patrick. 1974. *John Locke and medicine*. New York: Prometheus.
Schaffner, Kenneth F. 1993. *Discovery and explanation in biology and medicine*. Chicago: University of Chicago Press.
Stempsey, William E. 2000. *Disease and diagnosis: Value-dependent realism*. Dordrecht: Kluwer.
Stempsey, William E. 2004. The philosophy of medicine: Development of a discipline. *Medicine, Health Care and Philosophy* 7: 243–251.
The Complete Oxford English Dictionary. 1994. England: Clarendon Press.
Virchow, Rudolf. 1981 (1858). One hundred years of general pathology. In: *Concepts of health and disease,* ed. A.L. Caplan et al., 190–195. Reading: Addison-Wesley Publishing Company.
World Health Association. 2013. International classification of diseases information sheet, *Classifications*. http://www.who.int/classification/icd/factsheet/en/index.html. Accessed 13 Feb 2013.
Wulff, Henrik R. 1981. *Rational diagnosis and treatment: An introduction to clinical decision-making*, 2nd ed. Oxford: Blackwell Scientific Publications.
Wulff, Henrik R. 1986. *Philosophy of medicine: An introduction*. Oxford: Blackwell Scientific Publications.

Chapter 6
Sanctity of Life: A Study in Ambiguity and Confusion

Kevin Wm. Wildes

Bioethics is a field that has defined itself in moral controversies. For example, bioethics has emerged as a discipline in the attempts to resolve moral controversies surrounding medical practices and health care policies in areas such as experimentation and research, abortion, reproduction, and the allocation of resources in health care. One source of constant moral controversy has been the issues surrounding death and dying. There have been controversies about the definition of death, the extent of the obligation to treat the dying, the use of resources for the care of the dying, euthanasia, and assisted suicide. From the celebrated cases in the United States of Karen Ann Quinlan and Nancy Beth Cruzan to the ruling on assisted suicide in the state of Washington (see, Compassion in Dying) the controversies of death and dying have been part of bioethics and public policy.

In the debates over moral issues in death and dying terms like "sanctity of life," "respect for life," and "human dignity" have been deployed often. These terms have been appealed to as men and women have sought to create public policy and develop moral consensus in the controversies of death and dying. Such terms are often used as the final court of appeal to justify particular moral choices or public policy in health care. Yet such terms are often heterogeneous in their meaning. Each of these terms is filled with enough ambiguity so as to bring together a wide range of hopes, images, feelings, and values that make the meaning and use of the terms very different. As a consequence people can appeal to the same term and draw very different conclusions about what should be done. In the case of Nancy Beth Cruzan one finds

Reprinted from: Kazumasa Hoshino (ed.), *Japanese and Western Bioethics*, 89–10 1. © 1997 Kluwer Academic Publishers.

K.W. Wildes, SJ, Ph.D. (✉)
Loyola University New Orleans, Marquette Hall, Room 221, 6363 St. Charles Avenue, Campus Box 9, New Orleans, LA 70118, USA
e-mail: wildesk@loyno.edu

L.M. Rasmussen et al. (eds.), *At the Foundations of Bioethics and Biopolitics: Critical Essays on the Thought of H. Tristram Engelhardt, Jr.*, Philosophy and Medicine 125, DOI 10.1007/978-3-319-18965-9_6

different parties in the controversy holding very different opinions about the case and yet each appeals to the concept of "sanctity of life."

In this essay I want to examine, in a limited way, some of the different meanings of these terms. The first section of this paper will examine how terms such as sanctity of life are used within different religious traditions. In the second section I will outline how some have tried to use these terms in general, secular bioethics. It will be argued that as these terms are progressively excised from their religious or other specific context they become so ambiguous as to be meaningless. I will illustrate this point by examining the use of such terms in the case of Nancy Beth Cruzan. The principal argument of this presentation will be that terms like sanctity of life and respect for life can only be understood within the context of a moral framework and language. When excised from such frameworks the terms become incapable to direct health care policy in secular, morally pluralistic societies and they ought to be discarded.

6.1 Religious Perspectives and the Sanctity of Life

One finds terms such as "sanctity of life" and "respect for life" used in different religious traditions. The assumption is often made that the terms are more or less equivalent in meaning. However, after examining how either of these terms are defined within a tradition it becomes clear that they are not interchangeable. While they may have a family resemblance one will not, necessarily, draw the same conclusions from their different uses in different traditions. In the Christian view it is God who is the source of life's sanctity and human life has a unique role in the created world because of its relationship to God. This view, focused on human life, is quite different from the Buddhist view that all life is intrinsically sacred. Indeed one finds that even within different religious traditions the terms take on different meanings. A consequence of such ambiguity in meaning, within and between moral traditions, is that terms like "sanctity of life" often convey very different meanings and justify different choices of action.

6.1.1 A Christian Perspective

While different meanings can be associated with "sanctity" the meaning that seems central to the term sanctity of life is the meaning of "being hallowed or sacred." This definition conveys the notion of inviolability which is what the Latin root of sanctity ("sanctitas") means. In the Christian tradition the claims about the sanctity of life seem to communicate the supposition that *human* life has an inviolability or a sacredness. One function of the claim that human life is sacred is to direct our actions. This claim of sanctity put limits on what can be done (e.g., human life ought not be taken) and makes demands on what should be done (e.g., life ought to be preserved). It is true that a fundamental and traditional concern of Christianity has been a concern not to harm human life. For example, *The Didache*, which dates to the first century, enjoins Christians: "[T]hou shalt not procure abortion nor commit infanticide" (Didache I 1965). It also condemns the murder of children (Didache V).[1] While traditional Christianity has held specific prohibitions against killing, it is too much to claim that these prohibitions form a doctrine of the sanctity of life. Joseph Boyle has argued that the phrase "sanctity of life" represents a family of values that is rarely articulated carefully even in the Christian tradition (Boyle 1989). Two themes seem to be interwoven in the Christian uses of "sanctity" when talking about human life. The first theme is that such life is holy because life is a gift from the Holy. In Christianity the origin of human sanctity is God. Life is a gift from God, who is Life, to human beings. However, this alone does not explain the Christian tradition insofar as *all life* is a gift from God and yet human life has a unique place in the created world. A second theme, or necessary condition, is that human life must have a special relationship with the Divine that sets human life apart from other forms of created life.

The uniqueness of human life, for Latin Christianity, has been explored in reflections on man as the "imago Dei" (Genesis 1:26). That is, human beings are made in the image of God. The Fathers of the Church were influenced by the view of the Old Testament that emphasized God as both the beginning and the destiny of man. Since *all* life is created, what is it that distinguishes human life as "sacred?" Human life has a "unique" status in that God impresses onto the human person God's own image and resemblance and therefore makes the human being above other beings which are God's creatures but not mirrors of the Creator. Human beings are part of the creation but they are distinguished from the rest of the creation as they are to rule as God rules. All life, since it comes from God, has a sacredness about it and demands respect, for it belongs to an-Other. The special dignity and sanctity of human life comes from bearing the image of God and the responsibility to rule like God.

The divine is expressed in the world in the human. Irenaeus best captured this patristic sense when he wrote: "Gloriam enim Dei vivens homo, vita autem hominis visio Dei" (Irenaeus 4. 20. 7 1979). The glory of God is the living human and the life of the human is the vision of God. The "glory of the human" is not a modem sense of self-improvement but the expression of the Divine in the human. Human life is made "holy" and endowed with "sanctity" and "dignity" because of its relationship to God.

Throughout the history of Christian theology and spirituality the meaning of human sanctity and dignity has been developed in different ways. For example, in the reflections of twelfth and thirteenth century Latin theologians there is a search to identify the characteristics that distinguished human life from all other forms of bodily, created life.

The epoch of scholastic philosophy was characterized by an emphasis on the intellectual and rational dimension of God's image impressed onto the human. The

[1] For an excellent overview of the Christian tradition on abortion see Noonan 1970.

expression “imago in specula rationis,” commonly used by scholastic thinkers, manifests this view. The pivotal point in this interpretation is that human sanctity consists in the human ability to know himself and God. The faculties of intellect and will, unique to man in the embodied, created world, were seen as the most divine of human attributes.

In contrast, theologians in the twentieth century have sought a less rationalistic and more balanced, integrated view of the human person. Many theologians have sought to develop a Christian anthropology that situates the rational within the relational and social context of human life (Rahner 1968; Schotsmans 1991; Wojtyla 1979). In contemporary Latin theology one finds that these two conditions for sanctity of human life have often been blended together under a theme of “stewardship.” As life is a gift and since it is a gift with a purpose, one is called to be a good steward of the life given. The two conditions for the sanctity of human life lead to different moral imperatives. First there are proscriptions against the taking of innocent human life. For example, according to the Roman Catholic tradition innocent human life can not be directly, intentionally taken. Thus there are proscriptions against suicide, abortion, and murder. At the same time the tradition has reflections not only about the protection of life but that it should be used well. Archbishop Daniel Cronin writes: “Among the natural gifts with which the Most High God have favored man, there is none so excellent as that of life, because it is life that is the basis for all else that man has or can hope to attain” (Cronin 1958, 3). In this vein there are serious and prolonged reflections on the positive obligation to preserve life insofar as it is a basis for achieving other goods (see Cronin 1958).

As one examines particular moral questions surrounding death and dying one discovers that the term “sanctity of life” has contributed very little to their discussion in the Christian tradition. Issues about suicide and euthanasia have been treated by the prohibition against the direct, intentional taking of innocent human life. The questions about the obligation to preserve life by pursuit of life-sustaining treatments have been addressed by the distinction of ordinary and extraordinary means. Questions about the definition of death have been addressed by discussion of the metaphysical questions of hyelemorphic composition and whole brain death. In no one of these areas has the term “sanctity of life” contributed to the substantive discussion. Rather, the term seems to have emerged more in secular discussions of bioethics and in the abortion debate (see Brody 1975).

6.1.2 Buddhist Perspectives

The ambiguities in terms like “sanctity of life” are made ever more clear when one contrasts a Christian tradition with other religious traditions. In Buddhism the first of the Grave Precepts is to “affirm life; do not kill.” The principle of respect for life, as it has been called, is one of the foundations of Buddhist ethics (Fujii 1991). It has been the basis for a clear-cut position against practices such as abortion (Nolan 1993, 194; Stevens 1990, 138–139). However, in Buddhist thought the principle of respect for life must be understood within the context of other aspects of Buddhist teaching as well as the other precepts. Different traditions within Buddhism balance the concern for respect for life with concern with doing "the most compassionate action" (Nolan 1993, 194). So while there is a general prohibition against a practice such as abortion, for example, such particular decisions must be made within the context of the other elements of suffering and with a view toward compassion. A decision to violate the first precept is one that should not be taken lightly but to fail to violate it, when compassion demands, it is to generate negative karma (Kaplean 1981, 228; Rinpoche 1992, 376). One finds then that the precept of respect for life needs to be interpreted within the framework of the situation and other moral demands. One also finds that the different traditions of Buddhism will make different interpretations of how to deploy the Precept.

In addressing the questions of treatment decisions at the end of life one finds different interpretations of the first precept. For some Buddhists following the first precept means to utilize whatever means of treatment and recovery are available. The argument is that human life offers an opportunity to transcend suffering through enlightenment and there is the possibility for every disease to be cured as long as life continues (Ratanakul 1988, 310). However, the Buddhist discussion of the issue needs to balance the first precept to respect life with the demands of suffering and compassion. Treatment cannot be refused simply to escape suffering but one can refuse treatment for selfless and compassionate reasons. That is, a person may refuse treatment to act compassionately towards family and friends, or to relieve burdens (e.g., economic) on them.

One finds similar sets of issues and themes arising in discussions of active euthanasia.[2] Those who disapprove of forgoing treatment in that it has the character of suicide or "death-seeking" oppose active euthanasia (Nolan 1993, 199). Yet, as one might expect, those who allow an exception do so because they appeal to compassion. What is crucial for my argument is that the precept of "respect for life" is balanced with compassion and it is this balancing that helps determine the significance of the precept. This balancing only takes place within the context of a tradition.

6.1.3 Summary

This brief overview illustrates how terms like sanctity of life or respect for life have been understood within different moral narratives. In the Christian tradition human life is sacred because it bears the divine image, while in Buddhism human life is sacred because all life is sacred. However, what is perhaps most instructive for general, secular bioethics is that even within these narratives there are ambiguities in understanding these terms in light of particular moral choices. As one examines the

[2] There is in Buddhism the practice of self-immolation which has not been understood as an act of euthanasia or suicide. Rather it is understood as the practice of giving one's self over and merging one's self into transitory reality (see Fujii 1991).

use of such terms, these ambiguities offer a warning for secular bioethics. These terms, which are difficult to define within a moral tradition, will take on so many meanings in general, secular discourse as to become meaningless. The *Cruzan* case illustrates how people with very different views of moral controversies in death and dying can reach very different conclusions by appealing to the "sanctity of life."

6.2 Bioethics and Sanctity of Life

6.2.1 General, Secular Bioethics[3]

Discussions of sanctity of life seem to have entered the English bioethics literature in the early 1950s.[4] In 1957 one finds Glanville Williams using the term "sanctity of life" in some of his jurisprudential writings (Williams 1957). The use of the term initially centered around issues such as euthanasia and abortion. In criticizing the legalization of voluntary euthanasia John Bonnell argued that Christianity has emphasized "the sanctity of human life and the value of the individual, even the humblest and lowliest, including the afflicted in mind and body" (Bonnell 1951). The article by Bonnell was in part a response to an essay by Joseph Fletcher in which Fletcher argued for the centrality of persons over mere life. Fletcher argued that one would be better served to speak of the role that persons have in deciding for themselves rather than to appeal to principles like "sanctity of life" (Fletcher 1951).

In 1964 Norman St. John-Stevas argued that the Christian attitude toward euthanasia is based on "the principle of the sanctity of life" (St. John-Stevas 1964, 43). In these early uses of "sanctity of life" in bioethics there are clear religious (particularly Christian) presumptions (Ramsey 1967). Harmon Smith, in a commentary on Paul Ramsey, speaks of the religious framework within which the notion of sanctity of life was understood (Smith 1970). He writes:

> … the question of *when* sanctity attaches to human life is not religiously problematic at all (for Ramsey): "One grasps the religious outlook upon the sanctity of human life only if one sees that this life is asserted to be *surrounded* by sanctity that need not be in a man; that the most dignity a man ever possesses is a dignity alien to him … A man's dignity arises from God's dealings with him, and not primarily in anticipation of anything he will ever have it in him to be." (Smith 1970, 42)

Sanctity of human life for Ramsey and other Christians is derived from the fact that God values human beings.

[3] This discussion of general secular bioethics and sanctity of life has grown out of many long discussions with H.T. Engelhardt, Jr. and work that we have done together (see Engelhardt 1996).

[4] There are publications prior to 1950 exploring "the sanctity of life" (see Hillis 1921; Young 1932). In bioethics there was a renewal of some of the concerns of Albert Schweitzer for a reverence for life. In the *New England Journal of Medicine* William Sperry wrote that "reverence for life" is the ethical basis of both the profession of medicine and Christian ministry (Sperry 1948, 988).

There have been some attempts in bioethics to give the principle of the sanctity of life a less religious significance. One example is the work of Daniel Callahan in his exploration of abortion. Callahan understands the problem with using a religious term like sanctity of life in a secular society. In his attempt to give sanctity of life a secular meaning Callahan writes: "An affirmation of the sanctity of life which required that one accept a religious view of man's origin would provide a weak base upon which to build a consensus. One then would seem to be saying that there is nothing whatever upon which to ground the sanctity save that of religious belief…." (Callahan 1970, 315). Callahan attempts to use the content of religious, particularly Christian, views without their theological foundations. This analysis and transformation leads Callahan to understand the sanctity of life as an affirmation of a morality that affirms "the protection and preservation of human life, both actual and potential" (p. 343). Callahan articulates a diverse collection of rules gathered under the rubric of the term sanctity of life. These include: "(a) the survival and integrity of the human species, (b) the integrity of family lineages, (c) the integrity of bodily life, (d) the integrity of personal choice and self-determination, mental and emotional individuality and (e) the integrity of personal bodily individuality" (p. 327). While he recognizes the ambiguity of the term, Callahan still attempts to derive from it some useful direction and moral sense.

K. Danner Clouser, in 1972, criticized the term "sanctity of life" for this very ambiguity. He argued that all the different meanings, and their implications, are mixed together. Clouser wrote: "I find the sanctity of life concept to be impossibly vague and to be a concept that is inaccurate and misleading, whose positive points can be better handled by other well-established concepts" (Clouser 1973, 119). William Frankena, in 1975, sorted out the different meanings that are gathered under the term sanctity of life.

1. The sanctity of bodily human life should be distinguished from that of individuality or personality. The sanctity of human life (bodily) is relevant to the discussion of questions of shortening or preventing human life.
2. Mere life, whether that of a vegetable, animal, or human organism, has no moral sanctity as such, though it may have aesthetic and other kinds of nonmoral value, and may be a necessary condition of consciousness, rationality, or morality.
3. Life has moral sanctity, but only where it is a condition of something more, as it is in human fetuses.
4. There is something inherent – consciousness, feeling, reason – in such living beings.
5. Even if the moral sanctity of human life (bodily) is not absolute, it is considerable, at least from the moral point of view, but it is only *prima facie* or presumptive.
6. The only tenable view, then, is a derivative, qualified, and noncomprehensive ethics of respect for life (Frankena 1977, 58).

Frankena's work points out the ambiguity of terms such as "sanctity of life" and "respect for life." His work also points out the conceptual problems with deploying such terms in a general, secular bioethics. In the midst of all these different accounts of these terms there is no way, in a general secular context, to pick out which is the correct interpretation. General, secular bioethics runs the risk of speaking in babel when terms like "sanctity of life" and "respect for life" are used. Indeed the case of Nancy Beth Cruzan, and the general issues of death and dying, make this babel very clear.

6.2.2 Cruzan: An Example

Despite the ambiguities in the term it has played a crucial role in certain controversies in bioethics about decisions at the end of life. The Missouri Supreme Court, for example, in its decision in the *Cruzan* case asserted that the protection of the "sanctity of life" was a *state interest* such that the state should prevent the withdrawal of feeing and hydration (Cruzan v. Harmon).

The case involved a patient who had been in a persistent vegetative state since an automobile accident. As she was anoxic for 12–14 min, Mary Beth Cruzan suffered irremediable brain damage. Subsequent to the accident a gastrostomy feeding and hydration tube was placed with the consent of her husband. When it became clear to her parents that their daughter had no chance of recovery, they sought removal of the tube. The employees of the hospital refused to comply with the request. The Supreme Court of Missouri, in a divided opinion, denied the parents' request. The state argued that there was a state interest in the sanctity of life under the *parens patriae* doctrine of common law (Payton 1992). For this essay it is important to note that the State Supreme Court, as well as several opinions of the U.S. Supreme Court upheld "sanctity of life" as a value which trumps other values. The Missouri Supreme Court wrote:

> The State's interest in life embraces two separate concerns: an interest in the prolongation of the life of the individual patient and an interest in the *sanctity of life itself*. (*Cruzan* 1988, emphasis added)

In the view of the majority, these general interests are strong enough to foreclose any decision to refuse treatment for an incompetent person unless there exists clear and convincing evidence that the person previously had made such a choice.

While asserting sanctity of life as a value that orders other values, its meaning is never made clear nor is there an argument as to why this value should trump other values. The opinions of the justices in these decisions, in fact, reflect a pluralism of moral vision and language. Indeed, Justice Stevens, of the United States Supreme Court, in his dissenting opinion in the appeal of the case, pointed out that: "Life, particularly human life, is not commonly thought of as a merely physiological condition or function. Its sanctity is often thought to derive from the impossibility of any such reduction" (*Cruzan* 1990, Stevens' dissent, Part III). Stevens' dissent represents a very different interpretation of sanctity of life from the one deployed by the State Supreme Court. For Stevens, and others, the sanctity of human life is centered on the capacities of personhood. In a sense Stevens' view is not far from the view of Medieval theologians who focused on the rational capacities as the essence of the divine image in the human. The loss of these capacities ends the obligation to sustain life.

The different interpretations of the nature of the "sanctity" lead to very different outcomes of the case. Stevens captures the difficulty of using terms like "sanctity of life." He points out "the more precise constitutional significance of death is difficulty to describe; not much may be said with confidence about death *unless it is said from faith*, and this alone is reason enough to protect the freedom to conform choices about death to individual conscience" (*Cruzan* 1990; Stevens's dissent, Part III, emphasis added). Indeed one sees in the Cruzan case two different appeals to sanctity of life. One appeal interprets the term to require aggressive medical treatment so that she can be kept alive. The other interpretation concludes that the treatment should be withheld.

6.3 Conclusion

The argument of this paper has been that terms like "sanctity of life" are so ambiguous that they can support starkly contrasting choices in the treatment of the dying. Outside the context of a particular moral community or moral narrative the principle of sanctity of life can be interpreted in at least the following ways:

1. The principle of the sanctity of life requires one to save human life at all costs.
2. The principle of sanctity of life requires that one preserve the values associated with human life and these can be jeopardized if one tries to save mere biological life.

In the second interpretation sanctity of life is not to be achieved simply through biological life, but through a self-conscious moral life. This is the interpretation that lies behind the traditional Roman Catholic distinction of ordinary and extraordinary means (Wildes 1991). There is a recognition that if one makes the mere prolongation of life an overriding good, the place of other moral goods will be disturbed (Pope Pius XII 1958). The difficulties with appeals to terms like "sanctity of life," "respect for life," or "human dignity" is that they are more like slogans than principles outside of a particular context. I have argued that such terms can be understood within the language of a moral community. Excised from such communities such terms become ambiguous and useless. That is, they bring together a number of issues and attitudes more than set out foundations or rules for choices. People with very different views rally around them. It becomes impossible to establish a canonical interpretation of the principle of the sanctity of life. Indeed if one looks at the Latin Christian tradition it is clear that "sanctity of life" has not been understood as a moral principle. Rather, one might see it as a background assumption that shaped moral principles such as the prohibition against directly intending to take human life or the distinction of ordinary and extraordinary means. It is only in recent years, in debating issues of death and dying, that such terms have been invoked and used as principles. However, as I have argued, when such terms are taken from their basic role they yield very little. The result is more like babel than language.

References

Bonnell, J.S. 1951. The sanctity of human life. *Theology Today* 8: 194–201.

Boyle, J. 1989. Sanctity of life and suicide: Tensions and developments within common morality. In *Suicide and euthanasia*, ed. B.A. Brody, 221–250. Dordrecht: Kluwer Academic Publishers.

Brody, B.A. 1975. *Abortion and the sanctity of human life: A philosophical view*. Cambridge: MIT Press.

Callahan, D. 1970. *Abortion: Law, choice and morality*. New York: Macmillan.

Clouser, K.D. 1973. "The sanctity of life": An analysis of a concept. *Annals of Internal Medicine* 78: 119.

Compassion in Dying *et al.* v. The State of Washington, U.S. District Court, No C94-119R.

Cronin, D.A. 1958. *The moral law in regard to the ordinary and extraordinary means of conserving life*. Rome: Pontifical Gregorian University.

Cruzan v. Harmon, 760 S.W.2d 408 (MO.Banc 1988).

Didache. 1965. *The Apostolic Fathers*, vol. 1. Trans. K. Lake. Cambridge, MA: Harvard University Press.

Engelhardt Jr., H.T. 1996. *The foundations of bioethics*, 2nd ed. New York: Oxford University Press.

Fletcher, J. 1951. Our right to die. *Theology Today* 8: 202–212.

Frankena, W.K. 1977. The ethics of respect for life. In *Respect for life*, ed. O. Temkin, W.K. Frankena, and S.H. Kadish. Baltimore: Johns Hopkins Press.

Fujii, M. 1991. Buddhism and bioethics. In *Bioethics yearbook: Volume 1*, ed. B.A. Brody, B.A. Lustig, H.T. Engelhardt Jr., and L. McCullough. Dordrecht: Kluwer Academic Publishers.

Hillis, N.G. 1921. *The better American lectures*. New York: Better America Lecture Service.

Irenaeus. 1979. Against heresies. In *The Ante-Nicene fathers*, ed. A. Roberts and J. Donaldson. Grand Rapids: W. Eerdmans.

Kaplean, P. 1981. *To cherish all life: A Buddhist view of animal slaughter and meat eating*. Los Angeles: Zen Center Publications.

Nancy Beth Cruzan vs. Director of Missouri Department of Health, 111 L.Ed.2d 224, 110 S. Ct.284, 58 U.S.L.W. 4916 (D.S. June 26, 1990).

Nolan, K. 1993. Buddhism, Zen, and bioethics. In *Bioethics yearbook: Volume 3*, ed. B.A. Lustig, B.A. Brody, H.T. Engelhardt Jr., and L. McCullough. Dordrecht: Kluwer Academic Publishers.

Noonan, J. 1970. An almost absolute value in history. In *The morality of abortion*, 1–59. Cambridge, MA: Harvard University Press.

Payton, S. 1992. The concept of the person in the *parens patria* jurisdiction over previously competent persons. *The Journal of Medicine and Philosophy* 17: 605–645.

Pope Pius XII. 1958. The prolongation of life. *The Pope Speaks* 4: 393–398.

Rahner, K. 1968. *Spirit in the world*. New York: Herder and Herder.

Ramsey, P. 1967. The sanctity of life. *The Dublin Review* 241: 3–23.

Ratanakul, P. 1988. Bioethics in Thailand: The struggle for Buddhist solutions. *The Journal of Medicine and Philosophy* 13: 301–312.

Rinpoche, S. 1992. *The Tibetan book of living and dying*. San Francisco: Harper.

Schotsmans, P. 1991. When the dying person looks me in the face: The ethics of responsibility for dealing with the problem of the patient in a persistently vegetative state. In *Birth, suffering, and death: Catholic perspectives at the edges of life*, ed. K. Wildes, F. Abel, and J.C. Harvey. Dordrecht: Kluwer Academic Publishers.

Smith, H.L. 1970. *Ethics and the new medicine*. Nashville: Abingdon.
St. John-Stevas, N. 1964. *The right to life*. New York: Holt, Rinehart and Winston.
Stevens, J. 1990. *Lust for enlightenment*. Boston: Shambhata Publications, Inc.
Sperry, W.L. 1948. Moral problems in the practice of medicine. *New England J Med* 239: 985–990.
Wildes, K.Wm. 1991. Life as a good and our obligation to persistently vegetative patients. In *Birth, suffering, and death: Catholic perspectives at the edges of life*, ed. K.Wm. Wildes, F. Abel, and J.C. Harvey. Dordrecht: Kluwer Academic Publishers.
Williams, G.T. 1957. *The sanctity of life and the criminal law*. New York: Knopf.
Wojtyla, K. 1979. *The acting person*. Dordrecht: D. Reidel Publishing Company.
Young, D.R. 1932. *The sanctity of life: Secular and mid-week services*. London: Epworth Press.

Chapter 7
A Transcendental Argument for Agreement as the Sole Sufficient Basis of a Philosophical Ethic

George Khushf

7.1 Introduction

In his *Foundations of Bioethics*, H. Tristram Engelhardt seeks to develop a sparse ethic that can be used by moral strangers to peaceably interact with one another. Moral strangers are agents who do not share a common, content-full vision of what is right and good. This absence of a shared vision potentially undermines the possibility of sustaining an ethical framework that enables agents to resolve controversies without deploying force.

The role of "peaceable resolution of conflict" in framing his book and his ethic is difficult to specify. Sometimes "peaceable resolution of conflict" seems to be a postulated *end*, and his ethic for moral strangers seems to be worked out as a *means* for advancing that end. For example, Engelhardt states that his approach to ethics "*requires* only *a decision to resolve moral disputes in a manner other than fundamentally by force*" (1996, 68–69, emphasis added; Engelhardt emphasizes the "only" when making his argument, but he is advocating a decision none the less). Further down on the same page he summarizes:

> By appealing to *ethics as a means for peaceably negotiating moral disputes*, one discloses as a necessary and sufficient condition (sufficient when combined with the decision to collaborate) for a general secular ethics the requirement to respect the freedom of the participants in a moral controversy (i.e., in the sense of gaining their permission for using them) as a basis for common moral authority (i.e., from the permission of those collaborating). Since moral controversies can in principle encompass all moral agents (and, as we shall see, only moral agents), one has a means of characterizing the secular moral community as the possible intellectual standpoint of persons interested in resolving moral controversies in ways not fundamentally based on force. (1996, 69, emphasis added)

G. Khushf, Ph.D. (✉)
Department of Philosophy, University of South Carolina, Columbia, SC 29208, USA
e-mail: khushfg@mailbox.sc.edu

L.M. Rasmussen et al. (eds.), *At the Foundations of Bioethics and Biopolitics: Critical Essays on the Thought of H. Tristram Engelhardt, Jr.*, Philosophy and Medicine 125, DOI 10.1007/978-3-319-18965-9_7

From these claims, it seems reasonable to conclude that a "decision" to resolve disputes peaceably (without force) is made at the beginning, and Engelhardt's secular ethic is worked out "as a means" for realizing this end. The postulated end – peaceable resolution of conflict – imposes constraints on the ethic, which would serve as a means to attaining that end. Specifically, "peaceable resolution" has a meaning that, when unpacked in a transcendental exposition, implies that the ethic that provides the means must be an ethic that involves a commonly accepted moral authority, and the only possible candidate for such a commonly accepted moral authority is the agreement of those who will to collaborate.

When interpreted in such a way, Engelhardt's analysis requires postulation of some interest or end at the outset. Peaceable action is valued, posited as an end. Further, there are actual instances of such peaceable action. Peaceableness is a quality of some interactions among agents. The function of the "transcendental argument" is to clarify the conditions of the possibility of attaining that end, i.e., attaining such a quality of peaceableness in all agent interactions. Such an approach acknowledges many possible variants of ethics, and the ethic for moral strangers is simply that version of ethics we must have if we start by postulating an interest in "peaceable resolution of conflict." The advantage of approaching the foundations project in such a way is that we have a potentially clear criterion for determining which satisfaction states matter.[1] We are not just interested in any old kind of resolution of a conflict. We are interested in those resolutions that don't utilize force or appeal to force. We can thus start with an analysis of what force is and how we can tell when it is used. We then seek to develop an ethic that is instrumentally effective in attaining resolutions of controversies that don't utilize force or appeal to force. At the end, we can test the success of our project by examining whether the ethic that is provided does, in fact, provide a means that moral strangers might use to resolve controversies in such a way that their resolutions don't involve appeals to force. Two logics would intersect: one logic of force provides an independent basis for analyzing some resolution and determining whether it involves force. A different logic associated with the ethic provides a set of tools for resolving controversies in a specific way. We can compare the resolution accomplished by the ethic and the resolution required by the end and see if they are in accord.

[1] This interpretation is defined by two commitments: (1) A recognition that an independent account of force is required, and (2) explicit positing of peaceableness as a value, end, and good. When, for example, Engelhardt defines a person's innocence as "they have not used force against others without their consent" (1988, 387–388), he must assume some framework that enables him to determine whether force has been used. His ethic does not provide criteria for determining when force has been used. Some independent account is needed for determining that. In another early essay, Engelhardt presents his secular ethic as an effort to "sketch … the consequences of a commitment to a peaceable community" (1973, 1982, 65). Secular ethics "presumes only that one is interested in resolving issues in ways not directly grounded in unconsented force against the innocent. It also defines 'innocence' in terms of not engaging in force against the unconsenting who similarly eschew force" (67). In these accounts, peaceable community and notions of force enter into the very constitution of ethics.

Engelhardt should advance exactly such an interpretation, and he seems to do so in some of his earlier works. But he moves away from such an interpretation in the second edition of his *Foundations of Bioethics*. He seems to think that positing an end such as peaceableness would undermine the general contrast he wants to make between a purely procedural account (his contentless ethic for moral strangers) and a content-full ethic.[2] Just one page after the above-mentioned citation, Engelhardt takes back what he seemed to say about the role of peaceable resolution as an end, about the decision that frames the project, and about the way his ethic functions as a means to attaining that end. Here I quote at length the paragraph that explicitly excludes any interpretation that posits peaceableness as an interest, concern, commitment, decision, choice, or end:

> *This view of ethics* and bioethics [i.e., the ethic for moral strangers that Engelhardt presents] *is not grounded in a concern for peaceableness*. It is *not based on an interest in establishing peaceable community*. This view cannot be shown in general secular terms to be good, praiseworthy, or rationally to be desired. It should, *instead*, be recognized as *a disclosure*, to borrow a Kantian notion, *of a transcendental condition*, a necessary condition for the possibility of a general domain of human life and of the life of persons generally. *It is a disclosure of the minimum grammar involved in speaking of blame and praise with moral strangers, and for establishing a particular set of moral commitments with an authority other than through force*. (1996, 70)

[2] Engelhardt uses his contrast between a purely procedural ethic and a content-full ethic (1996, 126) to define and clarify several other contrasts: between a secular ethic for moral strangers and diverse forms of ethics for moral friends; a deontological ethic and a consequential ethic; a principle of permission and a principle of beneficence; the right and the good; society and community; Abstract Right and Sittlichkeit; negative rights and positive rights. These strata are not independent. Rather they involve co-implicated concepts that Engelhardt uses to clarify what is required for his ethic of moral strangers. But there are other ways to develop the contrasts that are in tension with his dominant interpretation. For example, one other way to present his contrast involves a hierarchy in the specification of the good. At the most general, abstract level, the good is specified in terms of the interests of a purely rational agent, who values contents in terms of their transparency to thought. The ethic for moral strangers provides the most general categorial structure integral to any other ethic. Any communal ethic completes the content by filling in the noncategorial, contingent strata until a complete account of the good is specified. In his earlier work, Engelhardt tended to such an interpretation. His secular ethic gave "the most general sense of what it means to speak of an ethical order" (1973, 1982, 67), and any more specific ethic of community needed to be worked out in ways that were consistent with this. This preserved a dialectical relation between the two (1973, 1982, 76), and reflected a Hegelian confidence in the capacity of reason (1973). (Engelhardt presents this as Hegel over Kierkegaard, 1973, 1982, 77; it is also Hegel over Kant 1987, 389; i.e., Sittlichkeit over Abstract Right. But here emphasis falls on the increasingly comprehensive categorial specificity of categories like community, society, and state, rather than the contingent givenness which is regarded as an indifferent negativity from the categorial standpoint. In his later work, the emphasis shifts to the contingent givenness, e.g., 1994b). This approach to secular ethics (as the most comprehensive, but also involving the most abstract specification of the end/good) assumes that the ethic for moral strangers does have some discursively available content that is included in any other ethic, including that of a robust religious community. This seems to be the reason he moves away from such an interpretation. However, while he moves away from viewing the most general ethic as a sparse specification of the good, i.e., the good of a pure rational agency, he at times seems to miss the way the valuation integral to such an intellectual standpoint informs his ethic and makes it more than "purely procedural."

Note a curious aspect of the final sentence: the transcendental exposition will disclose the "minimum grammar" for two things: (1) speaking blame and praise with moral strangers, and (2) establishing commitments with an authority other than through force. In the second clause, force again enters. If we interpret the final sentence as a kind of conjunctive constraint, then it is extremely difficult to see what Engelhardt means when he says that his view of an ethic for moral strangers "is not grounded in a concern for peaceableness." One can in a straightforward way take the two parts as nothing but an exposition of the properties an ethic must have if it is to be the required means for peaceably resolving differences.

We may, however, come up with an alternative interpretation. An analogy might be drawn from the domain of mathematics. Imagine an Ancient Greek who begins a discussion of geometry by first considering some practical problem that arises for surveyors mapping land. The problem is of such a character that it cannot be solved by using their intuitive sense of space, measurement, and their surveying tools. The Ancient Greek might ask: is there a way of developing geometry that is fully general and independent from any consideration of chains and fields? In this case, we get a statement that roughly tracks the one Engelhardt has for his ethic: the counterpart of a pure geometry is Engelhardt's grammar of speaking of blame and praise with moral strangers (his ethic for moral strangers), and the counterpart of the "without chains and fields" is Engelhardt's "without force." As geometry might be "applied" to surveying fields, so Engelhardt's ethic for moral strangers might be "applied" to resolving controversies without force. In this case, the account that is provided of ethics would in no way depend on any prior logic of force or peaceable resolution. While the ethic could be "used" as a means, it would not be derived as a means in a way that makes the derivation depend on the meaning of peaceable resolution or force.

This second interpretation is, in fact, Engelhardt's favored one. At several places, he clearly offers his ethic and his moral authority as something that can be worked out in a way that is completely independent from any valuation of or commitment to "peaceable community" and any logic of force. He insists that "even peaceable action is not to be valued" (1996, 126). No descriptive framework for determining when force is deployed is ever considered, and no final justification is ever given that the ethic and moral authority he provides does, in fact, provide a sufficient means for resolving disputes in ways that don't involve uses of force or appeals to force.[3]

While Engelhardt advances the second interpretation, I will argue that he needs the first. In this essay I attempt to reconstruct the transcendental argument Engelhardt

[3] Engelhardt acknowledges in several places that his ethic might not be sufficient. He seems to allow for two kinds of insufficiency. The first concerns possible tragic contexts where considerations related to the good are so great that they override the forebearance strand. The second concerns contexts where "problems stand indefinitely in a state of insolubility" (1996, 130). He presents both of these under the heading of "TEYKU problems" (129–131), but there might be forms of insolubility that only assume the forebearance strand; i.e., two agents committed to resolving disputes without appeal to force may find themselves in a controversy where use of force or appeal to force is unavoidable, and thus the interaction necessarily is not peaceable.

provides for his ethic. I argue that there is an unacknowledged ambiguity in his transcendental exposition that reflects divergent ways Kant and Hegel might interpret a transcendental argument (for an overview of this contrast, see Engelhardt 1994b, 2000, 80–96; Pinkard 1988, 111–151). Engelhardt fused these elements, but without sufficient consideration of the difficulties associated with aligning a categorial account of controversies with actual controversies. There is a discretion in an interpretive task faced by actual agents in controversies that allows agents to extend those controversies into meta-controversies regarding whether force is used in resolving a controversy and whether agreement and permission are attained. This leads to a regress. If the requirement of agreement is extended from an ideal agreement to an actual agreement, and if the actual agreement concerns the meta-controversies as well as the controversies, the resolve to act in ways that involve agreement can imply a host of very strong constraints on actual agents. Instead of a minimal ethic like that advocated by Engelhardt, his requirement can lead to a maximal ethic. The inversion arises because of an additional ambiguity associated with Engelhardt's moral standpoint: if an agent assumes the stance of a purely rational agent, this brackets away the contingent reasons agents have for acting as they do. If the postulated agreement is a universal basis, then this must be taken as an ideal resolve, not the brute factical resolve of actual agents. When an actual agent takes up that resolve, this commits the agent to the conditions necessary for assuring an actual agreement that can be mutually acknowledged by agents in the controversy as an agreement of the right kind. Since the moral standpoint requires the actual agent to bracket contingent commitments when assuming the stance, the contingent commitments of that actual agent don't have standing when the agent reflectively adjusts his commitments so these come into accord with the required resolve. This biases resolution of indifferent matters in the opposite way Engelhardt supposes: the minimal content at the categorial level leads to the maximal requirement that contingent matters must be reconstructed in ways that assure an actual agreement that is mutually acknowledged by agents as having the right kind of fit with the general agreement that is postulated as the sole basis. To avoid this conclusion, Engelhardt needs to define his ethic in relation to judgments about legitimate uses of force. This makes the "use" of the ethic for "peaceable resolution of conflict" constitutive of the way the game of blaming and praising is defined. This approach would require an independent account of force which has not yet been provided.

7.2 Disentangling Four Strands of Argumentation in Engelhardt's *Foundations of Bioethics*

Before I try to reconstruct some of the systematic parts of Engelhardt's secular ethic, I need to briefly address a challenge any interpreter of his work faces. One of the most beautiful but difficult aspects of his writings concerns the ways Engelhardt weaves together many strands of historical and philosophical analysis in mutually supporting ways. His secular ethic is situated in the context of a rich historical analysis of Enlightenment aspirations and their failure. Constructive parts of his secular ethic depend on creative reinterpretations of philosophers like Kant and Hegel, and many of his claims arise as a result of critical analysis of a host of contemporary works in ethical theory. At a yet broader level, Engelhardt works out his own meta-ethical language for mapping options, for example, his account of moral friends, affective friends, and moral strangers. As a result of these complex, overlapping strands of analysis, it becomes very difficult to determine exactly what evidence and which arguments are supporting which claims. There are usually a host of parallel strands of argumentation that converge to support any specific claim, with each supporting strand qualifying and conditioning all of the others. This, in turn, makes it difficult to isolate problematic features of his arguments and respond to them. I will refer to this as the "problem of entangled argumentation."

In this essay, I seek to disentangle Engelhardt's transcendental argument for his secular ethic and consider its merits in isolation from the other strands. He explicitly states that this part of his ethic is independent from the other parts and lexically prioritized (1996, 115). But there is no place where he fully isolates the argument in the way I will now suggest. I will try provide some arguments for why he needs to isolate the transcendental argument, but a full justification of my claims would require that I disentangle and independently evaluate the work done by all of the strands he weaves together. Only after this is done could I identify what work the transcendental argument must perform if Engelhardt is to sustain the full set of claims he makes. Such a critical appraisal would be as ambitious as Engelhardt's own project, and it would require a capacity to see how all of those strands of analysis align in just the right way. The one thing I have learned from many years of discussion with Professor Engelhardt is that he is always discerning and aligning many levels of analysis that I simply cannot assemble in one mental act. Whenever, with great effort, I strain to keep together the strands he presents, I usually find myself moving toward the claim he is making. Alternatively, when I think his argument has some gap, that usually indicates I have missed something. The gap is in my capacity to gather the strands, not in his argument, as I initially assumed. I think this feature of his work is one of the marks of a great philosopher. With most works, it is easy to quickly discern the problems. The more carefully you read, the worse the problems get. But a great work somehow remains ahead of you, so when you stretch a little further on your next reading, you find that parts link up even tighter rather than pulling apart. Eventually you hope you can get to a place where you might play your own little part in working the arguments forward.

Engelhardt is a great philosopher in just this way. I provide my criticisms as part of an attempt to properly understand the argument and reconstruct it in a stronger form. But the end of my effort might just as likely be that I will better see what I have missed in his argument. That too would be a good outcome, especially if others might also benefit from the clarification. Either way, I had to stretch many times to try to see the strands align. Now I try to isolate four of these strands, so I might tease apart what is essential for the transcendental argument that I want to consider in isolation.

The following four strands play a role in framing Engelhardt's account of his ethic for moral strangers:

A Meta-language for Describing Ethical Theories Engelhardt makes a distinction between robust accounts of the right and good that can provide guidance for how individuals should live their lives and a sparse, procedural ethic that might provide guidance for how moral strangers can resolve controversies without appeal to force (1996, 9). A full ethical theory provides a "content-full account of the right and good." He associates such accounts with "moral friends" (1996, 7). The commitments shared by moral friends enable them to resolve disputes they have with one another about what is right and good (1996, 24–25, note 13). In contrast, "moral strangers" lack such a common basis for conclusively resolving questions of the good. Moral strangers may pursue their ends in ways that bring them into open conflict. Where accidental alignments of interests might arise, moral strangers might collaborate to advance those interests. Engelhardt refers to the agents in such contingent associations as "affective friends" (1996, 7). Unlike moral friends, affective friends only have partial overlaps between their moral worlds. Where moral strangers come into conflict with one another, they may seek to peaceably resolve their differences, rather than resolve these by force. The question is then raised about whether it is possible to create conditions for peaceably resolving differences, i.e., whether an ethic for moral strangers might be developed that specifies a common basis that is recognized by opposing parties as having moral authority. He answers this affirmatively, and seeks to show how agreement provides the sole sufficient basis for such an ethic among moral strangers. With a stipulative definition worked out in relation to Hegel's Philosophy of Right, Engelhardt will say that moral friends are bound up with a community, while moral strangers are bound up with society (1996, 7, 74–78, 85).

Engelhardt uses his meta-language associated with moral strangers and moral friends to organize the various strands of his argumentation in *The Foundations of Bioethics* (my primary focus), and also in many of his other writings (e.g., 1991, 2000). With great oversimplification of his rich historical and philosophical analysis, we might now identify the following mutually supporting strands:

Historical Stages in the Development of a Post-modern Pluralism One motive for developing a rationally grounded ethic arose in the context of the wars among Catholics and Protestants during the Reformation. The different religious traditions provided alternate frameworks for resolving disputes among moral friends, but could not be used in ways that would be accepted as legitimate by those outside the religious tradition. By developing non-religious accounts, philosophers attempted to discover a universal basis for legitimating social uses of force (1996, ch 1; 2000, chs 1–4). This sets up the aspirations of modernity to work out a single, universally accessible basis to legitimate common social action (1996, 20–23, note 7 gives Engelhardt's historical stages and modernity). Over time, however, it became clear that the philosophical accounts advanced in the name of reason were just as particularist as the religious accounts.[4] In fact, as Engelhardt notes, "more people have been slaughtered in the cause of secular visions of justice, human dignity, ideological rectitude, historical progress, and purity than have ever been killed in religious wars" (15). Today, philosophical ideologies need to be viewed in the same way religious traditions were regarded when their advocates battled for state power. As modernist hopes in the prospects of an unbiased reason were shown to be false, we come to a post-modern context where there are again many competing traditions, only now the various competitors usually are advanced in the name of some kind of secular rationality (1996, 35–65). The net effect of this history is to overthrow any relevance for the distinction between philosophical and religious approaches to ethics. For Engelhardt, the crucial difference is between an ethic that involves contingent commitments that some people don't share (an ethic for affective or moral friends) and an ethic that would only assume what anyone must accept as a purely rational agent (an ethic for moral strangers). Nearly every ethical theory falls into the first category, and is thus no more or less fitting than any religious tradition as an ethic that moral strangers might use.

Demonstration that Contemporary Ethical Theories Should Be Taken as Ethics for Moral Friends, Rather Than as Ethics for Moral Strangers Engelhardt seeks to show why all attempts to justify a content-full ethic fail. He will argue that every available option appeals to a standard that is not itself universal. At this stage in his analysis, he moves from philosophical history to an analysis of contemporary ethical theories. He first presents what he regards as the available options for a standard or basis of an ethical theory. Here I quote his full statement of these options:

> Standards in ethics and bioethics can be sought (1) in the very content of ethical claims themselves, in intuitions, in what appears to declare itself to be self-evidently right or wrong, or at least (2) in a study of exemplar cases; (3) in the consequences of moral choices; (4) in a notion of unbiased choice, the ideal of an impartial observer, or a group of unbiased contractors; (5) in rational moral choice or discourse itself; (6) in game theoretical (or prisoner dilemma) account of the problems of social interaction; (7) in the character of reality

[4]In his book on secular humanism (Engelhardt 1991), he seeks to tease apart the philosophical strands and isolate the variant of philosophical ethics that is distinct from the rest. There is an interesting parallel between the way Engelhardt describes stages of Western history and the development of his own views about the prospects of philosophy. He notes in the first edition of his *Foundations of Bioethics* (1986, viii) that his recognition of the limits of philosophy came as an unhappy discovery: "I have endeavored to find grounds for establishing by reason a particular view of the good life and securing by general rational arguments the authority for its establishment. To my dismay and sorrow, such have not been available." He describes the transition from modernity to post-modernity in a similar way. It would be interesting in another study to explore where, exactly, this discovery came, since some aspects of his sparse ethic go very far back and reflect commitments that antedate his philosophical writings (e.g., his view of Texas, 1986, 46–49). A comparison of the first edition (1986) and second edition (1996) of the *Foundations of Bioethics* shows that the language of ontology was much more prominent in the first edition. This theme traces back to his dissertation (published as a book in 1973). But we also see a shift in more subtle uses of language; for example, he earlier would speak more of "insights" that "distinguish" Western philosophical ethics (as a privileged approach to ethics because of its generality (1982, 67). In his later works, he tends to speak of the "presumption" of such a Western tradition and regards most work on philosophical ethics as heretical versions of Christianity (2000).

> or nature; (8) in an appeal to middle-level principles; or (9) to some moral reference point that can canonically direct moral choice. More broadly, a standard can be sought in the content of moral thought (e.g., in intuitions), in the form of moral reasoning (e.g., in the idea of impartiality or rationality), or in some external objective reality (e.g., consequences of actions or the structure of reality) (Engelhardt, 1996, 41).

I will refer to this statement as Engelhardt's "disjunction of ethical standards." After presenting his disjunction, Engelhardt attempts to show how the most prominent ethical theories associated with each standard involves a tradition, contingent historical frame, or an arbitrary choice that disqualifies that theory and its associated standard from serving as a basis for an ethic of moral strangers (1996, 40–65). This analysis brings us to "the brink of nihilism" (1996, 65–67) and sets up his constructive, systematic project.

Transcendental Exposition of the Basis of an Ethic for Moral Strangers By means of a transcendental exposition of the conditions of possibility for sustaining ethical discourse among moral strangers, Engelhardt seeks to show that the sole sufficient basis for a secular ethic is agreement (1996, 67–72; 94–97, notes 82–88). This is the part of Engelhardt's argument I seek to reconstruct.

Before moving to the transcendental argument, I consider how the above-mentioned four strands of argumentation work together to justify the claims that Engelhardt makes. Specifically, I seek to show that the transcendental argument must be able to function in a modular way for Engelhardt's account to work. The transcendental argument provides THE foundation of the *Foundations of Bioethics* project. It is the heart of his secular ethic. Engelhardt's historical analysis (associated with the emergence of post-modern pluralism) and his demonstration that contemporary ethical theories are for moral friends (associated with his disjunction of ethical standards) can both only provide inductive support of his claims. These at best set up his transcendental argument. Further, Engelhardt needs an argument for why any content-full ethic is necessarily an ethic for moral friends and not moral strangers, and this argument must be different from the argument he bases on his disjunction of ethical standards. Finally, Engelhardt's meta-language for describing and categorizing ethical theories (associated with his account of moral friends and moral strangers) entails non-trivial commitments that are only justified if his transcendental argument is successful in isolating the conditions for the possibility of an ethic for moral strangers. Thus all the strands of Engelhardt's argumentation trace back to the transcendental foundation.

Why can't inductive, historical, or interpretive argumentation and analysis do the work Engelhardt needs? The answer relates to the modal status of both the positive and the negative claims Engelhardt makes. Negatively, he claims that the project of a Western, philosophical ethic has failed, and that no content-full ethic can be justified in a way that all rational agents must accept. This is a strong modal claim. It cannot be supported simply by surveying *some* efforts at developing a content-full philosophical ethic, and then showing that they fail. To justify the negative claim, some kind of analysis is needed that shows that given the way the task of philosophical ethics is posed it has conditions of possibility that cannot be satisfied. Only such an account could justify the modal claim about the impossibility of a content-full ethic for moral strangers. For Engelhardt, the negative claim is coupled with his positive claim, namely, that his ethic for moral strangers does provide a minimal satisfaction (his ethic for moral strangers) and his negative claim (that no content-full philosophical ethic can be provided) relate to the way he frames the project of a philosophical ethic in terms of a justification. While the project, as posed, and the associated "moral viewpoint" is "fashioned under the force of various sociohistorical influences," the project has a significance that is not "merely Western." For Engelhardt, this significance and this relevance beyond the Western context depends on the way the sought-for ethic hinges on justifications that compel "concurrence of persons as such." (All of the above quotes are from 1996, 38.) In this way, Engelhardt sets up the project of philosophical ethics in a way that requires a basis or moral authority that does not depend on anything that would distinguish one viewpoint from another. The account must be developed in a fully intersubjective way.

By linking the project of philosophical ethics to the demand for a universal justification, Engelhardt establishes his criterion for sifting the claims of specific ethical theories. If a claim or basis of an ethic involves some contingent content that is not accepted by all agents qua rational, then such an ethic would not count as the required ethic for moral strangers (his negative claim). To justify his ethic for moral strangers (the positive claim), he needs to show that there is at least one way of doing ethics, i.e., sustaining the game of blaming and praising, that does not make an appeal to some content that would not be universally accepted by anyone who plays that game. However, he doesn't imagine that his justification is one that will be accepted by everyone. He is not appealing to every actual agent, because many such agents will not consistently uphold the moral standpoint. Thus the justification for his secular ethic must take the form of an argument that shows that if one properly discerns the task of a philosophical ethic and assumes the viewpoint or moral standpoint of an agent who is providing a justification in the required way, then the sole condition for the possibility of such an activity is found in selection of just that basis he advances as the foundation of his ethic for moral strangers. An argument that justifies a modal claim about what is necessary for something else that is posited or given to be possible is a transcendental argument. Given the way Engelhardt has posed the project of a philosophical ethic, and given the strong modal claims he makes (both positive and negative), he thus needs a transcendental argument.

To see why Engelhardt's argument associated with his disjunction of ethical standards only provides inductive support and thus cannot support his negative modal claim that a content-full philosophical ethic is impossible, we must not be led astray by the deductive form of his argument. Engelhardt's argument against the possibility of a content-full secular ethic takes the form of a long proof by cases that is based on his disjunction of ethical standards (40–65). He attempts to show that the standard associated with each disjunct requires a contingent content that disqualifies an ethical theory using that standard from serving as an ethic that any moral strangers can use to resolve their disputes peaceably. After showing that each standard associated with the nine disjuncts requires some contingent content and is thus not universal in the required way, he concludes that no content-full ethical theory can provide an ethic for moral strangers. While I think there are some problems with the way he develops some of the branches in his argument, I'll allow that his argument demonstrates that all nine standards involve contingent commitments, and thus any ethical theory utilizing any of these standards involves some contingency.[5] Even if this is conceded, his argument is only sound if his disjunction of ethical standards provides an exhaustive set of possibilities, and if introduction of contingency into a standard used by an ethical theory demonstrates that the theory is insufficient for sustaining the game of blaming and praising among moral strangers. I'll consider each of these in turn.

First, Engelhardt has not shown that his disjunction of ethical standards provides an exhaustive set of possibilities. Since his ethic for moral strangers will count as an ethic, and since Engelhardt claims that ethic does not depend on any of the standards included in his disjunction, there is clearly at least one kind of ethic that is not

[5] Engelhardt's overview of diverse approaches to ethical theory does clearly and convincingly make one crucial point: that prominent philosophical approaches to ethics are just as dependent on contingent contents as the religious approaches they were supposed to replace. His criticisms of a Roman Catholic natural law ethic are of exactly the same kind as his criticisms of consequential, Rawlsian, and discourse ethics, and they make clear that none of these approaches are more or less fitting for a secular, pluralistic society. His overview calls into question the way civic discourse is normally framed. What usually is presented as a demand for ethical justifications that are universally accessible, i.e., based on reason, turn out to be an effort to impose a secular, naturalistic content-full vision (such as the quasi-religious versions of secular humanism he discusses in 1991) at the expense of religious content-full visions. It is a pity that Engelhardt has so closely tied his criticisms of philosophical ethics – i.e. that they are just as particularist as the religious traditions they are supposed to replace – to his strong modal claims about the impossibility of a content-full ethic for moral strangers. The general claims he makes about the contingent character of philosophical ethics depend on very different evidential criteria than the strong modal claim he makes about the impossibility of a content-full philosophical ethic. His arguments for the former are compelling and have significant, unappreciated implications for the way ethical discourse takes place in our society. His strong modal claim is far more difficult to justify. If we evaluate Engelhardt's arguments for that strong modal claim, we can notice two kinds of problems. First, when he shows that specific ethical theories utilizing each of the standards involve contingent, contestable content, he usually does not demonstrate that any ethical theory using such standards must involve contingent, contestable contents. For example, he first considers intuition (1996, 42–43). While the examples he gives do show how various appeals to intuition utilize contingent, contestable contents, they do not show that there is no way to draw on just those intuitions that all moral agents must share. To show that, he needs to provide an account of what is necessary and sufficient for a moral standpoint that must be assumed by anyone who properly plays the game of blaming and praising, and then show such a standpoint does not involve universally shared intuitions. In other words, the explication integral to his transcendental argument is actually a condition for making the strong modal claim he wants to make about the impossibility of a content-full philosophical ethic. The second problem with the argument based on his disjunction of ethical standards concerns the relations between the nine possibilities he considers. He presents them as independent, e.g., an ethical theory based on intuition is independent from one based on exemplary cases, and that from consequences, and so on. However, there might be an ethical theory that uses sparse, generally available intuitions, representative cases, and so on, and unites these in such a way that the combination yields a content that cannot be obtained when these standards are considered independently. In sum: even if his disjunction of ethical standards mapped the whole possibility space of options, Engelhardt's justification of a strong modal claim about the impossibility of a content-full, philosophical ethic for moral strangers would be at best problematic.

covered by his disjunction.[6] His argument that no content-full ethic can serve as an ethic that moral strangers use to peaceably resolve disputes then depends not just on the proof by cases associated with his disjunction, but also upon his transcendental argument providing the secular ethic together with something like a transcendental dialectic demonstrating that the ethic for moral strangers cannot be extended to a content-full ethic.[7] Even this extended argument would require that the options

---

[6]A tenth approach to ethics that would cover Engelhardt's argument might be called "ethical theories based on transcendental arguments." A recent appraisal of the role of transcendental arguments in grounding ethics is provided by Illies (2003). Alternatively, Engelhardt's ethic for moral strangers might be covered by one of his nine options; for example, his fifth standard (1996, 52–55) which concerns norms of rational discourse. Engelhardt's ethic for moral strangers might be regarded as an ethic that emphasizes norms of rationality, but restricts rationality to just that content and resolve which is necessary and sufficient for sustaining the game of blaming and praising. His dispute with Habermas and others would then relate to whether any content can be introduced into ethics, i.e., whether there are norms of rationality that are not just negative and thus go beyond Engelhardt's worry about rigorous justification (38), contradictions (105–108), and irrationality (104). When emphasizing the Kantian aspects of his work, Engelhardt highlights the will, rather than reason: "authority is derived not from reason … but from the bare will to have one authority moral strangers can share: permission. Secular moral authority is derived from a bare will to morality" (72). But when he emphasizes the Hegelian aspects of his work, general norms of rationality govern: concerns of finite knowing and willing are placed within a general categorial framework, and the norms of rationality are worked out in relation to a systematic ordering of the categories. These alternative ways of understanding his ethic for moral strangers – Kant vs Hegel – map to the alternative interpretations discussed in the opening section.

[7]Here I am using "transcendental dialectic" in Kant's sense as a "critique of dialectical illusion," an illusion that arises with efforts to use logic, as the form of understanding, as an organon for the production of objective assertions. Kant's transcendental dialectic thus involves "a critique of understanding and reason in respect of their hyperphysical employment" (e.g., Kant 1965 [1781/1787], 99, 100). This definition from Kant's first critique is qualified in relation to knowledge. In the second critique, Kant's dialectic concerns inappropriate inferences that move from the formal aspects of right to the supersensible conditions or highest good (Kant, 1956 [1788], Book II of Part I). In his third critique, Kant is concerned with reflective judgments that lack a given, determinate concept that can be used to subsume given contents, and thus uses of judgment where the faculty of judgment itself must provide the basis for subsumption. In this context, Kant brings together his account of knowledge and moral willing, e.g., by considering the supersensible basis (as a kind of supersensible community) of mechanical and teleological principles for understanding products of nature. Dialectic is given the negative role of blocking those inferences that lead to some kind of contradiction, e.g., when teleology is regarded as an objective, rather than subjective principle of nature (Kant 1987 [1790], sections 69–78). A fully general account of dialectic is given in Kant's Logic (1988 [1800], 19): it contains "the characteristics and rules by which we can tell that something does not agree with the formal criteria of truth, although it seems to. Dialectic in this sense would then have its good use as a cathartic of the understanding." Jointly, these uses of dialectic in Kant are especially important for Engelhardt's ethic because they clarify how rigorous contradictions (antinomies) arise with efforts to extend past the appropriate scope and limits of a given domain. Engelhardt will thus seek to specify the scope and limits of his ethic for moral strangers in ways that are patterned after Kant's mode of argumentation: first, providing an analytic and deduction of a restricted domain of valid judgments; second, providing a dialectic that exposes the irrationality that arises with efforts to extend past that restricted domain (1996, 105–108). However, he will also use "dialectic" in a more positive sense that is associated with Hegel's categorial ontology (1994b, esp. 212): a "categorial dialectic shows how conceptual problems that arise within one level can only be resolved at a higher conceptual or categorial level." In this same

associated with the disjunction plus the option associated with the ethic for moral strangers jointly provide an exhaustive mapping of possibilities. Any such account would itself depend on something like a transcendental argument which determines the conditions of the possibility of any content-full ethic. Without such an argument, the disjunction might only partially specify the possibility space that needs to be considered. As partial, the specification of the disjunction itself involves a kind of contingency. An ethic that depends on an argument based on the disjunction would be contingent in the same way as any of the theories are when they depend on standards with contingent contents. Specifically, if the argument based on the disjunction of standards is crucial to sustaining the conclusion that reason cannot provide a basis for an ethic used by moral strangers, then Engelhardt's postulation of agreement as the sole sufficient basis must itself have the same status as those other theories he discusses. An argument associated with such a disjunction at best shows that we have not worked out an ethic that does not partially beg the question of moral content. From this we cannot infer that it is impossible to advance a content-full ethic for resolving disputes among moral strangers.[8] Instead, we might infer that such an ethic would be difficult or unlikely or would require some novel standard that has not yet been identified. Whether we continue to try to work out such an ethic depends on a host of additional considerations, some of which relate to theoretical challenges associated with developing a universal grounding for an ethical

---

context, Engelhardt positively cites Klaus Hartmann's account of dialectic as "a means of linking categories with one another … to establish affirmative relations between the various levels of social formation that may be legitimate" (Hartmann 1984, 116). He even goes so far as to say that the dialectic (in a sense Engelhardt affirms) "may be regarded as a way of rationally reconstructing history from a particular content-full standpoint. It is most properly the method for organizing the categories in terms of reason's systematic needs" (1994a, 4–5; see also 17, note 16 where this reconstruction is associated with Hegel's *Philosophy of Right*). In the same overview of Hartmann and Hegel, he presents both his published dissertation (1973) and his *Foundations of Bioethics* (1996) as "realizations" of Hartmann's reading of Hegel, and takes that insight to be "the inability of secular reason to discover and justify moral content" (1994a, 17, note 17). He thinks his *Foundations of Bioethics* better advances this Hartmannian Hegel, presumably because his dissertation still exhibits too great a confidence in the capacity of reason to provide an ontology that maps real relations, e.g., between mind and body, and in such a way that establishes continuities between domains of fact and value. But this leaves unexplored how the "particular content-full standpoint" (1994a, 4) associated with the dialectical mapping of categorial relations (with its admittedly sparse and abstract content, but a content none the less) is related to the supposedly content-less "moral standpoint" and "bare will to morality" (1996, 72) of Engelhardt's ethic for moral strangers. The tensions between these Kantian and Hegelian strands are not reconciled in Engelhardt's secular ethic.

[8] In *The Grounds of Ethical Judgment: New Transcendental Arguments in Moral Philosophy*, Christian Illies (2003) explicitly considers what can be inferred from the failure of previous efforts at providing a rigorous justification for ethics. He notes that any inference from previous failures to the impossibility of a justification is itself a fundamental claim that requires a special kind of rigorous justification. "But how could this ever be provided, if the very claim is that there is no possibility of justifying first principles? Categorically to reject the possibility of any alternative method to achieve secure knowledge is therefore inconsistent. Hence we have no reason to assume that the search for a different method must be in vain – and with transcendental arguments we have further candidates for the task" (29).

theory, and some of which relate to practical concerns such as how greatly we value the goal and what effort we are willing to expend to attain it.

The second problem with Engelhardt's argument based on the disjunction of ethical standards is that it assumes that demonstration of contingency associated with a standard disqualifies any ethical theory using such a standard from serving as an ethic for moral strangers. But Engelhardt's transcendental analysis allows for some kinds of contingency to be introduced into conventions that are necessary for a secular ethic. This is especially apparent in his chapter on persons, property, and the state, where contingency in standards does not automatically disqualify some legal requirements (1996, ch 4). Simple examples include conventions like those that require we drive on the right side of the road and conventions that tell us what we must do to register property. At some higher level of abstraction, we can specify that it is necessary to have a contingent decision regarding the convention. While the specific decision involves contingency, the need for a decision is not contingent. In the same way, we might identify as necessary certain contingent specifications of standards. Since some kinds of contingency don't invalidate an ethic for resolving disputes among moral strangers, Engelhardt needs to provide criteria that enable us to distinguish which kinds of contingency are problematic. It is not enough to just review the standards and observe there is contingency integral to the specification of each of them.[9]

[9] Engelhardt acknowledges at least three legitimate roles for contingency in his ethic for moral strangers: (1) Contingency is required for specification of content; this leads to his contrast between a content-full ethic for moral friends (community) and a content-less ethic for moral strangers (society). This contrast is also put in terms of abstract right vs. Sittlichkeit: "Sittlichkeit is not just the category of social reality in which content is provided; it is most importantly the category within which contingency of moral content is recognized as necessary. None of the particular content is necessary. But its presence as contingent content is acknowledged to be necessary. … The problem is now to understand the unity of the diverse communities within which one comes to live a content-full, ethical life. The problem is to understand the unity of the diverse Gemeinshaften with their different Sittlichkeiten" (1994b, 216). (2) Contingency enters into selection of conventions that make peaceable interaction possible, e.g., while some convention regarding which side of the road to drive on is necessary (a physical necessity arising from the way embodied creatures like us circumnavigate using artifacts like cars), the convention to drive on the right side of roads is contingent, as are conventions for registering property, and so on. This form of contingency is central to the account Engelhardt gives of property, society, and state (1994b; 1996, ch. 4). It moves beyond the purely abstract, categorial specification in (1) to the *contingent necessities*, e.g., of giving content to a standard for driving. (3) Contingency also enters into the subsumption relations of the categories in so far as the systematic categorial order reflects something like inference to the best explanation rather than a strong logical or metaphysical necessity. On this score, we might notice the difference between Engelhardt's reconstruction of the relations between the Hegelian notions of an estate, community, and property (1994b, 218–221) and Pinkard's reconstruction (1994, 312–316). In Pinkard (following the real Hegel, rather than Engelhardt's reconstructed Hegel), there is a sharper criticism of religious communities (Pinkard 1994, 249–260) and a more positive appraisal of philosophy (261–268), and this leads to a sharper restriction of community by the state. Pinkard's account of the relations between these social categories (1994; 1988, chs 6–7) reflects commitments that Engelhardt held at an earlier stage (e.g., 1973, 22–27, and ch. iv). On the character of the "necessity" associated categorial order relations, see Pinkard on categories as "explanatory posits that are justified only by their explanatory value"

Since Engelhardt seeks to make fully general claims about what is and is not possible for any philosophical ethical theory, it is clear the core foundation must be provided by his transcendental argument. I now turn to that argument.

7.3 Engelhardt's Transcendental Argument for Agreement as the Necessary Basis of Any Ethic for Moral Strangers

Professor Engelhardt's ethic for moral strangers involves several strands that are built up layer by layer. At the foundation is a strand associated with forbearance and permission. This is developed in chapter 2 (pp. 67–72) of his *Foundations of Bioethics* immediately after he presents the argument based on his disjunction of ethical standards, and it is explicitly associated with a transcendental grounding. With Engelhardt (1996, 102), I will refer to this foundational strand as the "forbearance strand" and I will be concerned with the transcendental argument that is used to justify this strand. It is derived by considering the possible grounds for resolving a dispute among moral strangers with moral authority. The transcendental argument leads to a general constraint on how agents can interact with one another that is expressed in a general way as a "principle of permission" (1996, 103–108; 115–119; 121–123). This is a basic, lexically prioritized principle in Engelhardt's secular ethic, and it is independent of other strands.[10] He also considers a separate principle of beneficence, which guides the ways people collaborate in advancing their goods

(1988, chs 1–3, esp. 21–31; 1990 with Pippin 1990) and Engelhardt on the "must" of the dialectic and the *quid juris* (1973, 123–126). The crucial thing to note is how the categorial ordering itself involves contingency, and this ordering at the purely categorial level informs the way Engelhardt specifies the relations between the actual instances of community, society, and state. The contingencies associated with (2) and (3) lead to specification of the conditions necessary for applying Engelhardt's secular ethic to actual humans. If we were to review Engelhardt's ethic in the same way he reviews the standards integral to current ethical theories, then we would take (2) and (3) as evidence that his ethic of moral strangers likewise involves contingency. We might also add a fourth legitimate role for contingency to the three that Engelhardt explicitly recognizes; namely, the contingency associated with having an actual instance of any category. Deborah Chaffin (1994) discusses this in relation to the "concrete universal," and it is unavoidable in any application of an ethic by real world agents in real world contexts. This will be considered in relation to "reflectively aware agents" in Sect. 7.4.4 and 7.4.5.

[10] Engelhardt speaks of the "centrality of forbearance rights" (1996, 124–128). His principle of permission "marks the very boundary of all moral communities. … [T]he principle of permission is a principle of forbearance only. It is a negative principle. In general secular morality, the principle of permission is not beyond, but it is before, any concrete good or evil. It is only through the positive principle of beneficence that content is acquired for the moral life. Thus, not being beneficent is not to be an enemy of the moral community, but neither is it to be a member of the moral community. An individual who pursues his own solitary good, but without violating the rights of others, falls into a sort of moral limbo" (115). "In the context of general secular morality, the principle of permission *always trumps* the principle of beneficence" (1996, 127, my emphasis). What Engelhardt says here about an individual without beneficence not being a member of a moral community reflects his stipulated distinction between community and society (7), but it is in tension

(111–124). Jointly, these two principles provide the independent roots or headings of the ethic for moral strangers that is worked out in chapter 3 of the *Foundations.* In his fourth chapter, Engelhardt moves to a discussion of persons, property, and the state. These are for him co-implicated categories. However, in that chapter he makes clear that the logic of any legitimate use of force by a state would be the exact same as that logic of forbearance he initially developed at the end of chapter 2 in the context of two moral strangers (1996, 167; also 176 where he claims "states possess no special secular moral status"). In sum: the foundation of his whole ethic for moral strangers is provided by his transcendental argument for the logic of forbearance that he works out in relation to two agents who seek to resolve their controversy with a moral authority that is mutually acknowledged.

The development of Engelhardt's transcendental argument begins with a taxonomy of ways that controversies might be resolved. He identifies four bases or grounds to which agents might appeal: (1) force, (2) conversion, (3) rational argument, and (4) agreement (will).[11] Jointly these are taken to constitute the complete possibility space of options for moral authority that might ground any resolution. Engelhardt reviews each of the options in turn. After excluding force, conversion, and rational argument, he tries to establish agreement (will) as the sole sufficient ground of moral authority among moral strangers. Initially the argument seems to be presented as if it were a proof by cases, with the four bases providing the cases. Force seems obviously inadequate as a ground for a moral claim. Most people associate conversion with religion and also take it as obviously inadequate. Engelhardt extends conversion to cover ways someone might be persuaded or come over to some content-full notion of the right or good. Finally, the earlier extended proof by

---

with what he says in other places when he speaks of a "notion of a general secular moral community" that is associated with the "moral standpoint" (136–137).

[11] In the first edition of the *Foundations of Bioethics*, the fourth basis is presented as "agreed-to procedures" rather than "agreement," and "resolution of controversy according to a basis" is presented as an answer to "a rational question regarding why the controversy ought to be resolved in a particular fashion" (1986, 39). In the second edition, the fourth basis – his sole moral authority for a secular ethic – is changed to agreement, and much of the language about resolution of rational questions is eliminated. He also adds a lot of wording about the agreement as a brute resolve and he introduces a different kind of bare resolve associated with the moral standpoint: "Secular moral authority is derived from a bare will to morality" (1996, 67–72, with the citation on 72). This "bare will to morality" seems to be different from the bare resolve integral to an agreement which serves as a basis for resolving a controversy. I couldn't find any place where he discusses the relations between these different kinds of bare resolutions of will. Understanding the fourth basis as "agreed-to procedures" would avoid many of the problems of interpretation I consider regarding the nature of that agreement Engelhardt takes as the basis of a secular ethic (ideal vs surd content), because the agreement now only concerns the procedures. But that would shift the whole focus of his ethic, since the basis of the secular ethic would now be the procedures, not the agreement. This would raise even greater problems in relation to the kinds of claims Engelhardt wants to make about the nature of moral authority in secular, pluralistic contexts. We should thus take the second edition as the more fitting account of his crucial fourth basis. The first edition represents a transitional stage between the more robust categorial ontology of his dissertation (where the emphasis falls on intracategorial relations) and the ethic of his second edition (where the emphasis falls on actual agreements and the communal contents of moral life inaccessible to discursive thought).

cases based on his disjunction of standards might be taken as his proof that reason is not sufficient. He thus rules out the first three possible bases, and concludes that the fourth option – associated with agreement (will) – provides the only solution. But if this were the whole argument, we would not have moved past the problems earlier noted in the interpretive history. There would be no transcendental argument. To see where that enters, we need to more carefully consider how Engelhardt understands moral authority and how this is related to his "grammar" of moral discourse.

The arguments he provides are, in the end, quite tricky to specify. He presents his transcendental argument as an exposition of "the minimum grammar involved in speaking of blame and praise with moral strangers" (1996, 70). He follows Kant in arguing that any blame or praise implies there is an agent who has freedom to perform the action being evaluated: "That one cannot but think of oneself as free is a transcendental condition, not an empirical fact or metaphysical claim" (1996, 95, note 82, also 95–97, notes 83–87). Kant himself had two versions of this argument, and he didn't call either of these "transcendental." In one version, blame and praise implies there is an agent to whom an act can be attributed (the responsibility argument; Kant 2012 [1786]). In another version, blame and praise apply, first of all, to wills of agents, and Kant considers how agents must regard their own actions when they take themselves to be the spontaneous ground. Here freedom is a "postulate" of practical rationality of an agent (the freedom argument; Kant 1956 [1788]). But the arguments about the relation between blaming/praising and freedom/responsibility do not yet tell us anything about what is obligatory, permitted, or prohibited (deontic modality). This is where we come to the most difficult parts of Engelhardt's argument. Somehow he directly links moral authority in a secular context to "intellectual authority" (1996, 71), "intersubjectivity"(95, note 83), and a special kind of coherence that is possible to an agent who can sustain a kind of logical coherence integral to a given sort of practice (103–111). The forebearance rights he advances "are justified in terms of the perspective of moral agents in general" (127). He summarizes the task of developing a philosophical ethic as follows:

> in asking a question about morals as a philosophical question, one is seeking a rational reply that is, as far as possible, inescapable. One is seeking a clincher to a dispute concerning which of the possible ways one can live life … one ought to choose, where the sanction for violating the "ought" is not a threat of force or a feeling of guilt, but irrationality, worthiness of blame, or failure to realize the goods one wishes to achieve. In seeking a characterization of the fabric of morality and bioethics tied to the fabric of rationality, one is unlikely to secure content to characterize the concrete nature of the moral life. (Engelhardt, 1996, 104; the ellipsis simply excludes a reference to the practice of medicine)

In the section that immediately follows the one from which this quote was taken, Engelhardt considers how Kant smuggles in moral content by moving from the rigorous, purely rational notion of contradiction to a notion of contradiction in the will (105–108). Engelhardt wants to restrict the sanction of "irrationality" to the more rigorous notion of contradiction because only this can function as the required "clincher" in moral argumentation with strangers.

In his explicit account of the crucial transcendental argument, Engelhardt thus claims to work out how the grammar of moral discourse, regarded as a discourse of blaming and praising, works with a postulate of freedom as a necessary side constraint (105–106, 394–397). Moral strangers are then regarded as agents with minimal overlap in their content-full commitments. When regarded as ethical agents, they are taken as purely rational agents who avoid making moves that would directly undermine their capacity to consistently sustain the blaming and praising game.[12] We then find that force, conversion, and rational arguments based on contestable standards all fail to provide the required moral authority because putting them forward as authoritative has the net effect of giving up the whole endeavor of blaming and praising as a rational enterprise. Blaming and praising with standards that only one side recognizes as legitimate amounts to abandoning the whole enterprise of a philosophical ethic, since this enterprise requires rigorous justification of all claims.

7.4 To What Extent Can the Question of Moral Authority Be Disentangled from the Task Associated with Peaceably Resolving Conflicts (Resolutions Without Force)?

Engelhardt initially frames his whole investigation by characterizing the post-modern predicament as one of a potential war of all against all. Faced with this challenge, he asks whether there is a common moral authority that might serve as a basis for resolving disputes without force. When he attempts to provide an answer to this question, he tries to work out an account of moral authority and a kind of ethic that is completely independent from any questions about force or peaceable vs. non-peaceable resolution of conflicts. Instead of force, the sanction of his ethic is "blame." Blame, in turn, is associated with irrationality, and irrationality is associated with an incapacity of a rational agent to sustain the discourse of blaming and

[12] Engelhardt uses the term "moral stranger" for the flesh and blood agents who interact in society (1996, 24, note 13; 74–83). These agents are individuated in terms of contingent contents that differentiate each moral stranger from other moral strangers. However, when such moral strangers are regarded as ethical agents, emphasis falls upon the minimal content that they share qua rational agents with a resolve to interact in a way that sustains the game of blaming and praising. Here emphasis falls upon the "moral standpoint" as a kind of "intellectual standpoint" of a self-legislating agent in an intelligible world (Kant's *mundus intelligibilis*). What is shared by moral agents in a general secular community is the notion of "a community of entities who are self-conscious, rational, free to choose, and in possession of a sense of moral concern" (1996, 136). There is thus a shift that often takes place between individuated moral strangers where each is maximally differentiated from all others, and the general category for such moral strangers qua moral agents. The latter, general category includes only that sparse rationality and resolve that is found in any agent. Thus, when considering the sparse content shared by moral strangers, one has the maximally intersubjective content. This would be shared by moral friends as well. Viewed in this way, i.e., in terms of the shared, sparse content, moral friends and moral strangers are bound up in a common sparse community.

praising while at the same time resolving a given controversy in a sanctioned way. The inconsistency of the action and the general logic of the ethical discourse is what is conveyed in the moral sanction (the blame). Ethics is thus understood as a certain kind of discourse that can take place between rational agents who seek between themselves an account of authority which is prior to and independent of any uses of force. If such an account can be provided, then such agents may, in a second step, choose to restrict any use of force on one another to just those cases where the force itself might have moral legitimacy. For Engelhardt, this will only occur when force is used to counter the force of another who, by virtue of using that force, acts in such a way that a commonly acknowledged moral authorization can no longer be given for the action.

Engelhardt's fundamental justification of his ethic takes the form of a transcendental argument that explicates what is implied by a general discourse of blaming and praising that appeals to a basis that would be commonly acknowledged by moral strangers. His transcendental argument begins by presenting the categories of possible bases that might be taken by moral strangers as a commonly acknowledged ground or authority for resolving controversy.

To fully evaluate Engelhardt's transcendental argument we need to clarify the relations between (1) a controversy associated with an interaction among agents, (2) the nature and stance of the actual agents involved in the controversy, with their variable interests, reasons, and ends, (3) the stance of the "moral stranger" qua pure rational agent, (4) the moral authority (agreement, will) that provides the common basis, (5) the discourse of blaming and praising that requires an agent to take up a stance that can be reciprocally assumed by the other agent and relative to which an appeal must only be based on a common basis, and (6) the resolution of the conflict that may or may not be compatible with that stance an agent must assume when playing the game of blaming and praising. I now argue that all of these terms are insufficiently analyzed in the *Foundations of Bioethics*. More specifically, each term is not presented by Engelhardt in a fully general way. When we make explicit what remains implicit in his analysis, we find at least one other ethic of moral strangers that is fully consistent with the way he has characterized all of his terms. He presents a minimal ethic. Alternatively, we could posit a maximal ethic. As long as we posit the perspective of a pure, rational agency as that stance which must be assumed by an agent engaged in the practice of blaming and praising, no grounds can be given for selecting between the minimal and maximal ethic. However, the burden of proof associated with the maximal ethic is the exact opposite of Engelhardt's minimal ethic: wherever one agent has reasons for acting that cannot be made transparent to the other agent, then that agent must resolve to alter his action in such a way that the grounds of the action can be accessible to the other agent. This can be made into a maxim regarding how agents involved in a controversy must come to agreements if they are to sustain the game of ethical discourse as a game that equates sanction with irrationality. I refer to Engelhardt's minimal interpretation as the Libertarian Logic of Moral Strangers and will refer to the maximal interpretation as the Hobbesian Logic of Moral Strangers. I seek to show that if the only incompatibilities are the purely rational ones Engelhardt acknowledges, we arrive at an impasse. In fact, I take the Hobbesian logic as a fair account of what many philosophers demand, and think it is a better interpretation of the real Hegel's view, in contrast to Engelhardt's non-metaphysical Hegel. If Engelhardt can't show what is problematic with that Hobbesian logic, then one of the central goals of the Foundations project will fail. I then close by showing how we can get most of what Engelhardt wants back again if we slightly modify his project.

7.4.1 Meta-controversies About Controversies

Engelhardt's transcendental argument explicates the basis of moral authority that might be taken as a sufficient ground for an ethic that moral strangers can use to resolve controversies. What are controversies and why focus upon them? Unfortunately, Engelhardt does not provide us with an account to answer the two parts of this question (the what and the why), at least not in his *Foundations of Bioethics.*[13]

There are several ambiguities associated with the notion of a controversy. The word implies one thing is set against another. But what are the natures of the things that are set against one another? Are they topics or claims (mental thing) or agents (material creatures who might pick up clubs) or perhaps some combination of both agents and claims? Etymologically, controversy derives from the Latin, *controversus*, which in turn is from *contra-* "against" (with comparison) and *versus*, meaning a line of writing (originally from the context of plowing rows in a field). A controversy thus involves setting one line against another. This would imply that the two things that are opposed are claims, views, viewpoints, or positions in some kind of verbal or mental contest, and a resolution involves some kind of meta-statement regarding which claim or position was selected from the opposing options. Alternatively, a resolution might involve some higher order claim that somehow properly situates each of the lower order claims in a way that specifies their proper role, scope, and limits. In all cases, a controversy thus interpreted would involve a purely conceptual affair. The whole requirement for such an approach to controversies is that one abstract away all features of agent interactions that cannot be made transparent to thought.

[13]The background for some of Engelhardt's account of controversy might be found in a 3-year series of Hastings Center meetings on scientific controversies led by Engelhardt and Caplan (1987a). This set of conferences straddles the completion of the first edition of Engelhardt's *Foundations of Bioethics*, and he acknowledges his debt to conference participants, especially Tom Beauchamp (1987), in 1996 (92, note 74) and in his coauthored introduction for the controversies volume (Engelhardt and Caplan 1987b). But Engelhardt's interest in "resolving issues in ways not directly grounded in unconsented force" and his contrast between appeals to force, reason, and "peaceable manipulations" is a pervasive theme in earlier work (1973, 1982, 67). At best we might say that Engelhardt further refined the ideas in his *Foundations of Bioethics* while he was also actively thinking about the entanglement of questions of "knowledge, value, and political forces" in resolution of scientific controversies.

At times Engelhardt advances an account of controversies and transcendental argument of just this sort: the "claims" are categories and the function of a resolution is to specify the higher order categorial claim that properly subsumes and harmonizes the lower order claims.[14] But if this is what is meant, how are we to understand the use of force, conversion, reason, or agreement that might resolve the controversy? How might force be used to resolve a purely categorial claim? With the exception of "use of reason," all of these bases for resolving controversy shift attention from opposing claims to the processes, actions, and transformations used by or associated with agents involved in attaining a resolution. (Reason is ambiguous. It might refer to systematic ordering and order of the claims or the mental activity and communication of agents attempting to convince one another.) Engelhardt's four bases are kinds of actions, activities, or interactions (e.g., reasoning, force), events or processes (e.g., conversion, agreement), instrumental means used by agents (reason, force), or resolutions of the will or performatives arising from such resolution (conversion, agreement). In all cases they are deployed by and work on the agents, not on opposing categorial contents. We can have opposing propositions, categories, and so on without having any flesh and blood agents. For example, we might use a formal language and logic to represent two claims that yield a contradiction. We might introduce an algorithm that "resolves," for example, by having a set procedure for negating one of the claims and sustaining the other or for modifying the substructure of claims so they become compatible. We could implement this on a machine that uses automated procedures to make updates on asserted items. This would be one meta-level way of thinking about controversies. But it would not get us very far in understanding what Professor Engelhardt is concerned about when he speaks about things like force or conversion as possible bases of an ethic. We also can't get very far if we just focus on agents and ignore their claims, views, asserted positions, and so on. How might we even imagine making the opposition in a controversy into an opposition that is just between agents and not claims? We could perhaps imagine two agents standing next to each other, and say that they are "parties to a controversy" when both are staring at each other and yelling or when one is scratching at the eyes while the other is pulling on some hair. But this kind of bodily opposition is at best just one variant of what he has in mind.

To account for what Engelhardt means by a "controversy," we must bring both agents and their claims, positions, and viewpoints together. A controversy is then a clash of agents who each are acting in ways that make their actions into the kind of

[14] Specification of controversies in terms of categorial contents is a central theme in Engelhardt, 1973. This work can be taken as an effort to provide a set of increasingly refined specifications of general types of controversies and their general resolutions. Actual controversies might then be explained and a resolution suggested by identifying the general type that covers that controversy, and the general form of resolution it allows. Questions of controversy and resolution are quite explicit in his account of the dialectic, the categorial "must," and closure in relation to the *de jure* exhibition of categories that provide regressive justification (ch vi). Since conflict arises "progressively" and resolution is worked out "regressively," one might even find in his account of the dialectic a way to clarify what, specifically, a resolution of a controversy entails as a distinct step. Even when Engelhardt talks about real world controversies, his categorial account seems to hover in the background, e.g., in the way he associates ethics with a categorially specified moral standpoint.

interaction that is called "controversy." But the interaction is of a tricky, complex character, with ripple effects that somehow work back upon what the agents take to be important. Controversy involves two levels that are in play simultaneously: one involves two actual agents, another involves the specific patterns of action and agency that overlap in relation to contents that are partly material and partly mental. At this stage there seem to be many options for how we characterize a controversy, depending on the degree to which the conceptual content or the real, embodied, socially and historically situated agents enter into the specification of a controversy. There are also a host of variable qualities, such as whether a controversy is between two individuals or more than two, discrete or entangled, involving one isolated topic or many, and so on. Because of the special role agreement and permission plays in Engelhardt's ethic, my conceptual analysis will focus on controversies between just two agents. I will also use very simple examples. This will be enough to clarify some of the challenges associated with understanding the relation between the proposed bases for resolving controversies and the controversies. However, it should be noted that Engelhardt uses a broad range of examples and most of these cannot easily be assimilated to the two agent case.[15]

Consider, for example, a case where two individuals have filed a claim to the same property, but neither is aware of the claim made by the other. Are they involved in a controversy or is something only a controversy when the opposition is made explicit and two agents are actively aware that their claims are not compatible? If we require the explicit awareness of an incompatibility, then we will radically restrict what counts as a controversy. Many things an ethic for moral strangers should address would then be excluded, for example, how agents might make claims to property in public ways that prevent assertions of incompatible claims to property. In this case, the ethic would be deployed to avoid controversies, not resolve them.

[15] A more complete analysis of controversies might proceed by gathering examples of controversies from Engelhardt's works and categorizing them. One classification scheme might sort them from most abstract (emphasizing the thought content integral to a controversy) to most concrete (emphasizing the opposition of embodied, socially and historically situated agents). Such a taxonomy of controversies would range from the pure categorial oppositions and their dialectical resolutions (Engelhardt 1973), run through disputes about criteria for preferring one abstractly described possible world over another (Engelhardt 1996, 42–43), move to controversies about specific policies on topics such as the clinical determination of death (241–252), grade over toward general disputes between groups (e.g., Dominicans vs Orthodox on the nature of true conversion, 44), and finally move to specific, historical confrontations between named agents such as the 47 ronin and Kira Yoshinaka in 1703 Japan (44). Such a gradation of controversies in terms of the ingredients of thought and contingent givenness integral to specification of a controversy would clarify how a rank ordering of categories that cover the controversy type may alter the ways we understand bases for their resolution. It is not clear, for example, how force could ever be used to resolve a controversy about the criterion that determines which of two abstractly described possible worlds should be preferred. Any agent attempting to use force to compel another agent to accept World A rather than Worlds B or C (1996, 42–43) would simply demonstrate that s/he didn't understand the controversy properly. On the other hand, it is not clear whether political controversies or historical disputes between individuals such as the 47 ronin and Kira Yoshinaka could be properly understood in a way allows for a resolution that doesn't involve some use of force.

Let's imagine that two agents lay claim to the same property, for example, a lawn mower, and both are aware they are making incompatible claims.[16] The lawn mower is now in the garage of Fred. What would we mean if we say that Bob resolves this controversy by force? Does this mean that Bob goes to Fred's house and takes the lawn mower and puts it in his own garage? In that case the controversy seems to be in the exact same state with respect to resolution as it was previously, only now the lawn mower is at a different place. In both cases we still have an unresolved controversy. Alternatively, does resolution mean that one person gives up a claim to the lawn mower? Let's say Fred gives up openly making the claim and he even forgets they had a dispute, but the lawn mower stays in his garage. If Bob still asserts his claim, the controversy would not be resolved. If controversy remains whenever some lingering dissent remains, then the only way force can be successful in resolving a controversy is when some kind of conversion takes place. Every spontaneous resolution of will might be regarded as a kind of conversion. Alternatively, if force "compels" some change of mind (so we see the conversion as externally generated), conversion might be a step toward the spontaneous resolution of will associated with agreement. The conversion would simply be a middle step that alters the state of the agent in such a way that an agreement became possible. To distinguish this bad kind of agreement from the good kind Engelhardt advances, we need descriptions of controversies that enable us to see how force, conversion, reason, and agreement each work in relation to the agents, their actions, and any claims or views associated with them. We also need an account of how resolutions of controversies might be determined in a way that is independent from the means used for the resolution, so we can clearly distinguish which resolutions count as agreements and which do not.

[16] Engelhardt provides a rich account of property, ownerships, and persons in 1996, 154–166. He fully appreciates what an ethic must accomplish if it is to account for how a person gains ownership of a thing: "The difficulty is bringing things, objects, within a conceptual framework of possessions and possessors. Any particular system for speaking of possessions and possessors appears to be culturally relative and open to challenge by members of other communities with other conventions" (154–155). Engelhardt provides an account based on Locke and Hegel, and tries to root what he takes from them in his principle of permission. Objects owned are regarded as extensions of persons. But the crucial work required for understanding such an extension depends on an account of embodiment, and it runs up against a reality that Engelhardt clearly discerns in his early project on mind and body (1973): at any stage, the extension from person to possession of a thing (whether of one's body or other possessions) already reflects a long sequence of accommodations and resolutions of antecedent oppositions. We don't simply get persons and their bodies and then extension to things. Bodies are already communally configured products of mind, and any extension into other external bits of matter must somehow tie into the antecedent mediations of body and mind. This is problematic for an account that seeks to trace a notion of right back to a fully transparent basis. Even when general principles of ownership are worked out (as in 164–166), these reflect contingent, culturally conditioned conventions regarding ownership. A controversy over ownership thus cannot in a straightforward way be resolved by pointing to a principle of permission, as he seems to suggest. Here the problem is roughly like that in Hobbes's state of nature: different persons might lay claim to the same bits of matter. This is the kind of controversy for which a basis of resolution is sought.

Another ambiguity concerns how controversies are individuated, specifically, whether they must be individuated in just one way. When considering scientific controversies, Engelhardt has provided one of the most nuanced accounts available of how controversies about matters of fact often involve meta-level controversies about what is a matter of fact and what relates to interests, criteria for evidence, and so on.[17] With respect to whatever we mean by moral controversies, it seems we will in the same way have meta-level controversies about these controversies. I'll just focus on one kind of meta-level dispute: a dispute about whether, within the controversy, force is being used by one agent against the other. Let's consider again the

[17] The problems associated with individuation of controversies is explicitly addressed in Engelhardt and Caplan (1987b). There "'a' scientific controversy is defined by the existence of 'a' community of disputants who share common rules of evidence and of reasoning with evidence. If such rules are not shared, then the dispute is not a single controversy" (12). But this account of the individuation of a controversy cannot work for two individuals in an ethical controversy unless "existence of 'a' community" itself becomes completely indeterminate. In Sect. 7.4.3, I address the different strata of community of disputants by speaking of first through fifth community of disputants associated with a controversy. The conceptual analysis of controversies provided by Engelhardt and Caplan (esp. 1987b, 4–16) seems to indicate that the problem of individuation of controversies can be addressed by disaggregating controversies about fact, about values, and about collective action (politics), and isolating the relevant communities and criteria for resolution. Resolutions in terms of loss of interest, force, and consensus reflect dynamics of collective action, but are not correct or fair. Scientific controversies are about matters of fact, and are to be resolved by argument. Such resolutions have the quality of being correct. Ethical controversies are about matters of a practical rational nature and relate to values and choices, and these are to be resolved by negotiation. Such resolutions have the quality of being fair. When controversies involve entanglement of scientific, ethical, and political aspects, then resolutions will likely be partly correct (and attained by argument) and partly fair (attained by agreement). To show how the problem of individuation can be resolved, Engelhardt and Caplan consider disputes about Laetrile, and note how "at least two quite different controversies were intermingled in this dispute, one involving rules for establishing a scientific claim and the other involving political rules regarding access to particular medications. These two elements are, in principle, both distinguishable and separable. They are intermingled in one debate out of historical accident, owing to the drug control laws of the United States" (4). However, instead of viewing the entangled, complex controversy about Laetrile as a kind of false description – describing two distinct controversies as if they are one controversy – one might take the complex controversy as a first stage, and take the proposal by Engelhardt and Caplan (following contributions in Part II of 1987a, especially by Baruch Brody and Robert Schwartz) to partition the controversy into two subordinate controversies as one kind of a resolution of the original entangled controversy. The proposal to partition the controversy arises by deliberation and argument. If it is accepted by parties to the original controversy, and subsequent controversy neatly settles out into scientific and ethical/political strands, we might say that the complex controversy was resolved by argument. Nearly all (and perhaps all) real controversies are entangled. Resolutions often take the form of such refinements regarding what is, in fact, disputed. A resolution might take the form of differentiation of an antecedent controversy into multiple, simple controversies. Appreciation of these characteristics of so-called ethical controversies makes the problem of individuation of controversies more intractable and persistent than Engelhardt and Caplan suggest in their conceptual overview. Engelhardt deeply appreciates all of this, and I learned much of what I know about the entanglements of facts and values from his case studies and conceptual analysis. But his insights on the strata integral to actual controversies feed back upon meta-level descriptions of controversies in ways that problematize the way he wants to speak about ethical controversies and their resolution in his ethic for moral strangers.

property dispute between Fred and Bob, and assume the mower is now in the garage of Fred. From Fred's perspective no force is being used: he owns the lawn mower and Bob is making an unjustifiable claim. But from Bob's perspective, Fred is using force – for example, placing the barrier of a garage door – to prevent him from using his lawn mower. If use of force is interpreted broadly to include barriers to use or appeals to force, then one party may interpret actions as legitimate kinds of persuasion, business negotiation, and so on, which another party interprets as force or coercion. These kind of meta-level Gestalt shifts are as easy to generate for direct bodily actions as for disputes about property. The problems in all these cases is that a background web of rights is presumed to configure what counts as a use of force. However, acknowledging a web of rights as a condition for determining when force is used would invert the whole logic of Engelhardt's analysis. He is seeking to identify what might count as a common moral authority. If the argument for what counts as a common moral authority depends on presumption of a commonly recognized notion of right to determine what force is, then we face a problem of regress. I will refer to this as the problem of rights dependent descriptions of controversies.[18]

We can clearly see the problem of rights dependent description of a controversy in the dispute between Fred and Bob. Assume Bob is using the lawnmower, but temporarily leaves it on the public sidewalk in front of his house and goes into his house to satisfy his thirst. If Fred takes the mower and puts it in his garage while Bob is in his house getting a drink, this only counts as a use of force if the mower is Bob's, not Fred's. If Fred is using Bob's mower without permission and Bob goes out and tries to rend it from the hands of Fred, Bob is not attempting to resolve a controversy with a use of force. Rather, Bob is involved in a controversy where Fred is already using force to appropriate his property. Bob's effort to rend the mower from Fred's hands only involves a continuation of the controversy on the terms it has been advanced. If we view matters differently, this is only because we describe controversies with a meta-language that is informed by the contingent laws of our land. If we are to avoid a meta-level dispute about whether force is used, we need some independent account of a controversy and of what counts as a use of force, so we can go to the controversy and determine whether force is being used. If we cannot solve the problem of rights dependent description of a controversy, we cannot isolate the question of a moral basis, and then ground a sparse notion of right on the

[18] Engelhardt says that "one respects claims to ownership insofar as the entity owned has been brought within the sphere of the owner, such that violating that ownership would be a violation of the person of the owner" (1996, 164). But he leaves the meaning of "brought within the sphere of the owner" ambiguous. In the lawnmower example, "brought within the sphere of the owner" might mean "brought within the garage of Fred (or Bob)." If that's the case, when Fred has the lawnmower in his garage, Bob should respect his claim. When it is moved from one location to the other, then what the principle of permission would require shifts. This can't be right, since it would miss the way the controversy is about the right of ownership, and such a controversy remains unaltered by a shift in location of the thing. "Sphere of control" in such cases can't mean "power to directly use or dispose of the object." The question of right needs to be resolved prior to any application of the principle of permission, and that question of right raises the question of an ethical basis that is not addressed by Engelhardt's ethic for moral strangers.

moral basis that is regarded as the sole available moral authority that moral strangers would acknowledge. Such an independent account of what counts as a controversy and a use of force would seem to be one of the conditions for demonstrating the possibility of an ethic for moral strangers.

More generally, we might notice that the categories that Engelhardt uses to describe possible bases for *resolving* controversies – namely, force, conversion, reason, and agreement – might also be taken as meta-level categories for describing the nature of controversies in the first place. Agents interact in ways that involve explicit or threatened uses of force. They make various attempts to convert or convince, and these are directed toward attaining agreement. Certain kinds of agreement are required for many controversies to take the specific forms they take. At any stage, various moves and counter-moves have been made, and this sequence of moves constitutes the interactions as a controversy. These bring a controversy to a state where some possible actions might bring the controversy to closure while others might sustain it or escalate it. Any agreements are usually worked out in relation to this history of moves that might have involved uses of force, partial conversions, reasoning, and prior agreements. If at a given stage agreement resolves the controversy, this is partly an artifact of how we chose to describe the controversy. If we back up a little and look at the other actions that brought the controversy up to that stage, then we might say it was resolved by force or conversion or reasoning. Alternatively, if the controversy never involved any of the other things, how was it a controversy? What was opposed and in what way?

From the above considerations, we can conclude that moral controversies have the same character Engelhardt associates with scientific controversies: they are partly constructed, and the ways we construct them depend on the categories, languages, and logics of construction we choose to deploy for our descriptions.[19] We have a choice about these. Disputes about facts and their interpretation are just as much a part of ethical controversies as they are of scientific controversies. At any stage, questions of fact and questions of right are deeply entangled. Any capacity we have to isolate a controversy about a question of right already depends on a host of implicit background agreements we have about fact and right. The same prob-

[19] Engelhardt's account of controversies about disease (1996, ch 5) might be taken as representative of all controversies, including ethical controversies. When considering problems of medicalization he notes: "The problem is not simply to decide on the correct sick role or the correct staging or characterization of a disease, but whether to see a problem as a disease at all. The major social institutions offer competing construals of reality with competing costs and benefits. There are advantages and disadvantages in seeing disruptive behavior as a crime, a sin, a moral fault, or a disease. In some circumstances and from certain perspectives, it is more important, useful, and plausible to see individuals as responsible for their actions and in need of punishment or discipline, not treatment. In others it is more useful and plausible to see behavior as determined and open to technological manipulation" (224). This recognition, however, calls into question the way he implicitly frames the task of developing an ethic for moral strangers. Why be so worried about a fully explicit, universal, and authoritative moral basis for resolving controversies if contingency enters so directly into the way we "see individuals as responsible for their actions and in need of punishment or discipline"?

lems Engelhardt identifies at the level of resolving moral controversies might also characterize the meta-controversies about our controversies. But in this case, we can't just put resolutions in terms of agreement. This doesn't give enough content to get the project going. As with scientific descriptions, our descriptions of moral controversies are partly constructed and depend on initial stipulations we make about what a controversy ought to be. Background notions of right have already been deployed in the constitution of a controversy as a controversy. But controversies also depend on the ways things in the world are related and how they offer themselves to construction. While controversies have an inherent indeterminacy that makes them amenable to variant descriptions, and while descriptions selected by agents partly make controversies what they are, a controversy cannot be described in just any arbitrary way. It is never a matter of "just" construction. Descriptions must map in the right ways to the world.[20]

If we just focus on the agreement (the resolve of will), we get an unresolvable regress: we can have meta-meta-controversies about how we chose to resolve our meta-controversies, and so on ad infinitum. Alternatively, we might get the following simple but rather odd way of assuring all controversies are resolved in ways that moral strangers accept: we might simply stipulate that all controversies be the kind of thing that can be resolved by something that moral strangers regard as a morally authoritative basis. We can then develop rules for describing the controversies so the available basis for its resolution comes ready made in the description.[21] There will, of course, be many things that will not be subject to such a description. But we could just say the same thing about our approach as Engelhardt says about his ethic for moral strangers: while the postulated approach to controversies would make many of the real world struggles not amenable to our analysis, we simply have to

[20] When setting up questions about the standard or basis of an ethic, Engelhardt distinguishes ethical controversies from scientific controversies by highlighting the way the latter kind of controversy might be resolved by empirical considerations and secure intersubjectivity in relation to historically conditioned communities with a common sense. "Disputants can test competing accounts with reality and thus falsify some competitors and strengthen the claims of others. Some accounts will stand out as preferable in being simpler accounts of the facts with fewer anomalies and with fewer ad hoc assumptions. The 'facts' can cause difficulties for many conjectures concerning empirical reality" (1996, 39). Engelhardt thinks a similar basis for resolving moral controversies cannot be found. But the meta-controversies associated with description of a controversy have the same character as the scientific controversies. Standards of correctness in description of the controversy play as significant a role as those of fairness in resolution of a meta-controversy about the description.

[21] When Engelhardt and Caplan (1987b, 12) individuate a controversy by isolating a single community of disputants with common rules of evidence and reasoning, they might be taken as advancing a stipulative definition of just this kind. One need only add that the single community must proceed according to its common rules, and exclude rules that involve force or appeals to force. If at any stage some members of the community attempt to use alternate means for resolving the controversy, this would simply indicate that they are not in fact or no longer remain one community. The controversy is then redescribed as an entangled controversy. By definition, real controversies are the simple, resolvable kind. The messy, entangled kinds are not really controversies but bundles of many controversies falsely described as single controversies.

acknowledge the limits on what our philosophical analysis can provide. This will be a sparse approach to the ethics of controversies that only resolves some of the things we might hope to resolve, and it is closely tied to the conditions for sustaining a common discourse of blaming and praising among moral strangers. If, for example, we have two agents locked in a battle over their lawn mower, and if it is the case that we cannot discover any basis they jointly acknowledge as a common moral authority sufficient for resolving the problems associated with their unpleasant interaction, then we simply don't call this a "controversy." We might, instead, come up with some different term, say "intransigent dispute" or "surd opposition" or "agent opposition." We can then order our category of "controversy" in whatever way we like to our other general terms. For example, we might stipulate that any kind of negative interaction among agents will be called a conflict. Conflicts can be partitioned into controversies, intransigent disputes, and complex disputes. What makes a controversy a controversy is that it is simple, well-defined, and resolvable by a means other than force. We can then work out an ethic that only covers controversies. Disputes are subcategorized into intransigent and complex variants, and they are defined in such a way that disputants do not share a common moral basis for sustaining a univocal ethical discourse they can use to resolve their disputes. By means of such categorization of conflicts, we would solve the problem of a basis for ethics by the way we restrict it to our stipulatively defined controversies.

To block the artificial, stipulated way of "solving" the problem of a common moral authority, we need some solid account of what a controversy is and why it is the proper target of the secular ethic. We need to solve the problem of rights dependent descriptions of controversies in some way. When Engelhardt started with his question about peaceable interaction and legitimate uses of force, he had one strategy for solving this problem: the questions of his ethic could have been restricted to a consideration of when force may legitimately be used, and an independent account of force might have been developed to tell us when a problem comes within the scope of his ethic. To solve the problem of rights dependent notions of force, he would have had to invert the order of his topics. A categorial account of persons, property, and the state would need to precede any reflection on secular moral authority. An additional distinction would be needed between the kind of transcendental argument that is associated with the relative ranking of categories like subject, agent, person, property, community, society, and state, on one side, and the kind that involves exposition of conditions for actual agents to resolve to act in ways that are compatible with a resolve to continue playing the game associated with blaming and praising among moral strangers. But when he tries come up with an account of ethics and moral authority that is independent from any questions regarding the legitimacy of force, he loses both the semantic content and the syntactic structure of controversy that makes it the target of the ethical judgments made by the interacting moral strangers. Later I return to the approach he might have taken, and why this is required. But first, much more is needed before we can fully reconstruct his transcendental argument and see where it might go astray.

7.4.2 The Agents, Agency, and Action Associated with the Controversy

When Engelhardt asks about a basis moral strangers might jointly accept as authoritative, he is assuming some common thing – the controversy – within which actual agents are somehow inter-acting in an oppositional way. We might imagine the controversy as something that is described from some third person point of view, and we assume we have some third person description that is agreed upon by both agents. For the sake of argument, we thus imagine we have some controversy where there is some common description (thus no meta-level controversy) that is accepted by both agents, and the agreed-upon description is such that it does not make immediately apparent how the controversy is to be resolved. I'll call this the "common description condition," and it involves some solution of the problem of rights dependent descriptions of controversies. At this stage I will assume some simple controversy that satisfies this common description condition. Let's name the agents involved in this dispute Martha and Mary. In relation to this controversy and this agreed-upon description, what might Engelhardt mean by a "basis" that the agents in the dispute jointly accept as morally authoritative?

The only way I know to make sense of this notion of a *common basis* is to make mental moves something like the following: I first consider one of the two agents, say Martha, and some basis. Call the basis by the name B1. I ask: does Martha accept this basis as a moral authority? If I answer "yes," I then take this same basis and ask the question about Mary: does she accept this basis as a moral authority? If I likewise answer "yes," then I can assert that Martha and Mary both accept this basis as a moral authority for resolving their controversy. But when I do this, I make clear that there is another kind of relation that is even more primitive than the relation the agents have to one another in their controversy. This is a relation that each agent has to the basis. When we move to the perspective of one agent, then we might say a controversy consists of interactions wherein one agent says "no" (or has a surd opposition of will constituting a "no") to something another agent asserts or seeks to assert in relation to the first. This surd "no" from one or both sides constitutes the lack of agreement that makes the controversy a controversy. When a "no" is asserted in relation to some intrusive action of another agent (some criteria are needed here for conditions of intrusions that require permissions versus those that don't), then forbearance by that other is required. Alternatively, agreement involves a double "yes." When this double yes is provided in relation to some mutually conditioned, intrusive action, there is permission. When it is provided in relation to some commonly embraced end, there is a basis for beneficence.[22] One may now ask about the

[22] These statements about the "no" and "yes" scenarios might be taken as alternative ways of stating what Engelhardt means when he says "a difference between duties of forbearance and beneficence derives from the fact that another's refusal is sufficient to create an obligation of forbearance, whereas mutual agreement is required for a concrete duty of beneficence" (1996, 128). However, there is still an unclarified difference between the "no" that constitutes a controversy as a controversy, expressing an opposition or disagreement, and the "no" that constitutes a lack of permission

relation between a specific agent, the "yes" that constitutes that agent's agreement (permission/beneficence), and the determining conditions that serve as ground and end of the agent's resolution of will. This is a further determination of the general question about the basis or common moral authority required by an ethic for moral strangers.

Let us now further refine our question by asking: can the shared basis (the agreement) be some contingency integral to a specific controversy and its agreed upon description? Or must it be something any agent has access to generally, independent of any specific controversy that agent finds herself within? Further, how must the basis Mary and Martha have for resolving their controversy relate to the basis that other agents have for resolving their controversy? Let's now assume Fred and Bob are involved in a controversy that satisfies the same general conditions: namely, they agree upon a description of their controversy and thus have no meta-level controversy about their controversy, and they have some basis, B2, which they both acknowledge as a moral authority. Engelhardt's claim that moral strangers must have a common basis recognized as morally authoritative might mean two different things: First, it might mean that for any two agents involved in some specific controversy, there must be some common basis they have for resolving that controversy. In this case, B1 need not and probably would not be the same as B2. Second, Engelhardt's claim might mean there is some basis common to all agents involved in any controversy and, for any controversy, the two agents involved in that controversy must be able to appeal to that moral authority to which all of them have access. This would require B1 = B2. But both of these options would create problems for the way Engelhardt speaks about the common basis of his secular ethic.

The problem arises because of the way the two levels, associated with agents and claims, interact in a controversy. Associated with each level is a kind of ethic and a kind of transcendental argument. Kant's ethic focuses on the willing of agents, and he seeks a general form of willing that constrains and orients how the will is to spontaneously move itself. Hegel's ethic focuses on the ways higher order categories specify the scope and limits of lower order categories that might come into opposition. Each of these approaches leads to problems, but it is not clear how Engelhardt can avoid them if he stays at the fully general level. Consider the option where B1 need not equal B2, and where Engelhardt's claim simply requires that there be a common basis jointly acknowledged by agents in a controversy as sufficient for resolving the dispute. We are told that the general category for this common

and whose breach would constitute a blameworthy act. The first "no" does not require forbearance, the second does. The difference must relate to some mediating structure or criterion that specifies which oppositional resolutions in relation to which types of controversy establish the requirement for a common basis of resolution. In this case, the "common basis" cannot be the bare, surd movement of will that constitutes permission/agreement or its lack. It must be something that includes both the permission/agreement and the criterion or structure that enables an agent to determine when a "no" generates a duty of forbearance. A similar kind of mediating structure is needed to differentiate between the "yes" of permission and the "yes" of a collaboratively willed end associated with beneficence. To flesh out these mediating structures, more detail is needed regarding the underlying action theory than Engelhardt provides.

basis is "agreement" (also spoken of in terms of resolution of "will"). Let's now consider the controversy between Martha and Mary, the common basis they have, B1, and two stages of the controversy: a pre-resolution stage when the common basis is used and a post-resolution stage when the controversy has been resolved. For Mary and Martha, what might it mean to say that "agreement" is the common basis? When they have agreement, the controversy is presumably over. "Agreement" is then just a word for the mutual state of agents who are no longer in a controversy. If this is what is meant by "basis," we would get something like: resolving is the basis of a resolution. Perhaps this might have the net effect of moving the question of resolution from the level of the whole controversy down to each agent, who is supposed to "resolve." By saying the basis is common and shared, we then mean something like a joint resolution each makes with the other to resolve the controversy. This meaning wouldn't get anywhere close to the forbearance strand Engelhardt is concerned with in his ethic for moral strangers. Instead, it might get us to a maxim something like: whenever two agents are involved in a controversy, they should seek an agreement that is sufficient for sustaining the game of blaming and praising in relation to their interactions. This maxim may involve a host of non-trivial constraints. Hobbes uses a variant of it to generate his Leviathan,[23] and Locke uses a more limited variant to generate an account of civil society with positive law, an unbiased judiciary, and an executive power that have a legitimate scope of action that extends beyond the sparse content Engelhardt allows.[24] But should agents

[23] Variants of this maxim can be found throughout Hobbes's Leviathan. The wide scope he gives to it may be taken as a distinctive mark of his political theory. One statement of the maxim from an early chapter on reason and science reads as follows: "no one man's reason, nor the reason of any one number of men, makes the certainty; no more than an account is therefore well cast up, because a great many men have unanimously approved it. And therefore, as when there is a controversy in an account, the parties must by their own accord, set up, for right reason, the reason of some arbitrator, or judge, to whose sentence they will both stand or their controversy must either come to blows or be undecided, for want of a right reason constituted by nature; so is it also in all debates of what kind soever" (1962[1651], 42).

[24] In Locke (1988 [1690], Second Treatise) the full maxim needs to be generated in a sequence of steps. Since I will advance at the end of this essay a modification of Engelhardt's project that in net effect affirms Locke's variant of the maxim, and since Locke expresses what I think is a crucial insight about force, agreement, and conditions for sustaining a certain kind of blaming and praising game, it is helpful to briefly sketch his account. Within Locke's state of nature, each person has the power "to do whatsoever he thinks fit for the preservation of himself and others within the permission of the Law of Nature: by which Law common to them all, he and all the rest of Mankind are one Community, make up one Society distinct from all other Creatures. And were it not for the corruption, and vitiousness of degenerate Men, there would be no need of any other; no necessity that Men should separate from this great and natural Community, and by positive agreements combine into smaller and divided associations" (352). Since Locke, unlike Hobbes, has standards of right and wrong that govern people in this natural community, the maxim to establish conditions for sustaining the game of blaming and praising does not take the fully general, unconstrained form found in Hobbes. Instead, it is anchored in the natural standards of peaceable interaction, but requires elaboration that compensates for deficiencies and "wants" in the state of nature associated with inappropriate uses of force and forms of neglect that make people blind to the standards of right and wrong that they should recognize: e.g., in the state of nature "[t]here wants an establish'd, settled, known Law, received and allowed by common consent to be the Standard of Right and

involved in controversies accept such a maxim and seek agreements of that sort? Does the required basis lead to a resolve that wills the full set of conditions necessary for sustaining justifications in relation to actual agent interactions?

In Engelhardt's ethic of moral strangers, we have no imperative at all to seek resolutions of controversies. Nor do we have an imperative to establish whatever conditions are necessary to sustain the blaming and praising game or peaceable

---

Wrong, and the common measure to decide all Controversies between them. For though the Law of Nature be plain and intelligible to all rational Creatures; yet Men being biased by their Interests, as well as ignorant for want of study of it, are not apt to allow of it as a Law binding to them in the application of it to their particular Cases" (351). For this reason, people transfer their natural power to do whatever they see fit within the bounds of natural law (associated with peaceable interaction) to form a political society which promulgates positive law that in many cases confines liberty to a greater extent than the natural law confines it in the state of nature. People thus jointly establish conditions that constrain themselves in whatever ways they deem necessary for ongoing peaceable interaction. In addition to the concession of legislative authority to make binding positive law, and judicial authority to impartially judge cases according to such law, prerogative is also assigned to an executive power to enforce whatever it considers important for preservation of property (broadly construed to include life, liberty, and what is more narrowly construed as property): "Where the Legislative and Executive Power are in distinct hands, (as they are in all moderated Monarchies, and well-framed Governments) there the good of Society requires, that several things should be left to the discretion of him, that has the Executive Power. For Legislators not being able to foresee, and provide, by Laws, for all, that may be useful to the Community, the Executor of the Laws, having the power in his hands, has by the common Law of Nature, a right to make use of it, for the good of the Society, till the Legislative can conveniently be Assembled to provide for it" (374). "Prerogative can be nothing, but the Peoples permitting their Rulers, to do several things of their own free choice, where the Law was silent, and sometimes too against the direct Letter of the Law, for the public good; and their acquiescing in it when so done" (377). What distinguishes this from Engelhardt's "agreement" is the way consent of people lead to a political society with discretionary powers to make laws, judge, and execute law with the sanction of force over all who sojourn within its geographical jurisdiction. The blaming and praising are thus not fully general, but rather specified in relation to the legitimate uses of force. The crucial insight in Locke concerns the way the legislative, judicial, and executive powers of political society must track norms of peaceable interaction that are sparser and characterize a universal human community. These involve additional contingent content related to uses of force that are necessary to compensate for the ways actual agents deviate from ideal rational agents. The more a people lack an awareness of the norms of peaceable interaction, and the more they deviate from it in practice, the greater is the contingency that enters into positive law, judiciary, and executive use of its prerogative. For Engelhardt, the forbearance strand involves the sparsest content and is associated with the most universal society. For Locke, people withdraw from the sparsest universal community and form a "private, if I may so call it, or particular Political Society, and incorporates into any Commonwealth, separate from the rest of Mankind" (352). What Engelhardt says of people who irrevocably give up their property when entering a monastery (but not what he says of placing oneself absolutely under an abbott) is what Locke says happens when people with their property form a civil society (compare Engelhardt 1996, 159–160 and Locke 1988 [1690], 348). Since the initial transfer of land to a political society involves an irrevocable consent, once transferred, the property remains under the jurisdiction of the political society. The property, especially land, is permanently transferred and from that time forward comes with constraints on all who sojourn within the land. If this can be validly done for a monastery, why can't it be done for a whole geographical region in relation to a political society? What distinguishes Locke's political society is that the contingent content of "private," i.e., particular society is restricted by the logic of force that checks and compensates for gaps between ideal moral agents and actual agents.

interactions. Many controversies, e.g., philosophical ones about ethics, might go on endlessly, and from the perspective of an ethic for moral strangers there is nothing problematic about this. The controversies that need to be resolved are those involving unconsented uses of force. Engelhardt seems especially worried about the ways agents may resolve controversy. The core function of the forbearance strand is to block ways of resolving controversies that are inconsistent with the grammar of ethical discourse. If this is the case, there is no reason to worry about efforts to use reason or conversion when resolving a controversy. The only problem arises when force is used. But to get forbearance, a much more complex logical structure is required that links in a very special way each agent to herself, the other agent, the means each agent uses upon the other, and the controversy. Central to that forbearance is an act of categorization that partitions means into those that involve force and those that don't, and suspends such force as a means to attaining the agreement that resolves the controversy. One agent must be able to contemplate a possible action, discern how that action impinges on the agency of the other agent in the controversy, and determine whether the effect alters the conditions under which a resolution might count as an agreement. For this, we need some account of which resolutions count as the right kind of agreement, and this must be traced downward to specific agents, so we can know how they are to resolve controversies in ways that count as agreements.

One of the difficulties associated with understanding how agreement functions as a common moral authority might arise from two distinct ways we can imagine a controversy. We might imagine two agents who are already interacting, and whose interaction constitutes a controversy. "Agreement" is then a way to resolve the controversy that is underway. This is the context that raises difficulties. Alternatively, we might imagine agents who are not interacting at all. There is no controversy. But one or both of the agents is contemplating an interaction that might lead to a controversy. If we now consider "agreement" as a moral basis, we can introduce a forbearance constraint on how agents initiate interactions: they are to initiate interactions that are agreements. From this, we might come up with an account of controversies that are consistent with agreements. For example, two people might agree to wrestle, have a duel, or argue about some philosophical topic about which they disagree. We could notice that each of these interactions have oppositions between agents and claims they press. The agents agree to interact in this way, so such controversies are the good kind. But in this case, the agreement wouldn't be resolving the controversy. Rather, the agreement would be generating the controversy.

Let's now consider the option that requires a moral authority that is common for all agents, such that any two agents in any controversy reference that same common basis. This would require that B1 = B2, i.e., that the basis Mary and Martha utilize to resolve their controversy is the same basis that Fred and Bob use to resolve their conflict. If in all cases we categorize this common basis as "agreement," we cannot mean the contingent resolve and agreement made by two specific agents in a controversy. If the general category, "agreement," denoted the contingent, particular movements of the wills of specific agents in a specific controversy, then we would have to say B1 is not the same basis as B2, although it is the same type of common basis, namely, a joint resolution of the two agents involved in the dispute. If instead we say that the common basis is, in fact, common among all agents, then we would need to go with one of two options. Either we would need some kind of supra-individual "general will" that is somehow accessible within the awareness of particular agents involved in a dispute or we would need some purely formal, conceptual content that might be accessible to any agent qua rational agent. Engelhardt clearly doesn't go with the first option. He has an extensive body of research that seeks to work out the second option.

In the next section I'll more carefully consider the relation between this stance of an agent qua rational and the moral stance associated with Engelhardt's grammar of blaming and praising. In this section I've considered the relation between an agent in a controversy and the "common basis" that is categorized as "agreement." Here I want to make just one concluding remark about that actual agent and that general category used for the type of moral authority agents need: there is no way that the general category – "agreement" – can get us to the forbearance Engelhardt wants if we just focus on the contingent, actual agents without some mediating structure. When we took "agreement" to denote the particular resolutions of those actual agents embroiled in a controversy (the case where $B1 \neq B2$), we could at least see how some content might be addressed in relation to the category of the basis. While the category "agreement" is very sparse if considered in its generality, when that category is used in relation to specific agents, we get the facticity of those wills in their spontaneous action. However, when the category is deployed in that way, what is covered by a category in its use must be specific to the two agents in the controversy. In every controversy, the common basis will be different; namely, it will be the contingent resolve of just those agents in that controversy. When we move to the interpretation where "agreement" is regarded according to type, rather than instance, and when that is put forward as the common basis for resolving controversies with moral authority, we end up with the most vacuous of contents.[25] It is then not clear at all how we can get content back. "Agreement" might be taken as the outcome of an interaction, as the requirement for initiating an interaction, or as a constraint on any actions associated with an interaction. In each case, there are additional questions about the way an agent in an interaction is to contemplate that "agreement" when considering her own action.

[25] This is the problem of Abstract Right which Engelhardt, following Hegel (1967, 1991), contrasts with Sittlichkeit (1994b, 214–217). He presents this as a contrast between right and the good. But here we see the same problem in relation to that agreement that is supposed to be the basis of the ethic for moral strangers. This makes clear that there is another aspect of the problem of contingent content that is not addressed in terms of the intracategorial relations of Hegel's *Philosophy of Right*. This is the problem of the "concrete universal" discussed by Chaffin (1994), and which I discuss in relation to the "existential turn" from the pure categorial back to the ordinary domain of actual agents (Khushf 1994).

7.4.3 Why the Game of Blaming and Praising Requires the Stance of an Agent that Can Be Reciprocally Assumed by the Other Agent (and the Nature of the Self-Relation Between an Agent, Agency, and Action that Is Assumed by One Who Plays the Moral Game)

When Engelhardt speaks of a "common basis" and calls this basis "agreement," it is easy to imagine that he is somehow referring the "moral authority" back to the particular resolve of the actual agents involved in the controversy. In the last section, we saw why this cannot be the meaning. That particular resolve of actual agents is a brute, surd content. By itself, it is not a resolve for anything that is distinguished from anything else, nor is it a resolve that constrains what agents do in any way. While "agreement" might imply something more than just the sum of the two agent resolutions taken singly, this "more" – namely, that the brute resolve of each agent is somehow mutual – might be nothing but another instance of a brute, surd thing, for example, a resolve to mutual intransigence. To get a moral content and be able to call an "agreement" a "common moral basis," we need to introduce a non-trivial mediating structure. For Engelhardt, this mediating structure is a specific stance an actual agent might assume when deliberating about what to do and communicating with others about those possible actions. The possible stance an agent might assume is referred to by Engelhardt variably as "the moral standpoint" or "the standpoint of the purely rational agent."[26] Thus "agreement" only becomes a "common moral

[26] At times Engelhardt seems to use "intellectual standpoint," "moral standpoint," "secular moral community" and the "intellectual standpoint of persons" as all synonymous. "Since moral controversies can in principle encompass all moral agents (… and only moral agents), one has a means of characterizing the secular moral community as the possible intellectual standpoint of persons interested in resolving controversies in ways not fundamentally based on force" (1996, 69). "This concept of the person (as well as moral competence) is thus defined wholly within the practice of moral strangers resolving moral controversies by agreement, by giving and withholding morally authoritative permission. The very notion of a general secular moral community presumes a community of entities who are self-conscious, rational, free to choose, and in possession of a sense of moral concern. … Insofar as they wish to collaborate with common moral authority, they create the peaceable moral community. The peaceable secular moral community exists both actually and potentially. It exists potentially as a moral standpoint in terms of which self-conscious rational entities can speak of blame and praise, and through permission and agreement understand themselves as bound by their mutual authority. It is an intellectual standpoint … In terms of this possible moral standpoint, persons can at any time in any place conceive of themselves as belonging to, and being bound by the rules of the peaceable community. An examination of the moral language reveals a very important moral standpoint: the mundus intelligibilis of Kant" (136). Here the "accent" falls on a Hegelian categorial account that "avoids the Kantian difficulty of mediating between the sheer givenness of the object and the predicament of the finite knower. … [C]oncerns of the finite knower can be placed with a general categorial understanding" (95, note 83). Engelhardt is "seeking a characterization of the fabric of morality" that is "tied to the fabric of rationality"(105). A "will to the moral viewpoint" cannot just be an "inclination toward that viewpoint." It must be a "will to a moral fabric as general as the very concept of morality itself." By tying the "characterization of the moral fabric" to the "very enterprise of being a person," one

basis" when we assume another kind of common basis, namely, that of a shared rational agency. This shared rationality turns out to be the engine needed to get Engelhardt's forbearance structure into the agents who interact in a controversy.

While there are some places in the *Foundations of Bioethics* where Engelhardt provides some insight into what he means by this standpoint of the rational agent (usually in his endnotes), we really need to go to some of his other writings for a full account, especially his published dissertation, *Mind-Body: A Categorial Relation*, and several essays he has written on Kant and Hegel. I think the details of this account are important, especially when considering the relations between categories worked out at the transcendental level of the pure rational agent, on one side, and the actual agent who uses those categories and categorial orderings to specify what actions are necessary, possible, and impossible, on the other side. For Engelhardt, the kind of contradiction that must be avoided is the one that arises from the incompatibility between an actual agency and the assumption of the pure rational agency. Clarification of when incompatibility will arise thus requires we carefully specify what must be assumed by an agent when that agent takes up the stance of a rational agent, and when that agent qua rational engages in a practice of a specific sort, such as that associated with blaming and praising. However, for the purposes of this essay, I'm going to move as directly as possible to the core categorial features required by Engelhardt's transcendental argument.

How must we think about the agents who are involved in a controversy? First, we noticed that controversies were ambiguous things that involved at least two levels, one involving opposing agents, and another involving opposing claims, viewpoints, or stances. To stabilize our controversies, we needed a common description. We thus have two kinds of things that agents in a controversy must already share: first, they must share the fact of being in a controversy. This involves various ontic characteristics that make the two agents entangled in that manner called a controversy. I'll call this the "first community." Second, we have their shared description. I will refer to this as the "second community," and it is present whenever the common description condition is satisfied. Without this second community, there is a regress of meta-controversies. The generation of the shared description (and thus the second community) involves isolating and representing those ontic features that somehow do explanatory work for the agents, for example, features that enable agents to understand why they are in an oppositional stance, how they came to be in it, and

---

establishes the conditions for its rigorous justification (104). In all of this, there is a nontrivial categorial content and a special kind of rationally legislating will that is involved in making agreements. The agreements involve concomitant willing of the whole fabric associated with the moral standpoint. At other times, he emphasizes that "authority is derived not from reason but from the bare will to have the one authority moral strangers can share: permission. Secular authority is derived from a bare will to morality. Competence to give permission is the ability to so will" (72). Somehow the "will to the moral viewpoint" and the "bare will" associated with giving permission need to come together in an actual, embodied agent. His first book on mind-body as a categorial relation worked out the ways the complex strata of mind and body (with its "force") are integrated. But it is difficult to see how the rich sequence of mediating categorial relations associated with embodiment link up with both the "bare will" qua permission/agreement of an actual agent and the "will to a moral standpoint."

what possibilities are available for making moves, counter-moves, or escaping the conflict. The description introduces a stability into the controversy. By "stabilizing a controversy" I mean "makes the controversy explicit as the kind of controversy it is." (Since any controversy can also be viewed as a kind of instability, there is some strangeness to using the world "stability" in the above-mentioned way. My emphasis is upon a relative stability, and thus on the distinct gain associated with generation of a common description.) After stabilizing controversies in the second community, we asked how we might think of a "common basis." To make sense of this, we shifted first to one agent, asking what that agent might take as a basis. Next we shifted to the other agent, and asked if that other agent accepted the same basis as the first agent. As we reflected upon what might be meant by such a common moral authority, we saw that it split into a fully particular and a general kind. I'll refer to the particular, surd movements of will of actual agents (whether they be agreements or oppositions) in a controversy as the "third community." Agreements associated with the third community will be different for every controversy. It is the agreement of Mary and Martha, B1, that is different from the agreement of Fred and Bob, B2. $B1 \neq B2$. By speaking of first, second, and third community as "community" I seek to highlight how individuated controversies involve given, shared contents that might be appealed to by the agents involved in a controversy. Alternatively, we might consider such agreements according to their type, and postulate this type as one common basis for all moral strangers. Let's speak of this purely general concept of agreement as the "fourth community."

Now let us refer back to the two agents who we considered independently when we asked whether they shared a common basis, and let us reflect on the way we must regard those agents when they are in the third community and when they are in the fourth community. I initially named the two agents Mary and Martha. However, unlike the lawn mower dispute between Fred and Bob, I never said anything about Mary and Martha or their controversy. Mary could have been Martha and Martha could have been Mary. The only constraint was that Mary and Martha needed to be distinct. But if I now carefully consider the difference between the third and fourth community, I notice that even this distinctness is only required in the third community. There, the common agreement was the brute, mutual concordance of the distinct agents in the controversy. This required that I consider some actual controversy and actual agents who agree. But my controversy for Mary and Martha was nothing but the bare idea of a controversy, with the postulated opposition of different agents. The agents were "individuated" as distinct within the controversy, but the controversy itself was taken as fully general and thus not individuated in a way that made it distinguishable from any other abstractly regarded controversy.

To clarify what is meant by "individuation" of agents as distinct within a purely general type, we might draw on a distinction within the philosophy of language between type, token, and occurrence.[27] We can illustrate this with letters in words on a page. For sake of simplicity of this illustration, I just take words as identifiable word strings and thus ignore the semantics of the English words I consider. Consider the following line:

example example

In this line we find two token instances of the word string, “example.” (I regard the string according to mention, thus the quotes. But the use/mention convention breaks down here, because mention could be according to type or token.) I am concerned here with instances of a distinct pattern that is defined by the order of letters that make up the string. I say that I have an instance of the word when there is a string whose letters and order from left to right map one-one to the letters in the string “example.” If I consider the word according to type, there is only one such word, and the type is determined by the ordered string of letters that makes the word what it is. According to type, the word is an abstract object. However, if I consider the word as a token instance, then there are two word strings, not one. If we now consider the word according to type, we notice that the letter “e” is in the word twice, once at the start of the word and once at the end of the word. How do I speak about the double instantiation of this letter within the word type? If I focus on word string according to type, there are two uses of “e.” If I focus on word string according to tokens, for example, when I look at the two token instances of “example” in the above line, then we say there are four uses of the letter “e.” When speaking about the four uses, the “e” I speak about is the particular letter that is distinct from all other letters on the page. But when I speak about the two uses of “e” in the word string type, “example,” I am regarding the word as an abstract object. The letters thus can’t be tokens. But I still need the “e” at the beginning to be distinct from the

[27] A more detailed discussion of the type, token, occurrence distinction is provided in Wetzel (2009). Chapter 7 provides a defense of the notion of an occurrence. Three individuating parameters of occurrences are identified: “what expression is occurring, in what other expression it is occurring, and where it is occurring in the latter” (131) or “x occurs in y at position p” (132). I use this language as a shorthand for a different kind of logical language worked out in Hegel’s logic. Individuating conditions of occurrences are addressed in Hegel’s *Encyclopedia* by stating the order of a categorial occurrence within a higher order categorial whole. The individuation of the occurrence is as part within a whole, where the part-whole relation must itself have a place or position in a yet more comprehensive whole. Each higher category constitutes a “categorial place” that gives form and meaning to the notion of a position. Negativity, incompleteness, and inclusion relations between diverse categories are used to clarify the way individuation of an occurrence within a higher order occurrence changes as one progresses through the systematic ordering of the categories. Most significantly, the very structure of individuation of an occurrence in its higher order type changes, especially as one moves from the logics of being and essence to that of explicit concepts (the thinking of categorial thought determinations). This categorial logic is more complete and appropriate than the logic of types, tokens, and occurrences for elucidating the relations between agents, controversies, communities, and societies. It is worked out in *Hegel’s Encyclopedia Logic* (1991 [1830]), and summaries are provided in Engelhardt (1973) and Pinkard (1988). However, it was too difficult to clarify all of this for the current essay. By means of the notion of an occurrence I seek to highlight a kind of individuation that is different from that associated with concrete particulars, e.g., tokens. Once this difference is appreciated, we can move to a discussion of the relation between the ideal individuation of agent occurrences and the actual individuation of reflectively aware agents.

"e" at the end of the word. To speak of this special kind of doubling within a type, the word "occurrence" is often used. We can now summarize by saying that if we focus on the letter "e" according to its type as a letter, there is only one such letter, no matter how often that letter is presented in "letter" or "example" or any other word. If we focus on "e" as a token, then there are four such tokens in the two token instances of the word, "example" that is listed above. However, if we focus on the "occurrences" of the letter "e" within the word type "example," then there are two such occurrences, one at the leftmost part of the string and another at the rightmost part of the string. We have a superordinate type, categorized as "word," and a subordinate type, categorized as "letter." Within the superordinate type there may be more than one occurrence of the subordinate type.

We can now use something like this language of types, occurrences, and tokens to speak about agents integral to controversy. In a more complete account we would need to move from these linguistic terms to the proper categorial ones. But as a first approximation we could say that when we consider a controversy in the fully general way we did in the case of Martha and Mary, then we are considering the controversy according to type. Within this superordinate type categorized as "controversy," there are two occurrences of a subordinate type categorized as "agent." Within the third community, we imagine another kind of content that has the effect of individuating the occurrences within the type. When we just focus on controversies in a fully general way, as we are with Mary and Martha, we don't have the kind of otherness that makes particulars the token things they are. For example, the token "e" is made of black ink or involves an opacity on a computer screen that prevents light from being projected. Instead of any actual individuating substance like ink, the occurrence has a kind of ideal individuation which allows multiple instances of subordinate types to be individuated as distinct, but makes this a pure otherness without any token individuation. The occurrence of the letter "e" in the word string type, "example," has an analogue of individuation within the ideal. This ideal individuation takes place by a kind of postulated difference, negativity, or otherness of distinct occurrence of a subordinate type within its superordinate type. For Immanuel Kant (1965 [1781/1787]) these instantiations of occurrences within types took place in a medium of pure intuition and they were associated with schemata. For Engelhardt (1973, 1994a, b), these occur within the medium of pure thought, and involve sequences of thought determinations, which develop superordinate types, called "categories," as sublimations of subordinate types, which may themselves involve a host of antecedent steps in their constitution. Since our third community involved a postulated agreement of the actual agents, this "actuality" is now made into that of an instantiated occurrence, and it involves the way each occurrence of an agent "stands out" from the other within the categorial type called "controversy." The postulated otherness of that agreement is itself spread across the agent occurrences and thus partly "stands in" and "stands out of" each of the distinct occurring agents subsumed within the superordinate type, "controversy."

If we now focus on the fourth community, we notice that the "agreement" that is supposed to be shared is something that must be common to all agents. This must be that which any agents qua rational agents can take as a basis, and that by taking it as a basis, they will be utilizing exactly that basis that is used by any other agent.

Engelhardt signals this by calling the bases "intellectual" (e.g., the title of ch. 2, 1996). This shared basis and this agreement must be nothing other than that set of categorially specified determinations that constitutes the abstractly regarded controversy we had in the case of Martha and Mary. Now the abstractly regarded distinction between the third and fourth community must itself be taken as something that is included in the fourth community. In other words, this distinction between third and fourth community and the postulated relation between the distinct occurrences of agents within the superordinate controversy must all be taken as part of a general type that is available to each and every moral stranger qua rational agent. But if we do this, we lose the capacity to speak about an ambiguity in relation to the third community, since this might be the particular agreement of two actual agents in an actual controversy, for example, Fred and Bob, or this might be the postulated otherness of the occurrences within the type, and thus constitutive of what any agent qua rational agent can grasp about the fact of such a contingent, agent-based agreement.[28] To specify that distinctive kind of fourth community that arises when we postulate that categorial fusion of an abstractly regarded third community with the generality of the fourth community and regard this as a shared content associated with the purely rational agent, I will refer to this complex categorial content as the fifth community. We might say that in the fifth community, we eliminate all contingent individuating characteristics associated with the first and second community, and thereby equate them: now, according to type, the controversy is one with its general description. It is simply the abstractly specified category which is now "used" in relation to the subordinate occurrences that enter into the constitution of the general type. This includes agents, instantiated as occurrences of types, and also includes the postulated bonds of agreement that both "stand in" agents as that which distinguishes one agent from the other, and "stands out of" agents as that which unites them in that special resolution called "agreement." Within this general type, agents are symmetrical: each agent occurrence is symmetrical to the other agent within the higher order type, allowing agent perspectives to be flipped. Since both agent occurrences are within a general type that is itself included within any agent, this shared schema might be used when an agent deliberates about actions and controversies. Actual agents might click into one of the agent occurrences and regard the other agent in the alternate role. But this "clicking within" is something that occurs within the mental life of the agent who thus contemplates how he relates to another agent within a controversy.

With this categorial content associated with the fifth community, I think we can for the first time start to make sense of what Engelhardt might be doing when he provides his transcendental explication. From the start he was never considering an actual controversy at all. Instead, he was considering an ideal controversy, a categorially specified controversy in which there are two agent occurrences that are distinguished from one another by a postulated contingency of will. This is the

[28] This brings us to the crux of problems associated with Hartmann's non-metaphysical interpretation of Hegel. The analysis that follows involves further development of a criticism I originally presented in Khushf (1994). The ambiguities arise from the differences between intracategorially specified relations and relations between the actual instances covered by the categories.

moral stranger qua moral agent: an agent abstracted from all contingency and placed into a purely abstract relation with another agent similarly specified. We can now ask: how might we distinguish the resolution, with its postulated agreement, from the controversy with its postulated opposition? With this level of abstraction, the only difference between resolution and controversy is the category we use for the surd content distinguishing agents favorably oriented to one another within the higher order category. We imagine a kind of flipping of the wills from opposition ("no") to accord ("yes"). The flipping might come about in one of four ways. First, an agent or both agents might deploy external means upon one another, and thus compel a resolution. This involves a use of force or an appeal to force. Alternatively, agents might deploy some kind of internal means. This might involve one side moving the opposing side to accept some contingent basis that was not initially accepted. This is conversion. Agents might appeal to rational contents that are accessible as reasons to all agents, e.g., the type of rational content we find in the general category of the conflict. But the general category postulated no link between surd motion of the will and the categorial frame. Further, qua categorial frame, the two sides are symmetrical. There is no rational content that moves agents one way or the other. As a final alternative, agents might each resolve to not be in opposition to the other. This is agreement. It involves a symmetrical flip of the wills of the two agents from oppositional to concordant.

As the options were just presented, we did not yet link them to any presumed shared basis that they have for resolving the dispute. We only considered the categorial structure of the controversy itself as the basis, and we asked how some agent would contemplate the options for resolution. Some of the options for resolution depended on contingent contents that are not accessible to the agent who is considering these options from the purely general standpoint. Now we can move to the question Engelhardt asks when considering his ethic. He thinks ethics involves a commitment to a common basis that is then used in a discourse of blaming and praising. What distinguishes the moral standpoint from the purely rational standpoint is the commitment to a commonly accessible and accepted basis which is used to determine which courses of action are appropriate. The purely rational subject qua moral agent initially identified four possibilities of resolution – force, conversion, reason, and agreement. These are now sifted to see if they are compatible with the required commitment to a common basis. Force is not compatible, since none of the contingencies of power and its use are even accessible to the purely rational agent. Conversion requires one side to convert to the other side. But from a purely rational standpoint, the sides are symmetrical: there is nothing associated with one side that distinguishes one agent qua rational from the other. This leaves only agreement, namely, the mutual resolve of the agents integral to the controversy.

In a more complete discussion, some additional distinctions are needed. We should distinguish between the purely rational subject and the purely rational agent, and then define the purely rational agent as a purely rational subject who takes up the question of a practice and agency in a very specific way, i.e. as one engaged in a practice of blaming and praising which only appeals to what is universally accessible. Qua pure rational subject, there is no question of action or deliberation. Instead, one only brings into view the categories that must be taken as given for a subject to be rational. These categories inform any description. When the moral standpoint is assumed, this must arise as a resolution of will to deploy the categories of the rational subject in relation to a specified content to attain a specified end. If one seeks to specify the moral standpoint as a purely rational standpoint, then the contents on which agency works and the ends to be attained must each be fully transparent to the rational subject. Since categories are the contents transparent to agents, the agency must work on categories to attain a kind of categorial specificity in relation to those initially given categories. The result is something like Engelhardt's non-metaphysical interpretation of Hegel's *Philosophy of Right*. But in the *Foundations of Bioethics*, we have a much sparser categorial specification. Engelhardt's ethic for moral strangers might then be taken as the practice of an agent who assumes the stance of a purely rational subject and resolves to provide an account of actions that can be justified to another agent who likewise assumes that standpoint and justifies actions in that same way.

This is the best way I have been able to make Professor Engelhardt's transcendental argument work. To get to his conclusion, I need to abstract away all contingent contents that might separate me from any other agents. This yields the stance of a purely rational agent. If we take "moral stranger" as an agent who is considered simply in terms of the content that agent shares with all other agents, then the moral stranger is the same as the purely rational agent. If I now bring back all of the contingent contents and add them to the purely rational agency, then moral strangers might be taken as involving two strata, one strata of highly abstract contents that is shared and constitutes the rational agency of the moral stranger, and another strata of contingent contents that particularizes that moral stranger as the actual agent he or she is. When we come to Engelhardt's argument, we need to abstract away all of the contingent contents and contemplate available options from the purely categorial stance. We then consider available options for a common basis and how the logic of justification might work between moral strangers. We notice that if we bring in any of our contingent contents, we no longer have something that could be appealed to with any moral stranger. At any stage, if we want to know whether someone would accept the arguments we are advancing, we simply flip perspectives and imagine that someone is asking us to accept the basis we are proposing. The only general arguments that can work are those we could advance or another could advance to us when we assume this purely rational stance. If our argument requires the other side to accept a basis that we cannot motivate once we've eliminated all contingent contents then we have a contradiction with the basic premise of the ethical discourse among moral strangers, namely, the claim that a basis must be one that all accept as a moral authority. This basis is agreement.

We have now moved further toward our goal. But there are still problems.

7.4.4 How Are Actual Agents Related to Agent Occurrences Within Categorial Controversies?

In the end, the agreement that matters can't be the abstract agreement postulated as the sole sufficient basis for an ethic of moral strangers. The agreement that matters must be the agreement of actual agents involved in some controversy. Why does the agreement of actual agents matter? Engelhardt's answer seems to be that we can't answer this question in a fully general way, because different agents will have different reasons. When asking why things matter, answers must be given in relation to ends embraced by agents, and these differ in relation to the content-full commitments that they have. All that can be said generally is that the actual agreement matters for those who want to sustain the game of blaming and praising because it is a necessary condition for sustaining that game. The transcendental argument shows that the game breaks down when a basis is deployed that might involve dissent of the opposing agent. In the face of such dissent, there is no way to move forward in the ethics game. The game has been subverted by the way one agent does not abide by the constraints that enable those agents to sustain the game. If an agent chooses to suspend playing the ethics game, the other agent can try force, convert, or persuade that agent to again abide by the game rules. But all that forcing, converting, and persuading are working by a different game. It is no longer part of the ethics game played by moral strangers, even though it might count as some contingent kind of ethics game that moral friends play among themselves.

When Engelhardt characterizes how the game of blaming and praising breaks down, he is making a new move, one we have not yet considered.[29] Thus far, all of

[29] "A contradiction in will" can be associated with contingent contents of the moral life, for example advancement of freedom as a value rather than just freedom as a side constraint. Engelhardt seeks to distinguish such contradictions from the conceptual contradictions that are the mark of blame in his ethic for moral strangers: "affirming the notion of the peaceable community, in the sense of a community whose authority is not based on force, while at the same time deciding to use force against the unconsenting innocent, would involve a conceptual contradiction" (1996, 106; see also 103–108, 94–99, notes 81–89). If the agent directly and intentionally "decides to use force against the unconsenting innocent," then we do come to the strong, clear cases of contradiction. Even in these cases, the agent's understanding of intention would be informed by contingent, background norms of right that say when an action involves an unconsented use of force against the innocent. More significantly, many and perhaps most real controversies involve ambiguities which enable agents to engage in practices that others might legitimately interpret as using force, yet do this in a way that never requires an intention to use such force and in a way that can be interpreted by the agent as consistent with an ethic for moral strangers. We need a framework that enables reflectively aware agents to consistently interpret their actions in convergent ways. Engelhardt's notion of the "outlaw" is closely related to his account of conceptual contradiction: "if one rejects the principle of mutual respect, one cannot rationally protest when others respond with force. Since questions regarding the sanctions for immorality are intellectual, the sanctions are intellectual. They pronounce outlawry upon the offending individual, a charge against which that person cannot consistently protest as long as he continues in affirming the immoral action" (1996, 111; also 117, 133, note 12 and 137). First of all, Engelhardt interprets being an outlaw in terms of a failure of rationality: one loses a basis of justification for objecting when another would use force in return. Only secondarily does this provide an account of legitimate uses of force. In

the key moves in his transcendental argument have occurred with respect to purely rational contents and in relation to a special kind of rational practice called "ethics," where the contents of the ethic are restricted to those available to any agent qua rational. However, if this ethic is to be deployed by actual agents involved in real world controversies with all of their contingent messiness, we need some account of how the deliberations of the actual agents intersect with the sparse deliberations of that same agent when assuming the purely rational stance and engaging in the game of ethics. This requires guidance on how an agent moves from the intellectual stance qua pure self-legislating willing of a moral fabric back to the ordinary world where actual controversies with flesh and blood agents occurs.

I begin by noting an asymmetry between the kind of *abstracting action* that brings an agent to the standpoint of the purely rational agent and the *concretizing action* that brings an agent from that sparse, abstract space back into the stance of an actual agent involved in a real world controversy. To the degree that there are intersubjective, categorial contents that can be fully specified, then, to the same degree, when any two agents abstract from contingencies and articulate those categorial contents, they are moving to the same contents and the same perspective. The abstracting action thus brings actual agents to the same purely rational perspective. But when agents come back into their contingent contexts and deploy that awareness they have as purely rational agents there is an *unavoidable interpretive character* associated with the way categories and the abstract ethic are deployed. The asymmetry thus concerns this difference between the abstracting process which leads to a single perspective of pure rational agency and the concretizing interpretive process which potentially leads to multiple interpretations among multiple agents. Even if we accept all of Engelhardt's arguments qua transcendental expositions and justifications, this doesn't mean his ethic could be deployed by moral strangers involved in actual controversies. To assure that, Engelhardt needs to provide an account of how we get a single interpretive mapping back to actual agent interactions. More specifically, we need an account of what "agreement" among agents must entail, so actual agents can deploy these categories in mutually consistent ways. Without clear necessary and sufficient criteria for that agreement that is supposed to be the common moral authority, real agents would not be able to use the ethic to guide their practices. To "apply" his ethic for moral strangers in such a way that there could be a genuine contradiction of will, we need to see how an agent can

contrast, Locke (1988 [1698], 280, 284, 309) directly ties the notion of being such an outlaw to the legitimacy of using force (the executive power). By his transgression, the human "declares Himself to quite the Principles of Human Nature, and to be a noxious Creature" … "and therefore may be destroyed as a Lyon or a Tyger, one of those wild Savage Beasts, with whom Men can have no Society or Security" (273, 274). In Engelhardt's account, an extra step is needed from the inconsistency of an action with the justificatory game to a legitimate use of force against that person. In Locke, there is a direct link between the transgression and the legitimacy of force: by the act of violence, the transgressor demonstrates he "quit the principles of human nature" and thus placed himself outside the "natural society" of peaceable interaction that all humans have with one another. In such a case, it is "natural" to deal with the "wild beast" in the same way one would deal with any other wild beast.

bracket the contingent contents of his moral life and take up the moral standpoint while, at the same time, taking up the ordinary level stance where that agent is motivated by the contingent goods and situated within his moral community.

Following Klaus Hartmann (1966), I'll refer to the hyper-abstraction which brackets away all contingent contents as the "transcendental turn," and say an actual agent "takes the transcendental turn" when he or she engages in some kind of activity that step-by-step isolates and represents what any agent must conceptually grasp when engaging in some practice of a specified sort. Here "practice" will be interpreted broadly to include epistemic practices such as "seeing" or "experiencing." When we think about abstraction in this way, we notice that one agent might initially abstract from given contents in a different way from another, so they come into a dispute about what, in fact, must really be postulated by any agent engaged in an activity of the specified type. In the case of these controversies, since both agents are attempting to isolate what any agent must posit, they can agree that their controversies with one another must eventually be resolved by only appealing to just those things that they and anyone else might appeal to when explicating their contents. The whole project associated with taking the transcendental turn is oriented toward abstracting away whatever is not universal (in the sense of fully intersubjective), so the practices associated with attaining the transcendental turn must be in accord with the goal and outcome of the task. With this adjustment of practice to goal, we can imagine a long, complex sequence of moves and counter-moves where actual agents succeed in isolating and representing just those contents that they and any other agent must posit when considering a practice of a specified sort. The project is thus set up in such a way that all agents can converge on a common solution.[30] The project is only successful if the explicated contents can be systematically organized in such a way that it could be shown that any alternate account that satisfies the full set of conditions must be isomorphic to the one explicitly given.

In our above analysis, we made sense of Engelhardt's transcendental argument as one that assumes the transcendental turn and exhibits the categories that must be posited to jointly specify a practice of a specified kind. The practice in question is an ethic that resolves controversies by appealing to a basis that is mutually recognized as a moral authority. A controversy was then categorially specified in a way that involved postulated agent occurrences that were initially oppositional. The resolution took the form of an additional posit, which was to be the sufficient basis required by the postulated practice. The only posit that satisfied the transcendental constraint on both the ethic and controversy qua resolved was that of agreement. If we now more carefully consider the contents of Engelhardt's account of controversy and ethics, we notice that it simply prescribes what he presents more generally regarding the way any initially oppositional categories must be made to "agree" in some higher order category that subsumes them. If we now want to distinguish one kind of categorial agreement from another, we must give sufficient structure to the subsumed categories and the ways they are reconciled in the higher one, so we can

---

[30] Hartmann (1966) views Hegel's *Phenomenology of Spirit* as such an initiation into a common, purely rational standpoint. A summary is provided in Khushf (1994, 123–128).

clarify what this specific kind of agreement must entail. To do that we need a more explicit account of the agents subsumed within the controversy and of the determinate content that makes an accord of agents into that kind of agreement that qualifies as a common moral authority. To get this, we need to move from the abstracting process associated with the transcendental turn back to the concretizing, interpretive action of an actual agent who deploys the categories.

To further clarify the problems associated with the concretizing action, let's distinguish between three kinds of agents. First, we have what I will call "naïve agents." These are the agents who are ordinarily involved in the conflicts we are describing. They have categories for things like controversies, agents, force, resolutions, agreements, and so on. They use these categories when they think about the controversies, agents, agreements, and so on that they must think about when deliberating about their circumstances and their actions. Engelhardt follows Klaus Hartmann in referring to this as an "ordinary use" of categories. But these agents have not engaged in philosophical reflection upon their categories, and have never been tutored in Engelhardt's secular ethic. Any ethic they deploy will be one that Engelhardt will call content-full, and thus an ethic that incorporates a host of contingent commitments that make them unfit for sustaining an ethical discourse with others who do not have those commitments. Still, such agents periodically and perhaps frequently will find themselves acting and making agreements in ways that happen to conform to what Engelhardt's secular ethic requires. Further, they will often try to negotiate agreements and give reasons for their actions that other agents can understand, and they will try to create conditions that enable ongoing collaborations with those agents. But all of this depends on a host of contingencies that can only be worked out in relation to those agents, their contexts, and their controversies. (This poses ethical variants of Gettier problems.)

Second, we have what I will call "purely rational agents." These are agents who have learned to isolate the pure thought contents, the categories that are deployed by all agents. They can view themselves as "moral strangers," and thus disentangle the purely rational strands that must be shared by all other moral strangers from the contingent contents of their own actual agency and circumstance. Engelhardt's moral stance arises as a further determination of an agent who has assumed the purely rational stance, and who has, from within that stance, sought a basis for resolving controversy that any other agent qua purely rational agent must likewise accept. Engelhardt's moral stance is thus that of a pure rational agent with the added movement or spontaneous willing to resolve controversies with other agents on terms that can be mutually accepted when the agents involved in the controversy reciprocally assume the stance of an agent qua purely rational (and thus when they abstract away all of the contingent contents that distinguish them as actual agents). I assume that agents assuming the purely rational stance accept Engelhardt's claim that only agreement provides a common moral authority for resolving controversies.

Finally, we have what I will call "reflectively aware agents." These are like naïve agents in so far as they are moved by their contingent commitments and use their categories to make sense of their world and deliberate about their actions. But they also have the capacity to assume the perspective of the purely rational agent, and thus reflect upon the categories they deploy. The specific kind of reflectively aware agent I am considering can click into and out of the perspective of pure rational agency much like an agent can put on or take off a hat. By means of this ability to assume the purely rational perspective, such an agent can reflectively consider whether his actual agency is at any stage consistent with the commitment such an agent must assume when he takes up Engelhardt's moral standpoint, as a purely rational standpoint with the added commitment of resolving disputes with other agents in ways that sustain the game of blaming and praising among moral strangers. What distinguishes the reflectively aware agent from the naïve agent is thus the capacity to evaluate whether actual agency is consistent with the purely rational agency deployed with other moral strangers when engaged in ethical discourse. What distinguishes the reflectively aware agent from the purely rational agent is the contingent, content-full commitments that specify how he pursues what he as an actual agent takes to be right and good.

Note an ambiguity that has now arisen when we consider what it means to take up the moral standpoint: When an agent assumes the purely rational stance and from within this stance resolves to advance the game of blaming and praising, this resolve leads to the exposition of agreement as a necessary condition. Assume this moral standpoint simply involves the sparse resolve to play an abstract game in a certain way. But when the reflectively aware agent resolves to pursue his actions in ways that sustain the resolve associated with the moral standpoint, this involves a very different kind of resolve. Now the actual agent makes the rational moral resolve into a constraint upon how he pursues what he takes to be good. Engelhardt's emphasis (1996, 103–108) has been on the potential inconsistencies between an agent's actual resolve and the kind of resolve required to sustain the ethic for moral strangers. But we can't fully specify any of the rigorous kinds of contradictions he emphasizes unless we have a much clearer account of what is involved when a reflectively aware agent assumes the moral standpoint and evaluates whether his actions are consistent with it.

7.4.5 *How the Grammar of Blame Is Related to the Impossibility of Simultaneously Sustaining Two Incompatible Forms of Agency*

How is an actual agent to think about the accord between his actual resolve to pursue the good and right integral to his content-full ethic and the sparse resolve to sustain the discourse of blame and praise with other moral strangers? The answer to this question must depend on his contingent commitments, and on the way he maps the abstract content of the pure categorial realm back to the actual controversy he is entangled within. Since the contingent commitments will vary, the reflective accord (or contradiction) between the actual agency and the moral standpoint of the pure rational agent will likewise vary. This, in turn, means that one agent might consider a set of actions as consistent with the moral standpoint while another agent interprets those same actions as inconsistent. We can now ask: what should a reflective agent do when he becomes aware that another agent regards his actions as inconsistent with the moral standpoint that he regarded as consistent with that moral standpoint? Does this dispute of actual agents regarding the consistency of contingent agent actions with the moral standpoint have any standing for an agent who has resolved to act in ways that sustain the game of blaming and praising between moral strangers? The answer to this depends on whether the agent takes the consistency between the moral standpoint and the actual agency as a purely internal matter related to the form of that agent's willing or whether the agent takes the consistency as something that must be extended to an actual agreement between two agents involved in a controversy. Engelhardt's examples and explanations seem to indicate that the self-contradiction is purely an internal matter, and something that can be judged by an agent when contemplating his own actions in isolation from the ways other agents interpret those actions (1996, 69, 103–108). But if the game of blaming and praising requires that actual agents share a common basis for resolving their specific controversy, then the common basis must be something these actual agents share regarding whether their real actions and agreements are consistent. For this, all the excluded contingency comes tumbling back in, and must be accounted for in the interactions with one another.

We can put the argument for why contingency must be reintroduced as follows: If a reflectively aware agent assumes the resolve integral to the purely rational moral standpoint, then, by assuming this, the agent must accept all of the conditions necessary for sustaining that resolve. To the degree the conditions do not exist, the agent must productively will them, and resolve to bring them about. These conditions don't just relate to the categorial contents worked out at the level of the ethic for moral strangers. They also include conditions that enable actual agents involved in a real world controversy to recognize when they, as agents, come to agreements that are the kinds of agreements that involve common moral authority. Such agreements don't just concern results, but how the results were arrived at by the agents so actions of agents don't violate the general conditions. For example, Fred might threaten Bob. Out of fear for his life Bob might agree to move the lawn mower from his garage to Fred's. Further, Bob might decide that the lawn mower is not important enough to continue hostilities with Fred, perhaps also out of fear of his psychotic actions. We thus have a resolution of the controversy: Bob has agreed to Fred's terms, and no further conflict is present. Bob hopes to forget the whole affair, and Fred is likewise happy to forget. From Fred's perspective, his threat only made explicit a consequence that would arise were Bob not to cease using force to prevent him from using his own lawn mower. From Bob's perspective, the controversy was resolved by force, specifically by Fred's threat. The agreement was thus a forced agreement. If Engelhardt's postulated agreement is to be an actual agreement that has the kind of common moral authority he speaks about, then it must extend to the way actual agents describe all the actions integral to the controversy, not just their own actions. The resolve must extend to the whole social and political fabric that enables the agents to peaceably resolve their differences in a way that arrives at a mutually recognized agreement, permission, or forbearance. A shared interpretive framework is required for the reflectively aware agents to sustain agreements that are jointly recognized as the kind of agreement that satisfies the general kind of agreement that is postulated when each agent assumes the moral standpoint. This interpretive framework is partly constituted by the given, contingent strata that make up the shared life of actual agents involved in a controversy (constituting the first, second, and third community).

7.5 The Hobbesian Alternative to Engelhardt's Minimal Ethic for Moral Strangers

In sect. 7.4.1 I considered meta-controversies about controversies, and ended by simply assuming that agents would share a common description. Now we see that this cannot just be taken as a given. Rather, reflectively aware agents committed to Engelhardt's ethic for moral strangers must will the conditions that enable them to mutually recognize when an agreement of the right type has been attained in the right way. This brings us back to the maxim presented in sect. 7.4.2. The agreement must extend to any meta-controversy that might arise regarding the controversy or the concord that any agent actions may have with the moral standpoint. In this section I briefly note how the extension of this agreement to an actual, mutually recognized agreement can involve extensive constraint on the actions of moral strangers. In fact, the less the content available at the purely abstract level, the greater the imposition that might arise when an actual agent must assure that the other actual agent in a controversy judges all actions as concordant with the kind of resolution that counts as agreement. What was a minimal constraint when simply considered in terms of internal consistency now becomes a maximal constraint when considered in terms of a mutually acknowledged accord of an agreement with the postulated kind of agreement that involves common moral authority.

An analogy for the shift can be found in the standards for disclosure of information to assure an informed consent in health care ethics. Three standards of disclosure are commonly recognized.[31] A professional standard of disclosure requires that a clinician disclose what other similarly situated clinicians would disclose. The problem with this standard is that clinicians might generally disclose nothing. This standard is too sparse in what it requires of the clinician. It allows the practices of other clinicians to determine what is disclosed, rather than the needs of the patient

[31] Engelhardt's discussion of these three standards of disclosure is in 1996, 310–317. The parallel between his account of informed consent and his account of agreement as the basis of an ethic for moral strangers goes very deep. In earlier writings, his arguments about the conditions of a secular ethic were put directly in terms of consent, rather than permission: "the core of ethics is procedural in focusing on the acquisition of the free and informed consent of the individuals involved in a particular endeavor" (1973, 1982, 67).

qua rational agent collaborating in a practice. The professional standard only makes sense if emphasis is placed on the agency relation between physician and patient. With that emphasis, the physician is acting on behalf of the patient as that patient's agent, and informed consent is reduced to the bare, brute giving of ascent or permission (a patient "yes" to a physician action). If, after assent, the patient says "this is not what I gave you assent to do," there is no basis for resolving a controversy about redress for harm or battery.[32]

The second standard that might be used is variably referred to as the "objective" or "reasonable person" standard of disclosure. It says that a clinician should disclose what a reasonably prudent person would want disclosed. This standard seeks to introduce a balance between what patients need and want, on one side, and what physicians provide, on the other side. This faces the problem of determining what a reasonably prudent person would want or need. This of course is very tricky. In relation to specific medical procedures, this might be accomplished by bringing together a committee of clinicians, patients, and informed lay people and getting them to agree on what should be disclosed. This would reflect some sense of the community (*sensus communis*) regarding what disclosure requires.[33] It allows a particular

---

[32] The circumstances surrounding the professional standard of disclosure are analogous to making a contract or property transaction in a context where there are no laws (common or statutory) for how they become binding, are registered, and can be enforced if there is fraud or a breach of contract. The transaction works fine as long as there is no *post hoc* problem. If there is a problem, conditions for peaceably resolving the controversy may be lacking. The contingent, common standards of contract and property law aren't related to legislating content-full visions, but rather relate to establishing conditions for peaceably interacting and peaceably resolving disputes when they arise. In the same way, explicit, admittedly contingent standards for informed consent establish such conditions. In a way that is fully consistent with his purely internal account of consistency, Engelhardt states: "[u]nless individuals have taken steps to create special expectations and/or special requirements, the professional standard meets the principles of permission and beneficence" (312). He thinks requiring common social standards of disclosure involves a shift from "freedom as a side constraint" to "freedom as a value" (313). This response misses the way contingency and common, explicitly promulgated norms establish conditions for peaceable interaction, and thus make "freedom as a side constraint" something more than "abstract right." Since Engelhardt acknowledges that different physicians might be members of different medical communities and thus, by direct implication, might have different norms arising from "what other physicians similarly situated would disclose," his professional standard of disclosure is equivalent to allowing just one party of a contract the privilege of determining the standards for how contracts are duly authorized and enforced, for example, establishing as a rule that all and only the person providing services gets to determine how a contract for provision of services is drafted, registered, and enforced, as long as that provides standards that conform to those that would also be used by a self-selected peer group.

[33] Engelhardt criticizes this "objective standard" as follows: "This departure from using the judgment of reasonable members of the professional community as the standard to that of reasonable and prudent individuals can be justified if the medical profession is under the moral authority of society. However, in the absence of specific agreements and understandings, that is not the case. Members of a profession are as much entitled to their views regarding proper standards of disclosure as are members of the general public. Moreover, there is no socially or historically unconditioned, atemporal notion of a reasonable and prudent person. Such concepts are in fact always dependent on the particular values of a particular community. A reasonable compromise between the two viewpoints is to be achieved, if at all, through negotiation" (1996, 314). This criticism

physician and patient to have a more constrained set of obligations and expectations regarding what informed consent entails. Finally, there is what has been called the "subjective" standard of disclosure. This says that a clinician must disclose the information that the specific patient would want to disclosed. In this case, the problems are the flip side of those associated with the professional standard. A patient might want little or nothing, and in this case, the standard would require disclosure of no information at all. But the patient might also want the equivalent of a medical education, and require information that takes the physician hours to ascertain, for example, about the latest research results, information on prior probabilities, biases of ascertainment associated with the clinician's own patients, and so on. This could involve a tyranny of the patient over the physician.

The problem with Engelhardt's ethic as he has formulated it is that two reflectively aware agents, each resolving to sustain actions in ways that are consistent with the moral standpoint, may interpret such consistency in divergent ways similar to the divergent ways a standard of disclosure might be interpreted when assuring informed consent. At this stage we have not considered an analogue of the middle way associated with the reasonable person standard (sensus communis). But we do have variants quite similar to those associated with the professional and subjective standards. Engelhardt's chosen internal standard is very much like that of the professional standard of disclosure associated with informed consent. The clinician assumes that the thing called "informed consent" can be attained when the clinician discloses the information that other similarly situated clinicians would disclose. In Engelhardt's ethic, the standard is made even more internal to the agent, since there is no worry about what others of any kind would do. The notion of what others would do is captured by taking the pure categorial stance, a stance involving maximal intersubjectivity. Emphasis then falls on whether an agent can reflectively

---

nicely summarizes the whole sequence of moves he has made regarding an ethic of moral strangers. He is clearly correct that there is no atemporal notion of a reasonable and prudent person (although we might question whether his "moral standpoint" is supposed to provide such a notion). He is also correct that any positive standard established as binding for physicians requires that physicians be under the authority of the society establishing those positive standards. But all the same things are true for explicit standards about registering property. Again, the crucial thing is that contingency is required to have conditions for consistently determining which permissions, agreements, and transgressions among actual agents in real world circumstances count as the kinds that are abstractly specified within an ethic for moral strangers. This point is well addressed by Kant in his *Metaphysics of Morals* (1996 [1797]). A nice commentary on this work is provided by Byrd and Hrushka (2010). For example, when discussing legal requirements for property, they note that Kant "attaches legally significant consequences for our ownership rights to whether we performed or failed to perform some act of registration or documentation – an act of the same character as driving on the right (or left) side of the road, namely an act empty of moral content and, in the absence of positive law, legally irrelevant" (35). These otherwise empty actions are required to establish the external conditions of freedom, not just freedom as a value (although there is this in Kant), but also freedom as a side constraint. One wills the conditions for making actual what is otherwise a purely abstract right (note the maxim from section Sect. 7.4.1 and note 24 in the Lockean variant). My reference to the *sensus communis* is meant to align the *Metaphysics of Morals* with another one of Kant's later works where the *sensus communis* plays a prominent role, the *Critique of Judgment* (1987 [1790]).

regard his own actions as consistent with what the secular ethic requires. This is a purely internal standard of consistency, and allows the agent to ignore what other agents would say about how agents came to something they might call an agreement. However, as a purely internal standard, an agent can call anything he likes an agreement. All that is required is that the reflectively aware agent organize the contingent contents in such a way that when the contingencies are abstracted away, he has the right kind of sparse content left. However, if we move to the requirement that both agents acknowledge actions and agreements as satisfying the constraints on agreements that have a common moral authority, this makes one agent subject to the arbitrary demands of the other agent in the same way a clinician using a subjective standard of disclosure is subject to the arbitrary demands of a patient. Any controversy might be endlessly extended into meta-controversies that require for their resolutions that the reasons for actions be fully transparent to all agents. The more minimal the content directly included in the definition of a controversy and its resolution, the more maximal are the constraints imposed on real world agents when they seek to reach agreements that are mutually agreed upon as the kind of agreements that have that common authority required by Engelhardt's secular ethic.

At this stage, we seem to have reached an impasse. Engelhardt's ethic for moral strangers allows for at least two interpretations that might be used by reflectively aware agents who seek to act in ways that are consistent with that ethic. As a purely internal standard, his ethic yields a minimal content, as he suggests. But this interpretation yields a rather odd notion of common moral authority. Actual agents involved in real controversies wouldn't have such common moral authority unless they both assume the purely rational standpoint and only consider their controversy and their agency in so far as these conform to that abstract content. This shared basis reflects what both agents must assume as rational agents, and it might provide an abstract criterion of agreement as a condition for sustaining the ethic that is advanced at that purely rational level. But it tells nothing about how those actual agents should interpret their actions and controversies so they as reflectively aware agents have that kind of agreement that counts as a shared moral basis for sustaining their ethical discourse with one another. If those reflectively aware agents now are concerned with a shared basis for resolving their actual dispute, this requires a far stronger constraint, one that includes a mutually acknowledged consistency of actions with that ethic and an agreement that extends to all the potential levels of meta-controversy associated with the constitutions of their controversies. If we assume this shared moral basis must be present not just for this or that controversy, but for all controversies, then Engelhardt's standard leads to maximal constraint of a Hobbesian sort. Now all rights must be transferred to some kind of common authority that establishes sufficient conditions for such mutually recognized consistency under all scenarios.

From Engelhardt's minimal ethic we thus flip to a Hobbesian Leviathan.

7.6 On the Possibility of a Transcendental Argument that Explicates the Grammar of a Moral Discourse Restricted to Peaceable Resolution of Conflicts

I close with a tentative suggestion: we can use something like a reasonable person standard of agreement if we allow for some contingency in the initial way our project is framed. In the end, contingency is unavoidable. In addition to informing our standards, our arguments, and our agreements, it also plays a role in how we frame the whole project of ethics and thus our criteria for ascertaining success. Part of what a transcendental exposition should clarify is how any final success in applying our ethic arises by the way we constituted our project in the first place. (This is how Kant's transcendental arguments work.)

To make Engelhardt's argument work, we need to ask how reflectively aware agents who don't share a content-full vision of what is right and good might still identify a common moral authority that enables them to specify how controversies are to be resolved peaceably. Now a logic of force must enter in the framing of the task. Force, in turn, is related to a logic of enforcing, a logic of legitimate use of force to sustain a fabric of peaceable interaction. By introducing the question of force and peaceable resolution into the specification of the project of a secular ethic, we provide a criterion that can be used to address both the problem of rights dependent description and to specify which resolutions are excluded from being compatible with the project of blaming and praising that moral strangers seek to sustain. By raising the question in relation to reflectively aware agents, we make explicit something that Engelhardt generally assumes but didn't explicitly justify: namely, that actual agents have contingent reasons for acting that they take as valid AND other agents must recognize this if they are to properly make the kinds of agreements that have common moral authority. This, in turn, means actual agents have interiorities that are inaccessible to other agents.[34] This acknowledged interiority places a limit

[34] I take recognition of the inaccessible interiority of other agents as the crucial claim that is missing from the way Engelhardt explicitly develops his account of the moral standpoint. In ethics, this insight is closely associated with a principle of respect. Recognition of an inaccessible interiority of the other is a constitutive moment of the moral standpoint. In Kant, this is often associated with the supersensible. Many understand the supersensible in Kant as a purely negative concept, deployed to reconcile domains that would otherwise be irreconcilable (solve some kind of antinomy exhibited in a dialectic). As such a tool to reconcile opposites, it becomes the "negativity" in Hegel's dialectic. The crucial difference between the categorial and ordinary levels in Hegel hinges on how this otherness is regarded. Hegel, in net effect, brackets the surd content of the supersensible, regards it as nothing, a limit, an incompleteness. With this shift, the other can become nothing but a motive, a driver in the movement of the dialectic. But this reconstruction of the other as negativity involves a specific, highly problematic commitment. It reflects a distinctive valuation and value, namely that of a philosopher interested in the immanent ordering of thought determinations. As the early Engelhardt notes (1973, 166): "The task of thought is to comprehend the significance of being and then to comprehend the significance of the basic categories of being's significance for thought. Thought's task is to discover rationality – to come to terms with itself – to think of its own thinking, to reason out its own rationality, especially that which concerns the content of its world." This is how the task of justification is posed from within the categorial stance,

on the kind of descriptive agreement about the controversy that can be demanded by an opposing agent. This recognition is a non-trivial one, and it distinguishes a moral standpoint of a Lockean sort from that of a Hobbesian. Taking up the moral standpoint in relation to another agent involves giving space to that agent as one who has a set of inaccessible internal determinations. This is no longer the simple doubling of agent occurrences within a higher order type. Even at the categorial level, there are a set of additional internal determinations that break the symmetry between agents included in the controversy.

One of the ambiguities in Engelhardt's moral standpoint relates to the epistemic and valuational standing of contingent contents of the moral life of actual agents.

---

including the task of justification in ethics. Engelhardt's moral standpoint qua categorial stance has just this commitment. It is not neutral. The crucial difference between his work on mind/body and the *Foundations of Bioethics* involves the recognition that the categorial stance is not neutral, e.g., it leads to judgments about supersensible contents that are problematic. For example, in 1973 he explicitly states that "[m]ind embraces its other in itself by explicitly having the significance of its other integral to its own concreteness" (94) This embraced significance is understood in terms of value: the intracategorial affinities are said to express "not what is, but what should be" (97). In his own commitments, especially after his conversion to Orthodoxy, Engelhardt is fundamentally concerned with those features of reality that are, in fact, inaccessible to discursive thought. He favors the noetic over the discursive. First of all, he is concerned with knowing God: "Knowledge of God is not secured by philosophical arguments; in this Kant is right. We cannot with discursive arguments reach beyond the bonds of the possible empirical experience to an infinite God Who is by His nature beyond our sensory, spatiotemporal experience and discursive concepts. … Within the bounds of the immanent, the transcendent is always invisible, as Kierkegaard recognized" (2000, 164). But this insight regarding the need for a pause and respect before The Supersensible, God, is not just a feature of the holy, and it does not just reflect the content-full commitments of a religious community. It is also a constitutive feature of any moral standpoint. Without it, all are taken as pure rational agents, and anything that cannot be specified conceptually becomes indifferent and thus subject to the demands of thought thinking itself. Kant explicitly recognized the significance of the supersensuous, and thus did not simply reduce it to negativity. His notion of the supersensible was not just a tool to reconcile antinomies. It also reflected a sense of coming up against a depth and mystery, with a required pause and giving of place to the jurisdiction of the other, although at times it is a rather muted sense. He also continually used the concept of the supersensible to do positive work, for example, in organizing the notion of that distinctive whole/part dynamic associated with organic life (1987 [1790], secs 77–78, note how e.g. he takes "purposes as bases that make certain things possible," an insight that might feed back upon the way a transcendental investigation into the basis of an ethic is framed). This supersensible other is what, as given, requires the "fence" expressed in Locke's jurisdictional rights. Engelhardt also recognizes it as a pervasive feature of the moral life, and he implicitly requires its recognition by those who take up his moral standpoint. But he never explicitly introduces this, because of the way he is haunted by the valuation integral to the categorial standpoint. My critical reflections on Engelhardt's transcendental argument have been oriented toward exposing this feature and making it explicit. To this extent, my criticisms in this essay are a further development of those I made (Khushf 1994, 122–136) of Hartmann (1966, 1977) when he took the categorial ordering of Hegel's *Philosophy of Right* as a ranking of value. From a genuinely moral standpoint, the otherness, the inaccessible interiority, the supersensible ground of the other human agent encountered in a controversy is emphatically not nothing, negativity, insignificant. In its own valuations, the moral standpoint qua pure rational stance wills a transparent basis of justification. In dong this, it misses exactly what is most important. By its special bracketing, i.e., its "transcendental turn," it puts away exactly what requires explicit recognition; namely, that supersensible inaccessible interiority of the other agent.

When the transcendental turn is taken, contingent commitments of agents are abstracted away. In relation to the moral standpoint qua pure categorial stance, any way of handling the contingent commitments is as good as any other. Engelhardt assumes that these contingent commitments are the things that will be mutually recognized when reflectively aware agents consider the consistency of their actions with that stance of an agent that must be taken up if one is to play the game of moral blaming and praising. But this is only the case if the controversy doesn't degenerate into a meta-controversy. Once it does, the required agreement extends to the strands of the meta-controversy. At this stage, sustaining the blaming and praising game depends on sustaining the higher order descriptive categories that specify the conditions under which that game can be successful. That requires altering aspects of actions judged to be indifferent from the purely rational standpoint. Since nearly all matters of an actual agent's life are thus indifferent, the opposing agent's demands for stabilizing the controversy become part of the condition for attaining the kind of agreement that can count as the shared moral authority that is necessary to sustain the blaming and praising game.

The recommended changes to Engelhardt's argument would modify how minimal a state we get. Some tricky work would be needed to account for who has moral status, since moral agency would no longer be defined in terms of an agent pressing a claim, but rather in terms of which objects must be taken as the kinds of subjects that have the kind of inaccessible interiority that requires the pause integral to the moral standpoint. This too must trace back to a sense of community that has to extend to society as a whole, not just to the robust communities that Engelhardt associates with the content-full views of right and good. We now need an alignment of what was earlier called the first, second, and third communities (related to the contingent features of actual controversies and their descriptions, together with that surd aspect of actual agent agreement) and the fifth community, which was tied to the categorial structure of any controversy and its agent occurrences. When Engelhardt characterizes "the secular moral community as the possible intellectual standpoint of persons interested in resolving moral controversies in ways not fundamentally based on force" (p. 69), he already has been reflectively interpreting that community as one of "moral strangers" who include the specific strata of contingent contents that distinguish them from one another as actual individuals, not just as occurrences within a higher order type. I'm just recommending that he explicitly include it as a posit when the moral standpoint is specified. Either way, any blaming and praising game must be associated with a categorial structure that needs to be considered both from a pure categorial perspective and in relation to reflectively aware agents who must interpret that structure in the convergent ways that lead to agreements that are mutually acknowledged by actual agents as the kind of agreements that have moral authority. By loading greater content and constraints into the way the project is initially specified we can get less divergence of the interpretations that are regarded by reflectively aware agents as consistent with the categorial standard. As is true of Engelhardt's ethic, any actual agent might ask why he wants to play the game in the way it is initially formulated. The answer will depend on the actual commitments of the agent contemplating whether to play the game. But once they agree to peaceably resolve controversies in ways that recognize the standing each agent has to press reasons other agents cannot recognize as reasons, the transcendental justification of this game has the same form as that Engelhardt presents for his own variant. Only now we get an account of the moral standpoint that roughly tracks what most people in the community take it to be.

The content I recommend loading back into the moral standpoint is not just any arbitrary content. Rather it provides a valid interpretation of what assuming the moral standpoint requires. Some hermeneutic of that stance must be provided. A "hermeneutic of categories" advanced at a transcendental level cannot provide a sufficient specification of the moral standpoint, and it is emphatically not neutral. It explicitly disvalues exactly that inaccessible interiority of the other that is recognized when one actual agent takes up a moral stance toward another actual agent. Without some interpretation of the moral standpoint, a reflectively aware agent would not know how to see in the properly double way, qua actual agent with contingent commitments and qua rational agent resolving to justify in the manner required by ethical discourse. If someone chooses not to assume this richer moral standpoint, there are now independent grounds for criticizing the agent. Similarly, when states deploy force in ways that do not sustain this fabric of justification, there are now both internal and external grounds for criticism. Internally, the criticism traces the fabric of connections between the ethic and its application, and it clarifies the conditions for the possibility of making judgments about actual actions that are valid applications of the ethical apparatus and its categorially specified basis. Viewed in such a way, the transcendental argument justifying the ethic also provides the fitting basis for criticizing the application of the ethic to actions of agents in the world. The other basis for criticism traces externally to the convergence of interpretations of reflectively aware agents (the *sensus communis*) regarding the significance of respect (as a kind of pause and giving space to an other): an idea well captured by a Lockean jurisdictional right and an ethical principle of respect.

There is one other major advantage to the reformulated project: it now maps almost perfectly to the four interpretive strands that I initially identified in Sect. 7.2 of this chapter. These strands reflect the interpretation of the reflectively aware agent named H. Tristram Engelhardt. Now we have a clearer sense of why Engelhardt presented these other strands. He is not just doing a mapping project, as he sometimes suggests. His project is not just a modest, sparse one, clarifying how far a specific kind of justificatory game might be extended. Rather, Engelhardt is deeply concerned to make space for forms of reason and noetic insight that he takes as absolutely crucial to a good and holy life, yet which he believes are inaccessible to those who are outside his own community. All he needs to do is give up his claim that he has found a moral authority that is independent of his resolve to peaceably interact with other such agents, each with their own interiorities that likewise have standing.

At the beginning of this essay I noted a possible analogy between geometry and its application, on one side, and Engelhardt's secular ethic and its "use as a means" for peaceably resolving controversies. In this essay, I have argued that a transcendental argument that justifies agreement as the moral basis of a secular ethic must span and clarify the relations between the ethic, the agreement taken as the basis (with its bridging of the categorial unity and the diverse actual agreements), and the logic of force integral to any peaceable resolution of conflict. When we recognize how the transcendental argument bridges both the "theory" and the "application" sides, we might also notice an interesting parallel with the way Kant himself formulates his own transcendental arguments. Kant took the judgments of mathematics and science on the pure, theoretical side as the given, and he wanted to clarify how they were applicable to the world of experience. The key work of the transcendental argument was to explain how "applications" of a priori, synthetic judgments to our experienced world were possible. As Kant attempted to further work out these ideas and align theoretical and practical forms of rationality, he discovered the importance of a shared communal sense (the *sensus communis*) that played a mediating role in the reflectively aware agent. Engelhardt's transcendental argument for agreement as the sole sufficient basis of an ethic for moral strangers seems to point in the same direction.

References

Beauchamp, Tom. 1987. Ethical theory and the problem of closure. In *Scientific controversies: Case studies in the resolution and closure of disputes in science and technology*, ed. H. Tristram Engelhardt and Arthur Caplan, 27–48. Cambridge: Cambridge University Press.

Byrd, B.S., and J. Hruschka. 2010. *Kant's doctrine of right: A commentary*. Cambridge: Cambridge University Press.

Chaffin, D. 1994. The logic of contingency. In *Hegel reconsidered*, ed. H. Tristram Engelhardt Jr. and Terry Pinkard, 143–161. Dordrecht: Kluwer Academic Publishers.

Engelhardt, H. Tristram. 1973. *Mind-body: A categorial relation*. The Hague: Martinus Nijhoff.

Engelhardt, H. Tristram. 1982. Bioethics in pluralistic societies. *Perspectives in Biology and Medicine* 26(1): 64–78.

Engelhardt, H. Tristram. 1986. *Foundations of bioethics*. New York/Oxford: Oxford University Press.

Engelhardt, H. Tristram. 1987. The foundations of bioethics: The attempt to legitimate biomedical decisions and health care policy. *Revue de Métaphysique et de Morale* 3: 387–399.

Engelhardt, H. Tristram. 1988. Foundations, persons, and the battle for the millenium. *Journal of Medicine and Philosophy* 13: 387–391.

Engelhardt, H. Tristram. 1991. *Bioethics and secular humanism: The search for a common morality*. London/Philadelphia: SCM Press/Trinity Press International.

Engelhardt, H. Tristram. 1994a. Introduction. In *Hegel reconsidered*, ed. H. Tristram Engelhardt Jr. and Terry Pinkard, 1–18. Dordrecht: Kluwer Academic Publishers.

Engelhardt, H. Tristram. 1994b. Sittlichkeit and post-modernity: An Hegelian reconsideration of the state. In *Hegel reconsidered*, ed. H. Tristram Engelhardt Jr. and Terry Pinkard, 211–224. Dordrecht: Kluwer Academic Publishers.

Engelhardt, H. Tristram. 1996. *Foundations of bioethics*, 2nd ed. New York/Oxford: Oxford University Press.

Engelhardt, H. Tristram. 2000. *The foundations of Christian bioethics*. Lisse: Swets & Zeitlinger Publishers.

Engelhardt, H. Tristram, and Arthur Caplan (eds.). 1987a. *Scientific controversies: Case studies in the resolution and closure of disputes in science and technology*. Cambridge: Cambridge University Press.

Engelhardt, H. Tristram, and Arthur Caplan. 1987b. Introduction – Patterns of controversy and closure: The interplay of knowledge, values, and political forces. In *Scientific controversies: Case studies in the resolution and closure of disputes in science and technology*, ed. H. Tristram Engelhardt and Arthur Caplan, 1–23. Cambridge: Cambridge University Press.

Engelhardt, H. Tristram, and Terry Pinkard (eds.). 1994. *Hegel reconsidered*. Dordrecht: Kluwer Academic Publishers.

Hartmann, Klaus. 1966. On taking the transcendental turn. *Review of Metaphysics* xx(2): 223–249.

Hartmann, Klaus. 1977. Die Objectivitat der Freiheit. Hegels Rechtphilosophie heute. In *Freiheit. Theoretische und praktische Aspekte des Problems*, 315–337. Freiburg/Munchen: Alber.

Hegel, G.W.F. 1967 [1821]. *Hegel's Philosophy of Right*. Trans. T.M. Knox. London/Oxford/New York: Oxford University Press.

Hegel, G.W.F. 1991 [1830]. *The Encyclopedia Logic*. Trans. T.F. Geraets, W.A. Suchting, and H.S. Harris. Indianapolis/Cambridge: Hackett Publishing Company.

Hobbes, Thomas. 1962 [1651]. *Leviathan*, ed. Michael Oakeshott with intro. by Richard Peters. New York: Colliers Books.

Illies, C. 2003. *The grounds of ethical judgment: New transcendental arguments in moral philosophy*. Oxford: Clarendon.

Kant, Immanuel. 1956 [1788]. *Critique of Practical Reason*. Trans. Lewis White Beck. Indianapolis/New York: The Bobbs-Merrill Company.

Kant, Immanuel. 1965 [1781/1787]. *Critique of Pure Reason*. Trans. Norman Kemp Smith. New York: St Martin's Press.

Kant, Immanuel. 1987 [1790]. *Critique of Judgment*. Trans. Werner Pluhar, with forward by Mary Gregor. Indianapolis: Hackett Publishing Company.

Kant, Immanuel. 1988 [1800]. *Logic*, trans. with introduction by Robert S. Hartman and Wolfgang Schwarz. New York: Dover Publications.

Kant, Immanuel. 1996 [1797]. *The metaphysics of morals*, ed. Mary Gregor. Intro. by Roger Sullivan. New York: Cambridge University Press.

Kant, Immanuel. 2012 [1786]. *Groundwork of the metaphysics of morals, revised edition,* ed. Mary Gregor and Jens Timmerman, intro. by Christine Korsgaard. New York: Cambridge University Press.

Khushf, George. 1994. The meta-ontological option: On taking the existential turn. In *Hegel reconsidered*, ed. H. Tristram Engelhardt Jr. and Terry Pinkard, 119–142. Dordrecht: Kluwer Academic Publishers.

Locke, John. 1988 [1698]. *Two treatises of government*, ed. Peter Laslett. Cambridge: Cambridge University Press.

Pinkard, Terry. 1988. *Hegel's dialectic: The explanation of possibility*. Philadelphia: Temple University Press.

Pinkard, Terry. 1990. How Kantian was Hegel? *Review of Metaphysics* Xliii(4): 831–838.

Pinkard, Terry. 1994. *Hegel's phenomenology: The sociality of reason*. New York: Cambridge University Press.

Pippin, Robert. 1990. Hegel and category theory. *Review of Metaphysics* xliii(4): 839–848.

Wetzel, Linda. 2009. *Types & tokens: On abstract objects*. Cambridge, MA: The MIT Press.

Chapter 8
Equality Is Problematic: Engelhardt on Fair Equality of Opportunity, Health Care, and the Family

Ruiping Fan

8.1 Introduction

The requirement of fair equality of opportunity has become an indispensable element ingrained in contemporary liberal social democratic theories. It stands as a dominant ideology upheld by a great number of politicians and scholars to regulate the basic structure of society and to direct public policy formulation. In academic circles, it is fashionable to defend this principle in one way or another, but it is quite unpopular to argue against it. Unfortunately, the popularity and dominance of the principle do not lead to its tenability. In fact, the principle is entangled with a series of severe theoretical and practical problems. In his recent work, H. T. Engelhardt, Jr. has uniquely and sharply discerned those problems, which deserve our serious attention. In this chapter, I attempt, based on Engelhardt's insights and arguments, to provide a critical review of the principle. My focus will be on its disastrous implications for health care allocation as well as its baneful contribution to the erosion of the family.

8.2 The Principle of Fair Equality of Opportunity for Health Care

The principle of fair equality of opportunity does not require absolute equality in every respect of life, such as income or wealth. Liberal social democratic scholars admit that it would be squarely irrational to rule out the types of inequality that

R. Fan, Ph.D. (✉)
Department of Public Policy, City University of Hong Kong,
B7517, Academic Building 1, Kowloon, Hong Kong
e-mail: safan@cityu.edu.hk

L.M. Rasmussen et al. (eds.), *At the Foundations of Bioethics and Biopolitics: Critical Essays on the Thought of H. Tristram Engelhardt, Jr.*,
Philosophy and Medicine 125, DOI 10.1007/978-3-319-18965-9_8

would work to the advantage of everyone, especially the least well-off members of society. Accordingly, they allow certain inequalities in income and wealth to exist so as to benefit the least well-off individuals in society. However, since major rewards for individuals in contemporary society derive from jobs and offices, liberal scholars require that the competition for securing such positions must be "fairly" equal. Accordingly, as John Rawls argues in his seminal social democratic account of justice (Rawls 1971), the principle of fair equality of opportunity must be established and given lexical priority over any principle that permits inequalities in income and wealth.

What is fair equality of opportunity? From a Rawlsian account, it goes beyond the requirement of what Rawls calls formal equality of opportunity. While one's talents and skills are morally relevant factors for jobs and offices, there are also morally irrelevant features, such as race, religion, and sex. Society must impose necessary conditions on social institutions and markets, so that the influence of morally irrelevant features on individual opportunities will be ruled out. This is the requirement of formal equality of opportunity. This requirement, for Rawls, is not sufficient. As he sees it, while one's talents and skills are morally relevant features for jobs and offices, they are largely the result of one's genetic accident and/or a lucky family background – neither of which one deserves in the moral sense. Accordingly, Rawls argues, "it is necessary to impose further basic structural conditions on the social system…to recall the importance of preventing excessive accumulation of property and wealth and of maintaining equal opportunities of education for all" (Rawls 1971, 73). This is to say that in order to correct the influence of unequal family backgrounds and ensure "fair" equality of opportunity in society, the state must provide every child with suitable public institutions, such as early child care, childhood programs, and public education.

Liberal social democrats do not want to limit the application of the principle of fair equality of opportunity only to education. Instead, they also draw on this principle to regulate the area of health care. This chapter will primarily rely on Norman Daniels's work to illustrate how they seek to conduct this task. Daniels has, to my knowledge, offered the most systematic and influential arguments for "just" or "fair" health care in light of the requirement of fair equality of opportunity. Based on Christopher Boorse's work concerning species-typical normal functioning in evolution (see the next section for further detail), Daniels attempts, through the construction of a concept of the normal opportunity range open to individuals, to establish a list of "objective" health care needs so that they may reasonably be "fairly" protected by the state under the principle of fair equality. From his view, the normal opportunity range for a given society "is the array of life plans reasonable persons in it are likely to construct for themselves" (Daniels 1985, 33). Since disease or disability is, in his account, defined as that which impairs human species-typical normal functioning, it reduces the range of opportunity open to the individual, when compared to the hypothetical situation in which he is healthy (Daniels 1985, 26–31). Since an individual's fair share of the normal opportunity range is the array of life plans he may reasonably choose, "disease and disability shrinks his share from what is fair" (Daniels 1985, 34). Accordingly, fair equality of opportunity would, so Daniels argues, require that society meet everyone's health care needs, which are, from Daniels' view, "those things we need in order to maintain, restore, or provide functional equivalents (where possible) to normal species functioning" (Daniels 1985, 32). In particular, he states that health care needs include the following aspects:

1. Adequate nutrition, shelter
2. Sanitary, safe, unpolluted living and working conditions
3. Exercise, rest, and some other features of lifestyle
4. Preventive, curative, and rehabilitative personal medical services
5. Non-medical personal and social support services (Daniels 1985, 32)

It would be perfect if society could afford to satisfy all these needs for every individual. While Daniels stresses that this list indicates a legitimate functional relationship between responsible social institutions and various goods for individual health, he understands that people tend to think narrowly about personal medical services when they talk about health care (Daniels 1985, 32). In this regard, following his account of fair equality of opportunity for health care, Daniels, along with other social democratic scholars, has insistently argued for universal health care coverage and favored a single-payer form of financing the health care system in order equally to meet everyone's health care needs (Daniels and Sabin 2002, viii). In short, for Daniels, justice requires society to meet everyone's health care needs according to the requirement of fair equality of opportunity. Even if certain social and economic inequalities in welfare are tolerable due to variation in individual talents and skills, health inequality is unjust if it is the result of unequal distribution of the social determinants of health, especially access to health care (Daniels 2008). Accordingly, a right to health care becomes a special case of a right to equal opportunity that individuals must be granted in a proper social democratic society (Daniels and Sabin 2002, 17).

8.3 Value Pluralism in Health Care

Engelhardt has long recognized that the concept of disease is inevitably value-laden (Engelhardt 1974). On the other hand, Christopher Boorse attempts to distinguish between a concept of illness, which is acceptably value-laden, and a concept of disease, which is formed by appeal to species-typical functions established through evolution and is thereby independent of individual desires and societal expectations. Boorse argues that "a normal function of a part or process within members of the reference class is a statistically typical contribution by it to their individual survival and reproduction" (Boorse 1977, 562). Thus, on his account, "individual survival and reproduction" are taken as "natural" goals or values fitting into his view of species-typical functioning through evolution. A difficult problem, as Engelhardt points out, is that Boorse fails to takes into account the generally accepted notion of inclusive fitness in evolution – that is, what appears to be important in evolution is not whether a particular individual survives or reproduces, but whether that individual maximizes the chances of his genes being spread in the gene pool: a species may in fact be well adapted because of a balance among various contrasting individual traits (Engelhardt 1996, 200). Obviously, it is because of his generally individualistic, rather than species-inclusive, concerns that Boorse is so clear that homosexuality should count as a disease (Boorse 1975, 63). In contrast, Daniels seems to suggest that homosexuality, like masturbation, has been improperly classified as a disease, although he fails to explicate his reason (Daniels and Sabin 2002, 23, note 3). This disagreement between Daniels and Boorse indicates that although Daniels generally takes Boorse's species-typical function-identifying account as an "objective" account of health (and disease) and largely relies on it to construct his "fair opportunity" argument for a state-imposed single-payer system of health care distribution, Daniels may not have seriously shared the value judgments implicit in Boorse's strategy. It is obvious that in evolution, as Engelhardt points out, there is not a single design, but rather a number of designs. One cannot straightforwardly speak of either species design or species typicality without assuming some goals or values – such as individual survival and reproduction in Boorse's case – as "natural" or "universal" norms in evolution. Eventually, what Boorse has offered is a particular, though coherent in itself, individualistic view of human biological functioning and disease (as abnormal functioning) based on the values of individual survival and reproduction. On the other hand, Daniels' selective use of Boorse's account is at risk of incoherence.

The important point is that although Daniels attempts, by adopting Boorse's idea of "human species-typical functioning," to put forward his list of "objective" health needs so as to ground the application of his equality requirement in health care allocation, it is impossible to establish a non-evaluative view of disease or health, because, as Engelhardt argues, human species-typical functioning cannot be determined without specifying a set of values or goals in the first place (Engelhardt 1996, 201). While such values or goals are specified with overlaps across cultures or communities, there are also, undeniably, salient disagreements between cultures or communities. Indeed, no matter how enthusiastic Daniels is in denying the evident fact of value-relevance and value pluralism in the process of identifying species-typical function or disease in order to argue for his idea of an "objectively" fair medical system, he is unwilling to limit his "fair" medical system only to the work of curing diseases or maintaining normal functions in light of his understanding of species-typical functioning. On the one hand, as mentioned, he is unwilling to admit, as Boorse does, that homosexually is a disease that should be medically corrected. On the other hand, he stands ready to expect Medicaid to offer non-therapeutic abortions, although he concedes that such abortions are not health care needs because unwanted pregnancy is not a disease. Yet, they still should be offered, he argues, because it may well be the case that in addition to meeting health care needs, "Medicaid should serve other important goals, like ensuring that poor and well-off women can equally well control their bodies" (Daniels 1985, 32). These cases suggest that it is a series of social democratic norms of equality, rather than a descriptively discovered design of species-typical functioning for individual survival and reproduction, that are the genuine underlying values determining Daniels' health care decisions behind his apparently "objective" strategy of locating species-typical function or disease through evolution. It is a bit deceptive for social democratic scholars like Daniels to attempt to deny the fundamental value-relevance of their strategy in pursuing the social democratic ideas of equality in health care.

Engelhardt's lucid analysis of value-relevance and value pluralism in medical treatment and health care delivery in contemporary pluralistic society discloses the disingenuous tactic involved in the fundamental work of social democratic scholars, such as Daniels, for pursuing a "fair" health care system in light of the principle of fair equality of opportunity. The problem is not only that in seeking fair medical services people also care about values other than equality, such as free choice and efficiency. It is also that the social democratic understanding of equality is by no means the only proper understanding of equality for a fair health care system. For instance, in many European countries, a health care system does not merely treat disease or meet health care needs, but also heightens security and publicly reflects concerns with solidarity. In the United States, many Catholic believers contend that an equal health care package offered to its members in a Catholic institution should not offer contraceptives. This fact of value pluralism brings into question the legitimacy of the single-payer universal health care coverage proposed by social democratic bioethicists such as Daniels in light of their understanding of the principle of fair equality of opportunity. In other words, since every community in contemporary pluralist society is situated in its own unique cultural tradition, it is practically difficult and virtually impossible for all communities to agree on one particular understanding of equal opportunity as well as its proper realization. Just as the meaning of education as well as the way in which it relates to or contrasts with other goods varies from community to community, the significance of health care is inevitably communitarian.

8.4 Equal Input vs. Better Result

Value pluralism aside, it remains a fact that not all health care needs can be met, even if they can be specified in a value-neutral way. This is because health care resources are inevitably scarce in every society, including affluent liberal social democratic societies, such as the United States. As a result, if one argues for meeting health care needs in light of the fair equality of opportunity principle as Daniels does, one must indicate how to apply the principle under the circumstance of resource scarcity. Daniels attempts to appeal to Rawls' contract theory of justice to do the task. Among the various ways of extending Rawls' theory to health care in a society, he takes the most promising way to be including health care institutions and practices among the basic institutions that are involved in providing fair equality of opportunity for all individuals in the society. In this way, health care will not be taken as a primary good, just as things like food, clothing, shelter, or education are not. Instead, under his account, health care, just like education, will be provided adequately from the fair share of opportunity. Accordingly, it is opportunity, not health care, that is the primary social good to be used to establish suitable principles to regulate relevant basic health care institutions in a Rawlsian theory of justice. In this way, Daniels expects that his account will not have to face the resource-draining problem that inevitably results from any strategy of simply adding health care to the list of primary social goods in the application of Rawls' theory (Daniels 1985, 42–46).

Daniels' account is apparently appealing. First, it proposes to use Rawls' hypothetical contract theory by adopting a revised, thinner veil of ignorance than Rawls' original design (so that contractors know about some features of the society, such as resource limitations) to select principles to govern health care resource allocation decisions (Daniels 1985, 47). Second, it locates the support of health care with the support of equality. This will gain the support of social democratic citizens in contemporary society, given that equality has been taken by them as the paramount value in guiding health care allocation. Finally, this account, based on Rawls' equal opportunity principle, cannot be immediately doomed as practically impossible as it might if health care is taken to be governed by Rawls' difference principle. Indeed, some have attempted to place health care within the difference principle, so that inequalities in access to health care would be morally acceptable only if they redounded to the benefit of the least well-off class (Stern 1983). In this latter case, however, one difficulty would be intractable. As Engelhardt demonstrates, the contractors in the original position know that they might find themselves in the least well-off class; namely, those who would die early due to diseases. To deal with such predicaments, they would rationally require a massive transfer of health care resources to those types of treatment and research, such as pediatric treatment and research (rather than geriatric treatment and research), so as to ensure that children did not die young but could reach at least an average lifespan (Engelhardt 2012, 591). This would inevitably drain resources to the health care needs of these unfortunate members to the point where the rest of society is reduced to poverty. Accordingly, the strategy of using Rawls' different principle to handle health care allocation is practically impossible.

Can Daniels' account, without appealing to the difference principle, avoid this bottomless resource-draining problem while achieving "fair" equality of opportunity in society? Evidently, no matter what strategies are used, it is sadly difficult to achieve equality of result in health care because disease, disability, and early death render persons radically unequal. Daniels can pursue his health care ambition only by turning to a mechanism that emphasizes equal input, rather than equal result, in the organizing and directing of health care institutions. Indeed, on his account, the contractors in the Rawlsian original position will choose to set up at least three layers of health care institutions for ensuring just health care: preventive institutions that act to minimize the likelihood of departures from normal functioning; therapeutic institutions that deliver personal medical and rehabilitative services to restore normal functioning; and a third layer of institutions that provide more extended medical and social support services for the chronically ill and disabled and the frail elderly (Daniels 1985, 48). These layers, Daniels contends, do not imply that the contractors in the original position would establish a robust right to having all health care needs satisfied; rather, the contractors would have to be selective. Specifically, Daniels argues, the contractors, in deciding which needs are to be met and what resources are to be devoted to doing so, would have to carefully compare the importance of various health as well as non-health institutions affecting opportunity. They would also need to take into account the importance of institutions needed for basic social goods other than opportunity (Daniels 1985, 53–54). Frankly, given this imagined, complicated process of deliberation and balance, nobody would know whether it could really bring about a group of useful principles to direct health care institutions. But one thing is clear: if it could, then such principles would have to focus on equality of input rather than equality of result, because the latter is simply impossible to seek without creating a bottomless pit. That is to say, a "fair" system of health care that could turn out from this process would have to be a universally-imposed basic health care menu that everyone must use for one's health care services; and in order to ensure equality of opportunity, everyone must be prevented from receiving better or more basic health care than others.

From Engelhardt's view, this imposition of equal input on everyone is equivalent to turning to a position that has abandoned the original ground of many people's sensible concerns with equality:

> It would seem that in a Rawlsian account, the focus should be on equality of result... rather than on equality of access, because it is differences in health status that deprive such persons of equality of opportunity. However, because it is impossible to render all equal in terms of span of life, health, and freedom from morbidity, most of those who engage Rawls focus on the equality of input. They focus on assuring that none will receive better basic health care when most receive only basic health care. There is to be only one menu from which to order basic health care, so that one will be prohibited from acquiring better basic health care. The difficulty is that once one focuses on input, one will have abandoned the original ground for pursuing equality, namely, a concern with the circumscribed liberty and/or equality of opportunity of those who die young or who are radically disabled. (Engelhardt 2012, 594)

If I happen to have a rarely-occurring heart disease and its treatment, in part due to its prohibitively high cost, happens not to be covered in the universal health care package worked out through Daniels' "fair" account, my opportunity for life is by no means becoming more equal with those without the disease if I am only ensured that everybody with this disease will, like me, not receive the treatment. What is worse, as Engelhardt points out, this focus on equal input legitimates an egalitarianism of envy which is by no means reasonable or beneficial to the people in society, because an alternative focus on how we could as cheaply as possible decrease morbidity and mortality would be more useful to everyone. For Engelhardt, this focus on equal input is a type of envy egalitarianism because it implies that one is jealous of those who have access to better basic health care even when this circumstance will not harm those receiving only basic health care. Instead, this circumstance – in which one does not ensure equality of input in health care but instead seeks to reduce morbidity and mortality as far as possible – may indeed advantage the latter's health status through bringing more capital into the health care system (Engelhardt 2012, 594).

8.5 Democratic Ideology and Financial Unsustainability

Liberal social democratic scholars have gradually recognized that even if a group of general principles of distributive health care justice for fair equality of opportunity can be worked out, as Daniels did, from a philosophical mechanism like Rawls' original position behind a veil of ignorance, it is impossible, on such general principles, to determine concrete health care decisions in actual health care situations. The reason is easy to grasp. Actual health care decisions involve a series of important but mutually competing ethical concerns, such as whether we should give some priority to meeting the urgent claims of patients in last chance medical contexts, how we should appropriately manage collective resources, how much weight we should give to the work of generating scientific knowledge about the effectiveness of unproven drugs or therapies to benefit future patients, in what ways we should respect patient preferences about risk-taking, and so on. General principles cannot tell us how to weigh the relative claims of these different concerns and draw specific conclusions under the circumstance of resource scarcity. To do so we must rely on some fine-grained principles or context-relevant rules on which even the contractors in the original position cannot have consensus. Social democratic scholars have to admit that their preferred principles of distributive justice, no matter how important or necessary under their view, are indeterminate with actual health care situations. Accordingly, these scholars argue for the necessity of fair process to supplement appeals to general principles in order to arrive at determinate answers about specific allocation of health care in real-world contexts.

Daniels and Sabin have proposed the following four conditions to shape a fair process for making actual health care decisions under reasonable resource scarcity. The Publicity condition requires transparency with regard to the reasons for a decision. The Relevance condition constrains the kinds of reasons that can play a role in the process to those that are accepted as relevant by fair-minded people who are disposed to finding mutually justifiable terms of cooperation. The Revision condition demands mechanisms for challenge and dispute resolution as well as revision and improvement of policies in the light of new evidence or arguments. And the Regulative condition puts regulation on the process to ensure that the first three conditions are met (Daniels and Sabin 2002, 45–46). They expect that through fair processes regulated by these four conditions, specific health care decisions will be worked out and will be fair. They stress that the four conditions facilitate a public democratic process: "The four conditions compel decision makers in health plans or public agencies to contribute their deliberative capacities to whatever broader public deliberation is conducted through democratic institutions, formally or informally" (Daniels and Sabin 2002, 63). Indeed, they acknowledge that although they developed these conditions independently, they nevertheless fit well with the principles of publicity, reciprocity, and accountability governing democratic deliberation cited by Amy Gutmann and Dennis Thompson in their 1996 work, *Democracy and Disagreement* (Daniels and Sabin 2002, 66, note 2).

Through the adoption of this so-called "fair process" mechanism by imposing the four conditions, the issue of fairness seems to have been transformed into an issue of legitimacy – any decision that can legitimately be worked out through such a process will automatically be a fair decision. It is beyond the major concerns of this chapter to discuss whether this transformation is conceptually acceptable. However, their proposal does raise enormous concerns with its practical consequences. Surely these authors are primarily concerned with making health care allocations "fairly" under the reasonable scarcity of resources, including "fairly" setting limits in health services. Their argument does not address ways to maximize the offering of health care welfare or entitlements to meet all health care needs. Nevertheless, their proposal has clearly placed health care decisions into the sphere of social democratic politics in general, and welfare or entitlement politics in particular, expecting that the conditions they have proposed to regulate the democratic process will facilitate fair-minded people to make not only morally "fair," but also financially sustainable health care policies and decisions. Unfortunately, the practices of social democratic countries have encountered thorny problems regarding the financial sustainability of such processes. As Engelhardt has pointed out, in Western countries health care expenditures contribute to the financial crises of social-democratic polities, which have deep roots in the extensive state support of a wide range of entitlements, including health care entitlements, that now constitute a major portion of the budgets of Western European and North American states. For example, as Engelhardt notes, entitlements constituted 62 % of the 2010 US federal budget, in which 21 % was committed to Medicare and Medicaid (Engelhardt 2012, 585).

Such evidence indicates that in the democratic process, "fair-minded" people tend to opt for more welfare or entitlements (including health care welfare or entitlements) for themselves than what can financially be sustainable for their descendants. This is no surprise. From Engelhardt's view, all social insurance and similar entitlement-based approaches to the allocation of health-based resources in contemporary social democratic countries would involve three types of hazards, moral, demographic, and political:

> [T]here is a moral hazard in that, once an entitlement is established, it will tend to be exploited to its utmost. In addition, if a health care system depends on taxing those who are well and working to pay for those who are sick and/or not working, the system will face a demographic catastrophe if there is a significant decline in the percentage of the working well. The striking decrease in reproductive rates among Western Europeans and North Americans has been tied both to the effects of their welfare systems as well as to the emergence of an antenatal ethos in which it is no longer for granted that men and women will marry and have children. The result has been insufficient resources for (i.e., insufficient payers to support) generous pensions with early retirements, as well as for ample publically supported health care systems. Last, but surely not least, socially financed health care systems involve a political hazard, in that politicians have an incentive to promise better and more encompassing health care benefits in order to advance their political careers, even if the available resources are insufficient to pay for those benefits. These factors have combined to raise questions regarding the financial sustainability of social-democratic welfare systems. (Engelhardt 2012, 588)

This is to say, given these inevitably engaged hazards, even if the fair process proposed by social democratic scholars, like Daniels, could determine specific health care policies and decisions that will be "fair" in terms of the social democratic ideology of fair equality of opportunity, it will lead to a consequentially disastrous result in the long run: financial unsustainability.

8.6 Erosion of the Family

The most serious problem in implementing the social democratic ideal of fair equality of opportunity is the erosion of the family. Engelhardt has written a series of reflections on this problem. When Rawls constructed his account of justice, he seemed a bit ambivalent about the stand of the family in contemporary society. On the one hand, he appealed to the hypothetical contractors as heads of families to determine his general principles governing society. This apparently family-friendly consideration allowed him to give a ground for the contractors to be committed to the best interests of their descendants and thus to the just savings principle that is inevitably needed for his account of justice. As Rawls states, although it is not necessary for him to think of the parties in the original position as heads of families, this is still a useful motivational assumption to follow because it is essential that "each person in the original position should care about the well-being of some of those in the next generation" (Rawls 1971, 128). On the other hand, however, Rawls recognizes that a commitment to fair equality of opportunity brings the family itself into question. Indeed, he is morally suspicious regarding the traditional family. Like John Stuart Mill, he takes the traditional family to be a school for male despotism that inculcated habits of thought and ways of feeling and conduct incompatible with democracy. Accordingly, Rawls contends that "the principles of justice enjoining a reasonable constitutional democratic society can plainly be invoked to reform the family" (Rawls 1997, 791).

More fundamentally, as Engelhardt remarks, Rawls is critical of the family for the very reason that the family is useful in justifying nonegalitarian intergenerational obligations: families "advance the interests of their members in preference to the interests of others, thereby undermining fair equality of opportunity" (Engelhardt 2012, 596). Rawls admits that as long as the institution of the family exists, the principle of fair equality of opportunity can be only imperfectly carried out (Rawls 1971, 74). This is because, as he recognizes it, the family will inevitably lead to unequal chances between individuals. Accordingly, he acknowledges that the principle of fair equality of opportunity, "taken by itself and given a certain primacy," inclines in the direction of abolishing the family, although "there is much less urgency to take this course" within the context of his theory of justice in which inequalities must be arranged to benefit the least well-off class (Rawls 1971, 511–512). This is to say, as Engelhardt points out, given Rawls' commitment to fair equality of opportunity, he should have explicitly held that *ceteris paribus* the family should be abolished (Engelhardt 2012, 596).

Indeed, as Engelhardt discloses, although Rawls does attempt to interpret the contractors in the original position as heads of families for motivating savings necessary for future generations, the status of such contractors is taken "first as citizens, not as family members" (Engelhardt 2012, 596). They are actually conceived of as unencumbered individuals whose liberty, equality, and prosperity are realized in individualist, not familial, terms. Rawls denies to families an area of privacy within which family members could consent to nonegalitarian relationships: "If the private sphere is alleged to be a space exempt from justice, then there is no such thing...the equal rights of women and the basic rights of their children as future citizens are inalienable and protect them wherever they are" (Rawls 1997, 791). Accordingly, in a Rawlsian social democratic polity there is no family that has independent standing or authority between the individual and the state. The result is a Rawlsian social welfare system that must be established for the sake of supporting fair equality of opportunity. This system, as Engelhardt points out, will not only aim at aiding individuals apart from their location within families, but also structure itself in ways that enable individuals to leave their families easily. Accordingly, "[I]t is not an accident that the modern social welfare state is highly associated with the contemporary erosion of the family" (Engelhardt 2012, 596).

One result of the modern welfare state has been a substantive increase in the number of children born outside of traditional marriage. Engelhardt summarizes the situations of the United States and United Kingdom in the past little more than half a century as follows:

> The rate of out-of-wedlock births was probably at 4 % through most of American history and was such in 1950. The percentage had risen to 10 % in the United States in 1970 and 18 % in 1980. The rate in the United Kingdom for 1980 was 11.5 %. By 2000, the number of children born outside of a traditional marriage had risen to 33.2 % in the United States and 39.5 % in the United Kingdom. Nine years later, in 2009, the rate was 41 % in the United States and 46.3 % in the United Kingdom. (Engelhardt 2012, 589)

This trend continues in Western countries. Ironically, while the trend has been triggered to a great extent by the social democratic pursuit of fair equality of opportunity for every individual, the result – for most of those out-of-wedlock born children as well as the emergent single-mother families – is by no means the prospect of a promising or flourishing way of life (Engelhardt 2013). Engelhardt cites the researches recently made by social and political scientists to illustrate the problems. Evidently, "the unanticipated consequences of welfare programs like Aid to Families with Dependent Children (AFDC)" included those that "encouraged out-of-wedlock births and contributed to the culture of poverty" (Fukuyama 2006, 20). Families headed by single mothers are in most cases strongly associated with fewer resources, including health care resources, available for such families (Rector 2010). "Having two unmarried biological parents was associated with worse outcomes than having two married biological parents, and the outcomes were rarely better than those for children living with a single parent or in a 'cohabiting stepparent family'" (Murray 2012, 165).

Social democratic scholars committed to fair equality of opportunity tend to suggest the further enhancing of welfare programs in order to improve the culture of poverty associated with out-of-wedlock born children as well as single-parent families. This strategy will definitely hasten the speed of financial unsustainability and bring about more out-of-wedlock births and single-parent families. Nevertheless, the goal of accomplishing fair equality of opportunity can hardly be accomplished. Economists have long recognized a huge contrast of altruism between family interactions and market transactions: altruism is less common in market transactions and more common in families because altruism is less efficient in the marketplace and more efficient in families (Becker 1993, 299). Indeed, selfishness in market transactions has been assumed in virtually all subsequent discussion of the economic system, while altruism is generally recognized to be important within a family. The Rotten Kid Theorem states that "each beneficiary, no matter how selfish, maximizes the family income of his benefactor and thereby internalizes all effects of his actions on other beneficiaries" (Becker 1993, 288). Gary Becker shows that the theorem even has a surprising extension to envious behavior: "each beneficiary, no matter how envious of other beneficiaries or of his benefactor, maximizes the family income of the benefactor, and hence helps those envied" (Becker 1993, 288).

Why are traditional families most beneficial to the growth of children and their development? Following Engelhardt, "traditional family" is used here to identify a married man and woman who are the parents of the family's children (Engelhardt 2007). There have been appealing philosophical arguments for the traditional family's intrinsic orientation to childbearing and rearing (Girgis et al. 2011), the irreplaceable value of living with one's both biological parents for one's identity formation (Velleman 2005), as well as arguments based on liberal public reasons for the endorsement of traditional families (O'Brien 2012). Even if we put these philosophical or abstract arguments aside, there exists strong support of the traditional family for meeting the needs of children in light of utility-based pragmatic considerations. For example, from the study of Y. C. Richard Wong, a well-established economic scholar living in Hong Kong, we can summarize at least two important reasons. First, since children take a long time to grow up and must be cared for intensively, and since children in modern societies tend to take an even longer time to grow to independence as a result of the much larger investments in human capital parents make in them, the traditional family continues to be the best available institution for producing "quality" children. In contrast, children from non-traditional families, including those from out-of-wedlock births, single-parent families, or broken families, have a higher probability of being of lesser "quality" (the term "quality" is used in a very general and broad sense covering cognitive, behavioral, health, and other characteristics) (Wong 2013a, b). Second, while traditional families continually provide important rearing activities for their children, such activities in non-traditional families have increasingly been taken over by the state, including state support for single parents with dependent children in the form of education and other benefits. In this way, the state's expansion into the education, housing, and other services of these families has significantly reduced or replaced the role of the parents and other family members in providing for their children. The most significant effect is that such an aggressive state role reduces parental investment in children in these families (Wong 2013a, b).

In short, Wong has echoed Engelhardt to argue that undercutting parents' role in caring for their children is cultivating a disaster for generations to come. Such is, for Wong, the plight of the advanced industrial societies that have made this choice.

The public debts that have exploded in the US, UK, Europe and Japan today are the manifestation of a broken intergenerational compact. The crisis of the family, Wong warns, would speedily occur in the Pacific Rim, including mainland China, Hong Kong, and Taiwan, if they do not learn a grave lesson from the West (Wong 2013a, b).

8.7 Conclusion

Today's academics around the globe are lucky to have a wise man like Engelhardt to remind everyone of the unpleasant problems of promoting fair equality of opportunity as a cardinal principle. If the evidence and arguments provided in this chapter are sound, we should be persuaded to conclude that the fair equality of opportunity principle, no matter how attractive at first blush or how fashionable it is in the contemporary time, is not a suitable moral or political norm to govern society.

References

Becker, G.S. 1993. *A treatise on the family*, enlarged ed. Cambridge, MA: Harvard University Press.

Boorse, C. 1975. On the distinction between disease and illness. *Philosophy and Public Affairs* 5: 49–68.

Boorse, C. 1977. Health as a theoretical concept. *Philosophy of Science* 44: 542–573.

Daniels, N. 1985. *Just health care*. Cambridge: Cambridge University Press.

Daniels, N. 2008. *Just health: Meeting health needs fairly*. Cambridge: Cambridge University Press.

Daniels, N., and J. Sabin. 2002. *Setting limits fairly: Can we learn to share medical resources?* New York: Oxford University Press.

Engehardt Jr., H.T. 1974. Disease of masturbation: Values and the concept of disease. *Bulletin of the History of Medicine* 48(2): 234–248.

Engelhardt Jr., H.T. 1996. *The foundations of bioethics*, 2nd ed. New York: Oxford University Press.

Engelhardt Jr., H.T. 2007. Long-term care: The family, post-modernity, and conflicting moral life-worlds. *Journal of Medicine and Philosophy* 32: 519–536.

Engelhardt Jr., H.T. 2012. Fair equality of opportunity critically reexamined: The family and the sustainability of health care systems. *Journal of Medicine and Philosophy* 37: 583–602.

Engelhardt Jr., H.T. 2013. The family: Crucial to and divisive in bioethics. *International Journal of Chinese and Comparative Philosophy of Medicine* 11: 113–127.

Fukuyama, F. 2006. *America at the crossroads: Democracy, power, and the neoconservative legacy*. New Haven: Yale University Press.

Girgis, S., R.P. George, and R.T. Anderson. 2011. What is marriage? *Harvard Journal of Law & Public Policy* 34: 245–287.

Gutmann, A., and D. Thompson. 1996. *Democracy and disagreement*. Cambridge, MA: Harvard University Press.

Murray, C. 2012. *Coming apart: The state of white America 1960–2010*. New York: Crown Forum.

O'Brien, M.B. 2012. Why liberal neutrality prohibits same-sex marriage: Rawls, political liberalism and the family. *British Journal of American Legal Studies* 1: 411–466.

Rawls, J. 1971. *A theory of justice*. Cambridge, MA: Harvard University Press.

Rawls, J. 1997. The idea of public reason revisited. *The University of Chicago Law Review* 64: 765–807.

Rector, R. 2010. Marriage: America's greatest weapon against child poverty. *Backgrounder* 2465: 1–16.

Stern, L. 1983. Opportunity and health care: Criticism and suggestions. *Journal of Medicine and Philosophy* 8: 339–361.

Velleman, J.D. 2005. Family history. *Philosophical Papers* 34: 357–378.

Wong, Y.C.R. 2013a. *Economics of the family and same-sex marriages*. [Online] Available: http://www.wangyujian.com/?p=1997&lang=en. Accessed on 16 Mar 2013.

Wong, Y.C.R. 2013b. *Housing, retirement and the economy – Why family savings and investing in children matters?* [Online] Available: http://www.wangyujian.com/?p=2004&lang=en. Accessed on 16 Mar 2013.

Chapter 9
Bioethics After the Death of God – Reflections on an Engelhardtian Theme

Mark J. Cherry

> So farewel Hope, and with Hope farewel Fear,
> Farewel Remorse: all Good to me is lost;
> Evil be thou my Good...
> (Satan, John Milton, *Paradise Lost*, Book 4, lines 108–110).

9.1 So Farewell Hope: Man as the Measure of All Things

Absent a canonical point of view, a binding standpoint that is not already conditioned by a particular society, culture, or place in history, moral claims cannot be fully justified. It was for this reason that Immanuel Kant (1724–1804), who was very likely an atheist,[1] realized that if one did not accept the existence of God and the immortality of the soul as practical moral postulates, traditional Western morality would collapse. Kant recognized that unless God secured the moral project, morality would cease to possess definitive content or binding significance. God, or at least a fully objective God's-eye perspective on reality, is necessary to provide a foundation for morality in being-in-itself. Without such a foundation, all moral judgments become socially, historically, and culturally contingent; morality shatters into numerous potentially incommensurable perspectives. The recognition of this state of affairs acknowledges what Michael McCarthy describes as the "crisis of philosophy."[2]

[1] See Manfred Kuehn, who argues that "Kant did not really believe in God" (2001, 391–392).

[2] "In the present case, philosophers themselves are the severest critics of their own history. And these critics are thinkers of exceptional power and influence. Nietzsche, Dewey, Wittgenstein, Heidegger and Rorty, in their very different ways, have turned against the philosophical tradition in ethics, metaphysics, and epistemology. They have tried to delegitimatize the traditional philo-

M.J. Cherry, Ph.D. (✉)
Department of Philosophy, St. Edward's University, 3001 S. Congress Ave., Austin, TX, USA
e-mail: markc@stedwards.edu

L.M. Rasmussen et al. (eds.), *At the Foundations of Bioethics and Biopolitics: Critical Essays on the Thought of H. Tristram Engelhardt, Jr.*, Philosophy and Medicine 125, DOI 10.1007/978-3-319-18965-9_9

From G.W.F. Hegel (1770–1831) to Frederick Nietzsche (1844–1900) and Gianni Vattimo (1936–), a post-modernity has been embraced that radically transforms, indeed, demoralizes and deflates morality (Engelhardt 2010b).

This essay explores foundational and ontological (metaphysical, if you will) issues that lie at the foundations of such cardinal concerns. Throughout, I draw on H. Tristram Engelhardt, Jr.'s philosophical analysis to demonstrate why even in principle to talk about the possibility of a canonical morality to guide medical practice or to frame health care policy requires at least the equivalent of a God's-eye perspective, not out of religious but out of epistemic considerations. I argue that one must invoke God, or at least a God's-eye perspective, in order to preserve the unity and force of reality and morality. As Hegel was aware, nothing in particular follows from general concepts of the right, the good, the just, or the virtuous.[3] Even to sort useful information from noise, particular moral content must first be specified. Without such content, morality (and bioethics) cannot hope to provide definitive guidance for choosing among competing accounts of human flourishing, much less for properly evaluating how to proceed when the right and the good conflict.

In this paper, I argue in support of an Engelhartian diagnosis of our post-modern existential predicament: without God to secure a canonical perspective, morality and truth become plural. Reference to God and His unique knowledge of reality provides a perspective necessary to invoke the idea of unconditioned truth or unconditioned knowledge of reality, whether moral, social, or scientific. A God's-eye perspective provides knowledge of reality that is independent of all social construction; knowledge of reality as it is in itself. Such a perspective would, for example, provide a normative position from which to understand all of morality, as well as how and why one ought to act on considerations of morality when the right and the good conflict. Absent such an appeal, one loses a defining perspective that could, in principle, provide a canonical morality to set aside the great diversity of socially conditioned bioethical perspectives. Without God, the moral project confronts an irreducible plurality of competing moralities and moral narratives. Divergent

sophical disciplines by dissolving their defining problems and by rejecting the vocabulary and distinctions required for their formulation. Richard Rorty, in particular, has made an explicit appeal for a post-philosophical culture, a culture in which the concerns of Socrates, Aquinas, and Kant are no longer taken seriously. To indicate the magnitude of the change he desires, Rorty draws an explicit parallel between his project and the modern process of secularization. Over the course of five centuries, European culture gradually abandoned its preoccupation with religious and theological questions. Theology ceased to occupy the central position in Western universities and became an optional or marginal region of the intellectual landscape. Rorty hopes that a historic transformation has begun in which the Socratic concern for the good, the Thomist absorption with God, and the Kantian attachment to rigorous science will become equally marginal and optional. If Rorty's expectations are fulfilled, the crisis of traditional philosophy will end with its gradual disappearance" (McCarthy 1990, xii).

[3] As Hegel recognized, "Since action for itself requires a particular content and a determinate end, whereas duty in the abstract contains nothing of the kind, the question arises: what is duty? For this definition [Bestimmung], all that is available so far is this: to do right, and to promote welfare, one's own welfare and welfare in its universal determination, the welfare of others" (Hegel 1991 [1821], §134, 161).

accounts of morality also presume different rules of moral evidence and moral inference. As a result, a plurality of moral rationalities exist, grounded in various rankings of cardinal virtues, right making conditions, calculations of harms and benefits, claims to virtue or vice, and competing understandings of human flourishing. Without a God's-eye perspective from which to know the right, the good, and the virtuous, there is no definitive reason to choose any one particular moral perspective from among the various competing moral theories or sets of moral intuitions to guide medical decision-making. As a consequence, rather than providing an authoritative bioethics to guide medicine and frame public policy, secular bioethicists create their own potentially idiosyncratic criteria for veracity, rationality, and reliability, as well as rationalizations for ideologically driven political advocacy. Without God, or a God's-eye perspective, to secure moral content, all claims to moral truth are demoralized, reduced to socially and historically conditioned judgments, culturally relative virtues, personal intuitions, and particular political ideological commitments. As Protagoras (490–420 BC) so aptly put the point: "Man is the measure of all things."[4]

9.2 And with Hope, Farewell Fear: The Denial of Ultimate Foundations

> God is dead! God remains dead! And we have killed him! How shall we console ourselves, the most murderous of all murderers? (Nietzsche, *The Joyful Wisdom*, 1910, 168).

Without God to secure a canonical standpoint the very significance of epistemic claims regarding reality and morality changes. To illustrate, consider Nietzsche, who was well aware of the foundational implications of his diagnosis that "God is dead":

> The most important of more recent events – that "God is dead," that the belief in the Christian God has become unworthy of belief – already begins to cast its first shadows over Europe. …not to speak of many who already knew *what* had really taken place, and what must all collapse now that this belief had been undermined, – because so much was built upon it, so much rested on it, and had become one with it: for example, our entire European morality. This lengthy, vast and uninterrupted process of crumbling, destruction, ruin and overthrow which is now imminent… (1910, 275).

As Vattimo situates Nietzsche's analysis: "In Nietzsche, the death of God signifies the final dissolution of supreme values and metaphysical belief in an objective and eternal order of Being. That is nihilism in a nutshell" (2011, 58). Without God,

[4] "Man is the measure of all things, of those that are that they are, of those that are not that they are not." Quoted in Eduard Zeller (1980, 81). As Zeller notes: "that transfers the problem of knowledge from the object to the subject and thus made possible a real theory of knowledge. With 'things' we have not only to think of concrete things but also of abstract qualities; and what is more not only of sensual qualities such as warm and cold, sweet and bitter, etc., but also concepts like good and bad, beautiful and ugly, right and wrong" (81).

knowledge claims cannot be grounded in reality itself. Instead, all such claims are always situated within the taken-for-granted background assumptions of particular social and cultural frameworks.

Nietzsche's announcement "God is dead" succinctly captures the predicament of a secular culture committed to the non-existence of God: a culture ungrounded in being, trapped in immanence, and unable to secure its truth claims.

> But what I have in view will now be understood, namely, that it is always a *metaphysical belief* on which our belief in science rests, – and that even we knowing ones of to-day, the godless and anti-metaphysical, still take *our* fire from the conflagration kindled by a belief a millennium old, the Christian belief, which was also the belief of Plato, that God is truth, that the truth is divine. ... But what if this itself always becomes more untrustworthy, what if nothing any longer proves itself divine, except it be error, blindness, and falsehood; – what if God himself turns out to be our most persistent lie? (1910, 280).

Vattimo concludes that those who, like Nietzsche, are committed to living without God must also be content to exist – to engage in science, morality, and politics – without ultimate foundations.

> Nietzsche's announcement that 'God is dead' is not an atheistic thesis like 'God does not exist.' Nietzsche could not state a thesis like the nonexistence of God because the claim to its absolute truth would have to be upheld as a metaphysical principle, that is, as the true 'structure' of reality, having the same function as the traditional God of metaphysics. Wherever there is an absolute, even if it is the sheer affirmation of the nonexistence of God, metaphysics is always present in the form of the supreme principle, namely the God whose superfluousness Nietzsche believes to have discovered. In sum, for Nietzsche 'God is dead' means nothing else than the fact that there is no ultimate foundation (2002, 3).

There is no privileged position from which to obtain knowledge of reality as it is in itself. The unity and force of reality, whether scientific, historic, or moral, cannot, even in principle, be secured.[5]

Nietzsche's analysis echoes Hegel, who had already appreciated that without God to ground knowledge of reality as it is in itself, all knowledge claims become socially and historically conditioned. Hegel understood that the secular culture of his time had come to act as if God were dead; it was characterized by the "feeling that 'God Himself is dead'" (Hegel 1968a, b, 413; see also 1977, 190; 1968a, b, 414). As Engelhardt argues, "Even in his youth, Hegel understood that religion transformed by the Enlightenment no longer needed a transcendent God. The Enlightenment created a culture predicated on the irrelevance of a living, personal God" (2000, 96).[6] Neither a real belief in God nor even an in principle acceptance of a God's-eye perspective, as the regulative idea of an unconditioned perspective

[5] "Vattimo urges us to see that no interpretation, no particular disclosure of Being, is final or ultimate, thereby escaping a deeply inscribed contingency and provisionality. In the past, Enlightenment modernity (but surely not only the Enlightenment) has presented itself as the last word, the final truth; but the Enlightenment itself has been unmasked, as have all claims to privileged interpretation. What takes place over time is a continual 'rewriting' of the world. No one has access to universal structures, to absolute and immutable essences. One meaning of the 'death of God' is that there exists no final, encompassing vision of the world" (Guarino 2009, 27).

[6] See, for example, Hegel (1984, 104–165).

on reality, were any longer accepted as relevant as a cultural or moral point of reference (see, e.g., Engelhardt 2010a, b). As a result, for Hegel truth could no longer refer to correspondence with the "thing-in-itself," but rather thought thinking about being thinking itself. "Truth," he concluded, "… is only possible as a universe or totality of thought…" (Hegel 1892, 24). For Hegel, philosophical reflection replaced the transcendent perspective of God, albeit with a radically immanentized perspective, for philosophical reflection is itself always located within very particular social, historical, and cultural circumstances.[7] Hegel recognized that without access to a canonical standpoint, such as God would provide, one could no longer coherently speak of truth as correspondence with the thing-in-itself, but only of my subjective knowledge of the thing for me, and the hoped for coherence of my claims with the subjective judgments of others. Without God, or at least a God's-eye perspective, a fully objective account of being and knowledge of reality in itself simply cannot be secured. Cut off from a God's-eye perspective on reality, there can be no final, canonical perspective on either reality or the morality such a unique perspective would sustain. The implication is that as the categories affirmed by philosophical reflection change, so do reality and morality. The result, Hegel understood, is that philosophy (and bioethics) is always socially and historically conditioned.

9.3 Farewell Remorse: The Deflation of Epistemic and Moral Claims

Having rejected God, even as the regulative ideal of a privileged epistemological position, moral norms, and epistemic claims more generally, are thoroughly deflated. Without God, or at least access to a God's-eye perspective from which to know fully objective Truth, there ceases to be a socio-historically or culturally unconditioned position from which to know reality. Consequently, it is not possible, even in principle, to know whether our knowledge claims cohere with reality. All knowledge claims are already embedded within particular culturally and historically conditioned perspectives.[8] As a result, philosophical analysis and scientific exploration

[7] "For Hegel, philosophy, enjoying the coincidence of thought and an immanentized being, can reapproach its tasks, having finally put God in His place, free of true transcendence" (Engelhardt 2000, 96).

[8] In his *Treatise of Human Nature* (1739–1740), Hume argued against the possibility of discovering the ultimate characteristics of human nature: "And tho' we must endeavour to render all our principles as universal as possible, by tracing up our experiments to the utmost, and explaining all effects from the simplest and fewest causes, 'tis still certain we cannot go beyond experience; and any hypothesis, that pretends to discover the ultimate original qualities of human nature, ought at first to be rejected as presumptuous and chimerical" (Hume 1896, 11).

Experience, Hume argues, is our only guide concerning matters of fact; even though all experience is socio-historically and culturally conditioned, and subject to error: "Though experience be our only guide in reasoning concerning matters of fact; it must be acknowledged, that this guide is not altogether infallible, but in some cases is apt to lead us into errors. One, who in our climate, should expect better weather in any week of June than in one of December, would reason justly, and

moves away from considerations of "truth" per se, in the sense of a proposition's correspondence with the deep nature of reality, toward considerations of the more or less useful. The focus shifts to socially embedded judgments regarding utility, including how well our new judgments cohere with intuitions or worldviews we already affirm.

Consider, for example, Thomas Kuhn's complex idea of a "paradigm" from *The Structure of Scientific Revolutions* (1962). Kuhn engaged this concept in order to explain the complex ways in which various scientists, even from the same field, can be separated by different ways of understanding and experiencing reality. Divergent understandings of and approaches to reality separate Newtonian from Einsteinian scientists. Paradigms identify the "thought-styles" of communities of knowers and investigators; their approaches to reality, judgments of usefulness, and ways of resolving challenges.[9] In part, reasons for adopting a new paradigm (e.g., moving from Newtonian to Einsteinian physics) consider judgments of utility. In part also, they consider the relative ability of competing paradigms to solve the problems in which particular scientists have interest. Kuhn argued, however, that such considerations are only partly sufficient; other considerations that lead scientists to reject one paradigm in favor of another include: "… the arguments, rarely made entirely explicit, that appeal to the individual's sense of the appropriate or the aesthetic – the new theory is said to be 'neater,' 'more suitable', or 'simpler' than the old" (Kuhn 1962, 154). The history of science illustrates an on-going international undertaking that functions "…without benefit of a set goal, a permanent fixed scientific truth, of which each stage in the development of scientific knowledge is a better exemplar" (Kuhn 1962, 172). Kuhn cannot discern the availability of a God's-eye perspective from which to ascertain which scientific paradigm most closely approaches reality in itself. All that exists are a variety of scientific narratives, told from a diverse set of socially, culturally, and historically conditioned perspectives.

Medical reality and the expectations of clinical judgment similarly rely on conditioned social constructs. Diagnostic categories, for example, are not simply reflections of the underlying character of medical reality. They represent collective judgments regarding the modes of classification that will likely prove useful for

---

conformably to experience; but it is certain, that he may happen, in the event, to find himself mistaken. However, we may observe, that, in such a case, he would have no cause to complain of experience; because it commonly informs us beforehand of the uncertainty, by that contrariety of events, which we may learn from a diligent observation. All effects follow not with like certainty from their supposed causes. Some events are found, in all countries and all ages, to have been constantly conjoined together: Others are found to have been more variable, and sometimes to disappoint our expectations; so that, in our reasonings concerning matter of fact, there are all imaginable degrees of assurance, from the highest certainty to the lowest species of moral evidence" (Hume 1910, Section X, Of Miracles).

Hume repudiated theories of reality and human nature that purport to provide such ultimate principles.

[9] The term "thought-style" comes from Ludwik Fleck (1979), who uses it to identify a community of knowers who share metaphysical, epistemological, methodological, and axiological assumptions in their approach to both framing reality as well as assumptions regarding exemplar knowers and methodologies of discovery.

treating patients. As Engelhardt documents, in order for "… physicians to evaluate the efficacy of different therapies, collective endeavors such as the American Joint Committee on Cancer are organized to fashion rules for the use of descriptive terms in diagnostic categories. In such circumstances the social reality of the applied sciences is settled by votes within committees" (1996, 219). Moreover, diagnostic categories shift as alternative modes of classification prove more useful for achieving socially and culturally desired outcomes, however such outcomes are defined.

> Here one might think in particular of the ways in which cancers are staged. The classification and staging of cancer are fashioned 'to allow the physician to determine treatment for the patient more appropriately, to evaluate results of management more reliably, and to compare statistics reported from various institutions on a local, regional, and national basis more confidently.' Such classification and staging is the result of particular social organizations balancing the interests of various groups (Engelhardt 1996, 219).

Even the number of stages chosen to describe the cancer's progression ("Why, for instance, use three, four, or five stages for any particular carcinoma?" [Engelhardt 1996, 220]) presupposes particular cost-benefit calculations and judgments regarding the prudential treatment of patients. These are not purely independent scientific assessments:

> … they concern as well the proper balancing of benefits and harms in the organization of therapeutic choices around a particular number of stages of cancer. The choice to subdivide stages (e.g., stages IA and B) will reflect the decision that a more complex assessment of therapeutic options is appropriate. There are limitations to the number of subdivisions one can make and still have them remain useful (Engelhardt 1996, 220).

Medicine is an applied science as well as a social endeavor. Medical science and the treatment of patients are always set within particular cultures and human interests. As a result, concepts of health and disease, as well as the framing of proper health care policy, always intersects an overlapping set of culturally situated communities (scientific, moral, religious, and political), each striving to understand and to manipulate the world in ways that each finds socially useful, morally appropriate, aesthetically pleasing, or otherwise fitting.[10] Such taken-for-granted background conditions, in turn, impact clinical expectations as well as appreciation of bioethical obligations.

This idea of divergent ways of experiencing reality, given various judgments regarding simplicity, beauty, or usefulness, illustrates not just how scientists are separated from each other, but also how cultures, religions, and secular narratives are divided from each other by different understandings of and approaches to reality.[11] Such pluralism, however, goes deeper than simply differences in descrip-

---

[10] See, for example, the detailed analysis in Engelhardt (1980, 364–461); Engelhardt (1996, especially Chapter 5); as well as Engelhardt and Caplan (1987).

[11] Consider, Nelson Goodman's observation regarding the multiplicity of world versions: "Much more striking is the vast variety of versions and visions in the several sciences, in the works of different painters and writers, and in our perceptions as informed by these, by circumstances, and by our own insights, interests, and past experiences. Even with all illusory or wrong or dubious versions dropped, the rest exhibit new dimensions of disparity. … The dramatically contrasting versions of the world can of course be relativized: each is right under a given system — for a given science, a given artist, or a given perceiver and situation. Here again we turn from describing or

tion or depiction: every attempt to describe the world, absent an appeal to God, must do so given a particular socially, historically, and culturally conditioned frame of reference. As Engelhardt notes: "To speak of the nature of reality or of medical facts undistorted by historical and cultural contexts is to speak of a view that would be possessed by knowers in full possession of all information and unmoved by assumptions not grounded in reality" (Engelhardt 1996, 191). For finite knowers, however, such a perspective is impossible. Without God to secure objective being and objective knowledge of such being, each and every narrative may very well embody quite different socially conditioned interpretations of reality and morality.[12]

Cut off from God and His uniquely True perspective on reality, the so-called real world is, as Vattimo notes, like a story or a dream. Any particular account of science or morality is no more than one among the diversity of narratives that we tell each other:

> In a beautiful passage from *The Twilight of the Idols*, Nietzsche tells us how the real world has become a dream. It was the Platonic world of ideas that gave us the idea of the real world in the first place. Later, the real world was construed as the promised world after death (at least for the righteous). Still later, in the mind of Descartes, the thought of the real world was evidence of clear and distinct ideas (but only in mind). With positivism the real world became the world of experimental verified truths and then a product of the experimental scientist ... At this point, the so-called real world has become a story that we tell each other (Vattimo 2007, 39).[13]

Without God, there is no ultimate foundation; and, as a result, there are no facts, scientific or moral, only interpretations.[14] This does not mean that there is no underlying reality whatsoever; such a conclusion would similarly affirm an objective truth that could only be fully known from a God's-eye perspective. Rather, human knowledge and experience of the world, isolated from God, is always contextualized in terms of background human interests and judgments, including concerns

depicting 'the world' to talking of descriptions and depictions, but now without even the consolation of intertranslatability among or any evident organization of the several systems in question" (Goodman 1978, 3).

[12] "To be sure, there is no sense in purely and simply denying the world a 'unitary reality', in a kind of reprise of naïve empirical idealism. It makes more sense to recognize that what we call the 'reality of the world' is the 'context' for the multiplicity of 'fablings' – and the task and significance of the human sciences lie precisely in thematizing the world in these terms" (Vattimo 1992, 25).

[13] Similarly, in *The Transparent Society*, Vattimo argues that "Instead of moving towards self-transparency, the society of human sciences and generalized communication has moved towards what could, in general be called the 'fabling of the world.' The images of the world we receive from the media and the human sciences, albeit on different levels, are not simply different interpretations of a 'reality' that is 'given' regardless, but rather constitute the very objectivity of the world" (Vattimo 1992, 25).

[14] Nietzsche's famous conclusion against scientific positivism is appropriate: "No, facts are precisely what there is not, only interpretations. We cannot establish any fact 'in itself': perhaps it is folly to want to do such a thing" (Nietzsche 1968, 267). With regard to history, as Vattimo puts it: "... the idea of unilinear history ends up being dissolved. There is no single history, only images of the past projected from different points of view. It is illusory to think that there exists a supreme or comprehensive viewpoint capable of unifying all others..." (Vattimo 1992, 3; see also, Vattimo 1988, 7 ff.).

regarding morality and aesthetics, as well as the scientifically interesting, or ideologically useful.

All epistemic claims are deflated. Absent God, there exists no standpoint outside of particular cultural socio-historically conditioned perspectives from which to communicate any deeper perspective of reality or of the morality that such a perspective on reality would secure. Every attempt to describe the world is confined to explicitly or implicitly socially and historically conditioned ways of describing, given particular background scientific, moral, or political interests. As a result, persons come to see the world in terms of their expectations, backgrounds, and taken-for-granted assumptions.[15] All attempts to engage reality will necessarily bring to the encounter underlying ontological, metaphysical, epistemological, and moral assumptions through which the encounter is framed and understood. As Engelhardt concludes: "There is thus a tension between the universal aspirations of knowers and the particular context in which real individuals actually know and frame explanations" (1996, 218). Absent access to God to resolve conflicts among these various points of view, there will be numerous, competing, accounts of that reality – of the ontological, metaphysical, and epistemological assumptions that we bring to the understanding of reality as well as of the moral and utility expectations that shape our appreciation of reality. As Max Horkheimer (1895–1973) aptly summarized: "To seek to salvage an unconditional meaning without God is a futile undertaking" (quoted in Habermas 2002, 95).

9.4 All Good to Me Is Lost: Prudential Rationality vs. Moral Rationality

Without God, everything is approached as if all were without ultimate meaning. Given such circumstances, the very project of morality is radically transformed. The challenges here are significant. For example, how ought one to act when the right and the good conflict? No definitive guidance is forthcoming, because, absent God, in the end there can be no final purpose for moral striving. Here, consider Immanuel Kant (1724–1804), who in the first edition of the *Critique of Pure Reason* acknowledged the significant challenge at the foundation of the project of morality if right action requires great personal sacrifice. The potential for conflict in such a state of affairs is clear. Why should one act morally when immoral action will benefit oneself and those one loves, while acting morally will harm oneself and those about whom one cares? If there are no ultimate sanctions, it is decidedly unclear why

[15] To see truly one must already have an idea of what truth is and how to find it. For example, medical students have to be taught how to "read" the slides they examine under microscopes; researchers must be initiated into the assumptions and practices that guide the experimental methodology of their fields. Each must be initiated into a particular way of relating to and of experiencing reality.

moral norms bind beyond one's own personal prudential interests.[16] Kant noted: "Without a God and without a world invisible to us now but hoped for, the glorious ideas of morality are indeed objects of approval and admiration, but not springs of purpose and action" (1964, 640, A813 = B841). Kant recognized that God, and reference to a God's-eye point of view, offers the in principle possibility of unconditioned and categorical moral claims. While it would still be necessary to access such moral truths, and to address questions regarding human understanding of transcendent knowledge, God, Himself, would function as the epistemological gold standard: knowledge of moral reality as it is in itself.

Without God there is no reason to believe that it would always be rational to act morally, rather than prudentially to pursue one's own advantage, and the advantage of those whom one loves. That is, even were it possible definitively to establish secular moral norms, absent God it is unclear why one would act in accordance with such norms, unless moral action could be independently demonstrated to coincide with one's own prudential interests. Imagine, for example, that a new and deadly version of swine flu is sweeping through the country, killing one-fifteenth of all inhabitants, and that the lab in which one works discovers a cure. Unfortunately, the lab will be unable to mass produce the cure quickly enough to make it generally available for everyone. The government plans to distribute the limited quantities of the drug according to some lottery procedure it judges to be fair and just. As a lab worker, you can either steal the drug, protecting yourself and your loved ones, or subject yourself and your family to chance in the lottery. In such cases, prudence would appear to be at odds with what are usually taken to be the impartial demands of morality. Given such circumstances, what motivation would one have to forgo prudential action for moral virtue? Knowledge that one "acted rightly" might provide cold comfort at best given the better alternatives. Consequently, it is unclear why it would necessarily be rational (or even reasonable) to act in ways that conform with a disinterested commitment to right-making conditions, especially if such commitments conflicted with a considered judgment of one's own and one's family's prudential well-being. What motivation would one have to act morally in such cases?

Or, consider the seventh commandment, forbidding adultery. What if one finds oneself in circumstances in which one is very tempted to commit adultery, will be protected against pregnancy and disease, and will certainly avoid being caught? Or, what if one's spouse approves of extramarital sexual experimentation? How should one act? Secular morality can at most conclude that adultery is the breaking of a promise to be faithful, presuming that one has made such a promise. It cannot articulate the meaning of adultery as a wrongdoing except in terms of the prudential impact of the broken promise on those persons intimately involved. If you want to

[16] There is an entire genre of post-modern entertainment devoted to exploiting this particular insight. In such category, I would place various television dramas, such as "Weeds" and "Breaking Bad," for example. Each television drama is the exploration of an individual, otherwise "good" person, who given difficult circumstances decides to deal illicit drugs. The harms to others are self-evident, as are the financial benefits to themselves and their families. Without ultimate sanctions, the prudential calculations overwhelm any moral considerations.

have a good marriage and if your spouse does not approve of adultery, then it would be prudent to avoid being caught committing adultery. Secular morality at most reflects particular contingent and prudential considerations.[17]

Kant concluded that one must affirm God's existence so as coherently to act holding that moral rationality should always trump prudential rationality. Given such practical concerns, Kant argued that the existence of God and the immortality of the soul had rationally to be affirmed as postulates of practical reason so as to secure the moral project. God as the author of morality offers the hope for harmony between the right and the good, as well as a resolution to the tension between moral rationality and prudential rationality. Reference to God offers the unification of the origins and justification of morality, an account of the sanctions for immorality, while providing motivation to be moral (see, e.g., Kant, *Critique of Practical Reason*, 1956, 130, 134; AK 125, 129–130).[18] Affirmation of God's existence is necessary to secure the rationality of always acting morally. Still, while Kant affirmed his rationalistic God as the foundation for a unique canonical morality, he also sought to ground morality in reason itself, to vindicate a canonical secular morality through moral philosophy.[19]

Hegel appreciated that Kant's moral-philosophical project could not succeed. Hegel understood that reason could not establish an ahistorical anchor for morality in being, nor could it disclose a unique and unconditioned canonical content for morality. Without particular content, morality is empty:

> Whereas we earlier emphasized that the point of view of Kant's philosophy is sublime inasmuch as it asserts the conformity of duty and reason, it must be pointed out here that this point of view is defective in that it lacks all articulation. For the proposition 'Consider whether your maxim can be asserted as a universal principle' would be all very well if we already had determinate principles concerning how to act. In other words, if we demand of a principle that it should also be able to serve as the determinant of a universal legislation, this presupposes that it already has a content; and if this content were present, it would be easy to apply the principle. But, in this case, the principle itself is not yet available, and the criterion that there should be no contradiction is non-productive – for where there is nothing, there can be no contradiction either (Hegel 1991, §135, Additions, 163).

---

[17] Elizabeth Anscombe similarly appreciated that without God there are no ultimate sanctions for immorality, and thus no final motivation to be moral, unless moral action coincides with one's prudential advantage. Without God, the concept of "immoral action" is analogous to the concept of "illegal action" in the absence of any police, courts or prisons, even in principle, to identify and punish illegal acts (Anscombe 1958, 6).

[18] Here, one might also consider Habermas's reflections on the ways in which secular morality relies on Christian theology to secure the motivation to behave morally. "A critical theory that sees itself as the 'successor' to theology finds itself in this unhappy predicament because everything to do with morality ultimately derives from theology" (Habermas 2002, 99).

[19] Kant in *The Critique of Practical Reason* (1788) contends that morality is a fact of reason: "The consciousness of this fundamental law may be called a fact of reason, since one cannot ferret it out from antecedent data of reason, such as the consciousness of freedom" (1956, AK V. 31, 31).

Philosophical reason, in and of itself, cannot provide definitive moral content. Moral content must come from some specific social, cultural, and historical perspective.[20] The price of such particular content, however, is always the loss of universality. Different and incommensurable moral perspectives are always possible.

The result, Hegel appreciated, is that moral norms are always situated within particular social, cultural, and historical circumstances. Without God, morality is no more than what humans make of it; its content is contingent and socially and historically conditioned. Moral judgments, intuitions, and dispositions express no more than particular accounts of being-for-us. Any specific ideal of human flourishing or human good is revealed as only one particular set of moral intuitions among others, and unable, as Vattimo puts it, "...without violence, to obtain as the true essence of man, of all men" (1992, 4). Hegel recognized that without God's uniquely objective understanding of reality, moral norms can never be more than particular socially and historically conditioned human creations. Cut off from God, all secular moralities – all such claims regarding the right, the good, and the virtuous – are no more than particular socio-historically conditioned narratives sustaining idiosyncratic clusters of moral intuitions and ethical inclinations. Secular reason cannot even definitively indicate why one account of the moral life is better *ceteris paribus,* more reasonable or more rationally to be preferred, over another. Consequently, secular morality (and secular bioethics) does not and cannot identify a single unified account of the right, the good, and the virtuous. As Hegel puts it, philosophy simply cannot know how the world must be.[21] Or, as Engelhardt captures a similar analytic point: bioethics is a plural noun.[22]

---

[20] Hegel criticized Kant for encapsulating duty as an empty formalism and morality as empty rhetoric: "However essential it may be to emphasize the pure and unconditional self-determination of the will as the root of duty ... to cling on to a merely moral point of view without making the transition to the concept of ethics reduces this gain to an *empty formalism*, and moral science to an empty rhetoric of *duty for duty's sake*. From this point of view, no immanent theory of duties is possible. One may indeed bring in material *from outside* and thereby arrive at *particular* duties, but it is impossible to make the transition to the determination of particular duties from the above determination of duty as *absence of contradiction*, as *formal correspondence with itself*, ... and even if such a particular content for action is taken into consideration, there is no criterion within that principle for deciding whether or not this content is a duty" (Hegel 1991, §135, 162).

[21] "As the *thought* of the world, it appears only at a time when actuality has gone through its formative process and attained its complete state. This lesson of the concept is necessarily also apparent from history, namely that it is only when actuality has reached maturity that the ideal appears opposite the real and reconstructs this real world, which it has grasped in its substance, in the shape of an intellectual realm. When philosophy paints its grey in grey, a shape of life has grown old, and it cannot be rejuvenated, but only recognized, by the grey in grey of philosophy; the owl of Minerva begins its flight only with the onset of dusk" (Hegel 1991, 23).

[22] See, for example, Engelhardt's discussion of bioethics as a plural noun in *The Foundations of Bioethics* (1996, Chapter One).

9.5 To Reign Is Worth Ambition: Secular Bioethics as Conceptual Political Ideology

> We shall be free; th' Almighty hath not built
> Here for his envy, will not drive us hence:
> Here we may reign secure, and in my choyce
> To reign is worth ambition though in Hell:
> Better to reign in Hell, than serve in Heav'n.
> (Satan, John Milton, *Paradise Lost*, Book 1, lines 259–263).

Secular modernity, Gianni Vattimo rightly recognizes, is a Babel. Because for secular culture God is dead, that culture lacks any ultimate foundation or final point of focus. "The West, indeed, is a synonym for consumerism, hedonism, a Babel-like pluralism of cultures, loss of center, and obliviousness to any reference to 'natural' law" (Vattimo 2002, 70). Vattimo recognizes that without God there is no in-principle standpoint from which to provide a unique canonical perspective on morality. As a result, all attempts politically to provide such a perspective through the state are inherently coercive:

> According to Popper, the enemies of the open society are all those theorists, starting with the philosophers in Plato's *Republic* who have emerged from the cave in which ordinary people dwell and have contemplated directly the eternal ideas of things (the truth of Being, not just its shadows). These Platonic philosophers have the right and the duty to go back down into the world and lead their fellows, or compel them if necessary, to recognize the truth (Vattimo 2011, 1).

As I have argued, however, philosophy fails to provide secure foundations for particular claims about welfare rights and health care policy, social justice and human dignity. Consequently, as Engelhardt concludes, all such moralistic assertions "… have become at best covert political slogans masquerading as moral truths" (Engelhardt 2011, 140). The cluster of intuitions that guide contemporary secular bioethics are only those judgments and proclivities that the current prevailing secular culture assumes to be reasonable and rational.

The mark of being cut off from God, and His perfect knowledge of reality, is that one is also cut off from being-in-itself. Without God, all that one can know with certainty is the experiential phenomenological world. However, even the phenomenological world is fragmented into numerous, diverse phenomenologically experienced life-worlds as well as personal accounts of reality, each of which is necessarily socially, culturally, and historically conditioned. One is left without a definitive way to choose among potentially radically divergent accounts of reality. Cut off from God, and a God's-eye perspective, one is left floating between reality (being-in-itself) and the living subjective phenomenological experience of that reality (being-for-me). As noted, absent access to a perfectly positioned, fully objective God's-eye perspective, it is not even possible coherently to speak of objective scientific truth as correspondence with reality – as correspondence with the thing-in-itself. Claims to moral truth run aground in much the same fashion. We encounter numerous moral points of view; worldviews; moral narratives; incommensurable theories and normative accounts; claims to justice, virtue, and human good; each of which is socio-historically conditioned, all without access to a unique, canonical standard on the basis of which to choose amongst the various possibilities.

Despite the lack of foundations, ideologically driven secular worldviews may falsely claim to be able to establish how one ought in secular terms to act *sub specie aeternitatis*. The proper compass of public policy – indeed, the normative force of secular worldviews (e.g., human rights discourse and social justice[23], particular secular philosophers [e.g., John Rawls 1999], and claims regarding the "common morality" [e.g. Beauchamp and Childress[24]]) – often rhetorically give the impression that their views bind moral agents as such. However, though such robust claims regarding the capacities of reason lie at the foundations of Western morality from the Scholasticism of the later thirteenth century to the moral philosophy of Immanuel Kant, they are unjustifiable in general secular terms. Where the mark of a traditional religious moral understanding is a discourse regarding one's obligations to God, including commands regarding duties to one's neighbor, secular morality attempts to secure a discourse of human rights, human dignity, and social justice, a project which has failed (Engelhardt 2000).

It was Hegel's recognition of this failure, Engelhardt argues, that drove him to substitute the state (politics) for God.

> It is through the state, through politics, that one morality is established at law and finds enforcement through state power. State power rather than God realizes as far as possible in a particular society at a particular time a harmony between the right and the good, as well as the realization of an appropriate relationship between worthiness of happiness and happiness (i.e., the state punishes and rewards). As Hegel understands, the state is "the march of God in the world [*Er ist der Gang Gottes in der Weld, dass der Staat ist*]"; the state is the "actual God [*der wirkliche Gott*]" (Hegel 1991, 279, Sect. 258, Zusatz). Hegel also recognized that without a transcendent God secular morality cannot be a canonical or a common morality but only that dominant morality established through law and public policy in a particular jurisdiction (Engelhardt 2011, 140).

The political has replaced the moral; the state has been substituted for God. This is why secular bioethics routinely functions at the level of conceptual ideology.

Moreover, as Engelhardt has documented, there is a background functional assumption that bioethics as a conceptual ideology will politically support a progressive liberal social-democratic agenda.[25] The role of secular bioethics has

---

[23] "'Human rights' are those moral rights that all people possess. International human rights documents like the Universal Declaration of Human Rights, reflect the efforts of the international community to make the protection of these rights binding under international law. As the rights of all persons as such, human rights are importantly egalitarian. …human rights are universal rights, in the sense that they 'are held by people simply as people'" (Hessler and Buchanan 2009, 204).

[24] "The common morality is the set of norms shared by all persons committed to morality. The common morality is not merely *a* morality, in contrast to other moralities. The common morality is applicable to all persons in all places, and we rightly judge all human conduct by its standards" (Beauchamp and Childress 2009, 3).

[25] "There was a deep consanguinity between the cultural and educational movements and a political movement to secure at law patient rights and to forward what was generally a social-democratic political vision. It was no accident that Sargent Shriver and the Kennedy family found it quite

come to be appreciated in terms of its ability to provide rationalizations to support the political imposition of a particular ideological point of view: "Under the cover of supporting a particular set of moral and bioethical commitments, they function as partisans for a particular legal and public policy vision, as supporters of a particular political vision" (Engelhardt 2011, 144). Given the limits of philosophy and the deep plurality of secular bioethics, in the absence of an authoritative appeal to God, however, secular bioethics as bio-politics simply assumes its own particular moral vision. To rephrase John Milton's particular idiom: secular bioethicists see themselves as standing in the place of God, with the authority to declare the right, the good, and the virtuous for themselves.[26] Such advocates wish to define morality as well as to impose their very particular moral vision through state force, to "…lead their fellows, or compel them if necessary, to recognize the truth …" (Vattimo 2011, 1). Insofar as Engelhardt's diagnosis of our post-modern moral and existential context is correct, however, much will have to be rethought regarding the limits of secular bioethics and its appropriate role in the crafting of public health care policy.

References

Anscombe, E. 1958. Modern moral philosophy. *Philosophy* 33(124): 1–19.

Beauchamp, T., and J. Childress. 2009. *Principles of bioethics*, 6th ed. New York: Oxford University Press.

Engelhardt Jr., H.T. 1980. Philosophy of medicine. In *A guide to the culture of science, technology, and medicine*, ed. P.T. Durbin, 364–461. New York: The Free Press.

Engelhardt Jr., H.T. 1996. *The foundations of bioethics*, 2nd ed. New York: Oxford University Press.

Engelhardt Jr., H.T. 2000. *The foundations of Christian bioethics*. Lisse: Swetz & Zeitlinger.

Engelhardt Jr., H.T. 2009. Credentialing strategically ambiguous and heterogeneous social skills: The emperor without clothes. *HEC Forum* 21(3): 293–306.

Engelhardt Jr., H.T. 2010a. Kant, Hegel, and Habermas: Reflections on "Glauben und Wissen,". *The Review of Metaphysics* 63: 871–903.

appropriate to support the Kennedy Institute of Ethics, along with its Center for Bioethics. The language of rights to health care and concerns with equality in health care had important resonances with Senator Ted Kennedy's long time support of a substantial restructuring of American health care. …this ideological orientation expressed itself in tensions manifest in reactions against the President's Council on Bioethics chaired by Leon Kass for President George W. Bush. The Council had a quite different ideological framework from that underlying Bill Clinton's National Bioethics Advisory Commission. There had been a general silent background assumption that bioethics would support social-democratic agendas. The contrast between the President's Council on Bioethics and the National Bioethics Advisory Commission made salient the extent to which bioethics is also bio-politics" (Engelhardt 2009, 296–297).

[26] In his recent interview in *La Repubblica*, Pope Francis I appears to situate himself within this category as a post-modernist: "Each of us has a vision of good and of evil. We have to encourage people to move towards what they think is Good. … And I repeat it here. Everyone has his own idea of good and evil and must choose to follow the good and fight evil as he conceives them." Francis I, interview with Eugenio Scalfari. "The Pope: how the Church will change", *La Repubblica*, October 01, 2013. http://www.repubblica.it/cultura/2013/10/01/news/pope_s_conversation_with_scalfari_english-67643118/

Engelhardt Jr., H.T. 2010b. Moral obligation after the death of God: Critical reflections on concerns from Immanuel Kant, G.W.F. Hegel and Elizabeth Anscombe. *Social Philosophy and Policy* 27(2): 318–340.

Engelhardt Jr., H.T. 2011. Core competencies for health care ethics consultants: In search of professional status in a post-modern world. *HEC Forum* 23(3): 129–146.

Engelhardt Jr., H.T., and A.L. Caplan (eds.). 1987. *Scientific controversies: Case studies in the resolution and closure of disputes in science and technology*. Cambridge: Cambridge University Press.

Fleck, L. 1979. In *The genesis and development of a scientific fact*, ed. T.J. Trenn and R.K. Merton. Chicago: University of Chicago Press.

Francis I, Pope. 2013. Interview with Eugenio Scalfari. "The Pope: How the church will change". *La Repubblica*, October 01.

Goodman, N. 1978. *Ways of worldmaking*. Indianapolis: Hackett Publishing Company.

Guarino, T.G. 2009. *Vattimo and Theology*. New York: T&T Clark.

Habermas, J. 2002. In *Religion and rationality: Essays on reason, God, and modernity*, ed. E. Mendieta. Cambridge: The MIT Press.

Hegel, G.W.F. 1892. *The logic of Hegel*, 2nd ed, Trans. W. Wallace. Oxford: Clarendon Press.

Hegel, G.W.F. 1968a. Glauben und Wissen. In *Gesammelte Werke*, vol. 4: *Jenaer kritische Schriften*. Hamburg: Meiner.

Hegel, G.W.F. 1968b. Jenaer kritische Schriften. In *Gesammelte Werke*, vol. 4. Hamberg: Meiner.

Hegel, G.W.F. 1977. *Faith & knowledge*. Trans. W. Cerf and H.S. Harris. Albany: State University of New York Press.

Hegel, G.W.F. 1984. *Das Leben Jesu* ("The life of Jesus" in *Three Essays*, 1793–1795), ed. and Trans. P. Fuss and J. Dobbins. Notre Dame: University of Notre Dame Press.

Hegel, G.W.F. 1991. *Elements of the philosophy of right*, ed. A.W. Wood, Trans. H.B. Nisbet. Cambridge: Cambridge University Press.

Hessler, K., and A. Buchanan. 2009. Specifying the content of the human right to health care. In *Justice & health care: Selected essays*, ed. A. Buchanan. New York: Oxford University Press.

Hume, D. 1896. *A treatise of human nature by David Hume*, Reprinted from the original edition in three volumes, ed. Selby-Bigge, L.A. Oxford: Clarendon Press. This material originally appeared in the Portable Library of Liberty published by Liberty Fund, Inc. http://oll.libertyfund.org/title/222 and PLL v5 (generated 22 January, 2010).

Hume, D. 1910. *An enquiry concerning human understanding*, Harvard Classics volume 37. New York: P.F. Collier & Son. On-line edition: 18th.eserver.org/hume-enquiry.html#10.

Kant, I. 1956 [1788]. *The critique of practical reason*. Trans. L.W. Beck. Indianapolis: Library of Liberal Arts.

Kant, I. 1964 [1781/1787]. *The critique of pure reason*. Trans. N.K. Smith. New York: St. Martin's Press.

Kuehn, M. 2001. *Kant: A biography*. Cambridge: Cambridge University Press.

Kuhn, T.S. 1962. *The structure of scientific revolutions*. Chicago: University of Chicago Press.

McCarthy, M.H. 1990. *The crisis of philosophy*. Albany: State University of New York Press.

Milton, J. 1674. Paradise lost. In *The Milton reading room*, ed. H.T. Luxon http://www.dartmouth.edu/~milton, October, 2013.

Nietzsche, F. 1910. *The joyful wisdom (La Gaya Scienza)*. In *The Complete Works of Friedrich Nietzsche*, vol 10, ed. O. Levy. Trans. T. Common. London: T.N. Foulis.

Nietzsche, F. 1968. *The will to power*. Trans. W. Kaufmann and R.J. Hollingdale. New York: Vintage Books.

Rawls, J. 1999. *A theory of justice*, rev ed. Cambridge: Belknap Press.

Vattimo, G. 1988. *The end of modernity*. Trans. J.R. Synder. Baltimore: The Johns Hopkins University Press.

Vattimo, G. 1992. *The transparent society*. Trans. D. Webb. Baltimore: The Johns Hopkins University Press.
Vattimo, G. 2002. *After Christianity*. Trans. L. D'Isanto. New York: Columbia University Press.
Vattimo, G. 2007. Toward a nonreligious Christianity. In *After the death of God*, ed. J.D. Caputo, G. Vattimo, and J.W. Robbins. New York: Columbia University Press.
Vattimo, G. 2011. *A farewell to truth*. Trans. W. McCuaig. New York: Columbia University Press.
Zeller, E. 1980. *Outlines of the history of Greek philosophy*, 13 ed. Revised by W. Nestle, Trans. L.R. Palmer. New York: Dover Publications, Inc.

Chapter 10
The Ethical Conduct of Research: The Legacy of the Three Principles

Ana Smith Iltis

> *Engelhardt has been the enfant terrible of bioethics: irrepressible, irreverent, unpredictable, but ever insightful and brilliant*
>
> (Jonsen 1998, 82).

10.1 Ethical Principles and the Principles of Research Ethics

> Ethics, as a philosophical enterprise, is best conceived as an attempt to negotiate diverse moral intuitions. Ethics is the logic of a pluralism in the sense that ethics is an attempt to find the most general grounds or bases for judging the rightness and wrongness of conduct. Unlike religious ethics, or particular legal traditions, philosophical ethics hopes for general principles of conduct discoverable by disinterested reflection, apart from either grace or cultural prejudice. Though such a disinterested perspective cannot be attained, one can move towards such a vantage point by attempting to lay out ever more clearly general principles of moral conduct (Engelhardt 1978, 1).

H. Tristram Engelhardt, Jr. wrote those words in 1978 as part of a paper commissioned by the National Commission for the Protection of Human Subjects of Biomedical and Behavioral Research (hereafter the National Commission) to help it accomplish its congressionally mandated task of identifying the principles for the ethical conduct of human research. Engelhardt spent much of his career defining the "general principles of moral conduct" in bioethics, not only in human research. In the face of irreducible moral pluralism and given our inability to secure a truly

A.S. Iltis, Ph.D. (✉)
Department of Philosophy, Center for Bioethics, Health and Society, Wake Forest University, P.O. Box 7332, Winston-Salem, NC 27109, USA
e-mail: Iltisas@wfu.edu

L.M. Rasmussen et al. (eds.), *At the Foundations of Bioethics and Biopolitics: Critical Essays on the Thought of H. Tristram Engelhardt, Jr.*, Philosophy and Medicine 125, DOI 10.1007/978-3-319-18965-9_10

"disinterested perspective," what general bioethics principles might obtain? This project culminated in the second edition of the *Foundations of Bioethics* (1996), where Engelhardt proposes two general principles for bioethics:

> The Principle of Permission: Authority for actions involving others in a secular pluralist society is derived from their permission. As a consequence,
>
> i. Without such permission or consent there is no authority.
> ii. Actions against such authority are blameworthy in the sense of placing a violator outside the moral community in general, and making licit (but not obligatory) retaliatory, defensive, or punitive force....(122)
>
> The Principle of Beneficence. The goal of moral action is the achievement of goods and the avoidance of harms. In a secular pluralist society, however, no particular account or ordering of goods and harms can be established as canonical. As a result, within the bounds of respecting autonomy, no particular content-full moral vision can be established over competing senses (at least within a peaceable secular pluralist society). Still, a commitment to beneficence characterizes the undertaking of morality, because without a commitment to beneficence the moral life has no content...(123).

Engelhardt also dedicated much of his career to articulating the substantive moral commitments of Christians regarding issues in bioethics. Although this remains a focal point of his career today, the *Foundations of Christian Bioethics* (2000) was a major contribution to this discussion. This essay examines (1) the early principles Engehardt proposed to judge "the rightness or wrongness of conduct in the area of human research" in a pluralistic society and (2) his discussion of the relationship among those three principles to (3) argue that contemporary research ethics suffers from the failure to address clearly what the relationship is among the principles for the ethical conduct of research on human subjects set forth in the *Belmont Report.*

The principles for the ethical conduct of human research that appear in the *Belmont Report* grew out of the work of Engelhardt and others, most notably Tom Beauchamp (Jonsen 1998, 103), as well as discussions among the commissioners and their reading of other commissioned papers. The *Belmont Report* offers the following principles:

1. "Respect for Persons. – Respect for persons incorporates at least two ethical convictions: first, that individuals should be treated as autonomous agents, and second, that persons with diminished autonomy are entitled to protection. The principle of respect for persons thus divides into two separate moral requirements: the requirement to acknowledge autonomy and the requirement to protect those with diminished autonomy...."
2. "Beneficence. – Persons are treated in an ethical manner not only by respecting their decisions and protecting them from harm, but also by making efforts to secure their well-being. Such treatment falls under the principle of beneficence. The term 'beneficence' is often understood to cover acts of kindness or charity that go beyond strict obligation. In this document, beneficence is understood in a stronger sense, as an obligation. Two general rules have been formulated as complementary expressions of beneficent actions in this sense: (1) do not harm and (2) maximize possible benefits and minimize possible harms."

3. "Justice. – Who ought to receive the benefits of research and bear its burdens? This is a question of justice, in the sense of 'fairness in distribution' or 'what is deserved.' An injustice occurs when some benefit to which a person is entitled is denied without good reason or when some burden is imposed unduly. Another way of conceiving the principle of justice is that equals ought to be treated equally. However, this statement requires explication. Who is equal and who is unequal? What considerations justify departure from equal distribution? … It is necessary, then, to explain in what respects people should be treated equally. …"

> [C]onceptions of justice are relevant to research involving human subjects. For example, the selection of research subjects needs to be scrutinized in order to determine whether some classes (e.g., welfare patients, particular racial and ethnic minorities, or persons confined to institutions) are being systematically selected simply because of their easy availability, their compromised position, or their manipulability, rather than for reasons directly related to the problem being studied. Finally, whenever research supported by public funds leads to the development of therapeutic devices and procedures, justice demands both that these not provide advantages only to those who can afford them and that such research should not unduly involve persons from groups unlikely to be among the beneficiaries of subsequent applications of the research. (National Commission 1978)

The relationship among these three principles is unclear, and debate about this relationship has been ignored to a large extent. Yet the relationship among the principles is of paramount importance for understanding and evaluating numerous questions in research ethics. The contemporary research ethics literature ought to recognize the importance of understanding the relationship among the principles and return to this debate.

10.2 The National Commission for the Protection of Human Subjects of Biomedical and Behavioral Research

In the paper Engelhardt wrote for the National Commission in 1978, he proposed three principles for the ethical conduct of research:

> (A) One should respect human subjects as free agents out of a duty to such subjects to acknowledge their right to respect as free agents.
> (B) One should foster the best interest of individual human subjects.
> (C) One should have concern to maximize the benefits accruable to society from research involving human subjects, taking into particular regard interest in values such as (1) the amelioration of the human condition through advances in the biomedical and behavioral sciences and technologies; (2) preservation of human autonomy as a general value; (3) increase in knowledge apart from any consideration of its application to the amelioration of human condition; (4) the personal satisfaction of human subjects derived from their feeling of having contributed to the common good or to the advancement of human knowledge by participation in research. (Engelhardt 1978, pp. 5–6)[1]

[1]The first principle laid out here became more and more the focus of Engelhardt's secular philosophical project. The need to start off with a basic sense of respect for the status of another person as a free agent and to respect that agency applies not only to human subjects but to all persons. If we do not recognize persons as free agents and respect them as such, we have stepped outside the

As Jonsen (1998, 130) notes, the National Commission "took Engelhardt's first two principles and Beauchamp's principle of distribute justice and crafted 'crisp' principles: respect for persons, beneficence, and justice."[2] Although the Belmont principles are not identical to Engelhardt's principles, Engelhardt had a profound effect on research ethics and his principles bear significant resemblance to Belmont principles. This resemblance, and their historical connection, make it instructive to consider Engelhardt's discussion of the relationship among his principles in trying to understand the relationship among the Belmont principles.

Engelhardt's first principle reflected a basic premise of what it is to engage in the work of ethics and to even discuss questions of moral responsibility: it "concerns respect for persons as a logical condition for morality. Such respect for persons is not a value among other values. It is rather the basis for our sense of moral responsibility, and it is considered apart from any interest we might have in respecting other persons" (Engelhardt 1978, 5). One of the conditions for the possibility of talking about morality is that we must assume persons are free agents and treat them as such. Otherwise, it is meaningless to examine moral responsibility. The second and third principles invoke values and goals, but only in a generic fashion. He distinguishes the first principle from the other two. The first "is a deontological one in the sense of focusing on a consideration of rights and duties independently of any issue of goods and values. It is an appeal to respect for the freedom of persons whether or not such respect would in the long run contribute to the benefit of society. With regard to this principle, experimentation upon unwilling human subjects should be regarded as immoral, even if the results of such experimentation would be of considerable general quality" (6). The other two principles are teleological and concern goods to be achieved. Engelhardt demonstrates that these principles yield four major types of procedural maxims: "(1) obtaining the free and informed consent of the human subject involved in the research under consideration; (2) obtaining a proxy consent from individuals unable to consent in order to protect the best interests of those subjects; (3) avoiding coercion which would unduly direct by threat the consent of a human subject…; (4) one ought to weigh the benefits versus the risk involved in the use of human subjects in order to be assured that the costs of such research will not outweigh its possible benefits" (1978, 13–14).

Engelhardt addresses directly the relationship among the three principles. He says:

> Since we both respect persons and have interest in goods and values, we must allocate scarce resources as best we can to forward our commonly embraced goals within the bounds of never treating other persons as means merely. One may never treat others as means *merely*. But, one may, within the bounds of moral probity, treat others as means. Informed consent of the other one is the difference between using another as a means, and using another as a means merely. Which is to say, ethical principles (B) and (C), bearing on the best interests of the individual subjects and the best interests of society, find their locus

---

project of philosophical ethics and into something else. That something else might be a project of religious ethics, but it might also be a project of ethics grounded in particular, rather than diverse, moral intuitions.

[2] Jonsen is referring to the principle Beauchamp discussed in Beauchamp 1978.

> within the bounds of respect for human subjects as free agents (A). In the end, what distinguishes the moral problem of experimentation with human subjects from the moral problems of experimentation with subjects of other species is that only persons can make absolute claims to respect (1978, 38).

The first principle (A) must be satisfied before we can move to promote particular goals, such as the best interests of an individual or society. The three principles are not equal, unranked principles. Tradeoffs in respect for persons are not permitted, even in the name of social goals or even of doing what we believe is in the person's interest, because the first principle is a necessary condition for ethics. This view of the relationship among the three principles is very different from one that holds that the three principles are three equally important, unranked principles to be balanced against one another.

No doubt when the *Belmont Report* was written, the commissioners were aware of possible conflict among the principles, but it seems that they had different views of how to manage such conflict. The debate about how the National Commission understood, and how we should understand, the relationship among the principles and the question of whether it is permissible to violate some principles to fulfill other principles in the context of human research continues (see Ackerman 1996; Jonsen 1996; Veatch 1996, 2005). There are three main possibilities. First, the principles are equal, independent principles, all of which must be fulfilled for research to be ethical. Research that requires us to compromise on one principle to fulfill another principle is impermissible. Veatch has called this the simultaneity approach (2005). The second approach is that they are ranked or ordered principles. The first principle must be fulfilled before we may consider the second or third. Engelhardt held that the first principle of respect for persons was a necessary condition for the ethical conduct of research and that concerns about achieving the good could not justify failing to respect persons as free and autonomous agents. Veatch describes the ordering approach this way: "[t]he morally superior course is the one that moves furthest down the list without violating any higher-ranked principles" (2005, 185). This approach is similar in many ways to the simultaneity approach in that it does not permit tradeoffs of respect for persons for the sake of potential benefit to subjects or society or for the sake of justice. The third possibility is that the three principles are equal, unranked principles and that we should balance them when they conflict or compete with each other. Engelhardt rejected this approach in his paper for the National Commission (1978).

Many people treat the principles as *prima facie* duties that can be balanced to allow for violations of some duties in the name of others. Albert Jonsen and Terrence Ackerman defend different accounts of balancing. Ackerman argues that researchers have ethical obligations to research subjects and to society (to promote general welfare). These obligations "represent competing moral demands that must be balanced against one another" (Ackerman 1996, 83). Ackerman argues that the National Commission did not explicitly articulate and apply a weighting and balancing strategy, but such an approach is implicit in the National Commission's work and in the federal research regulations (88). Evidence of such a strategy includes the decision to allow for the alteration of consent procedures and waivers of informed consent (89–90). Ackerman concludes that "while pursuit of the general welfare should partly constrain satisfaction of the duties to subjects, substantial compromises in subject-related protections should not be tolerated. Compromises in these protections should be permitted only when they are necessary for the conduct of research that may generate knowledge useful in removing or preventing important harms to others. Moreover, these compromises must involve no more than minor infringements on the requirements of duties to subjects" (99). He says the reports and federal regulation call for balancing: "the reports of the National Commission and the federal regulations accept the proposition that there are no moral interests so important that their satisfaction may not be modestly compromised when counterbalanced by tremendous gains along other moral dimensions" (101).

Jonsen (1996) also argues that the National Commission saw the principles as equal, unranked principles that could be balanced against each in cases of conflict. He argues that the National Commission implies that conflict should be solved by making moral judgment after assessing the principles, values, circumstances, and consequences. Our social and cultural traditions, systems of moral reflection, and the particular circumstances of each situation give weight to the principles. The principles of respect for persons, beneficence, and justice are absolute in one sense, namely that they are always relevant. But they are not absolute in that they admit of exceptions. For example, "the weighty principle of autonomy dominates all our deliberations about the ethics of human research because both our cultural heritage and our critical reflection refuse to sacrifice individuals to the common welfare. Yet, we can appreciate situations in which quite specific harms to many may force us to contemplate exposing some to risks so that all might be saved. Exceptions are painful and specific, and the reasons that justify them never ought to be generalized nor routinized" (1996, 79–80). Difficult cases require that virtuous persons deliberate and make judgments in the face of conflicts among the principles (81). To make judgments, they must assess all the relevant circumstances, values, and consequences that apply in a particular decision. Despite claims of balancing, Jonsen also suggests that respect for persons is "the most significant of the principles" (2004, 10) and was especially important for the National Commission because "in the early literature about research ethics, there is a very strong attempt to repudiate the kind of crude social utilitarianism that seemed? – that was frequently used to justify research, that deprived human subjects of some of their rights" (*sic*, p. 10). It has a "pride of place that's unquestionable" (10), though he maintains that it may be balanced against other principles to resolve particular conflicts (11).

Despite support for balancing, we saw in Engelhardt's paper a different account of the relationship. To jettison the first principle for the sake of securing the overall good of society or the good of individuals is to violate the basic boundaries of ethical conduct. Veatch defends a combination of the simultaneity and ranked order views to reject the balancing view: "[I]n one sense it seems obvious that the principle of utility (the National Commission's 'beneficence') must be satisfied. No amount of respect for autonomy and justice in a protocol would justify research that was expected to do more harm than good. Thus, whatever we say about simultaneity and the other principles, this formulation of utility must be satisfied. On the other hand, if the principle of utility is formulated, as it sometimes is in utilitarian theories, to mean that, for research design to be moral, it must maximize utility, then almost no protocol that concerns itself with autonomy or justice would satisfy utility" (2005, 185). He notes that the *Belmont Report* and the Common Rule enumerate requirements for the ethical conduct of human research and, he claims, "The most reasonable reading of that requirement is that they all must be satisfied before research is ethically acceptable. Hence, the report on children permits only minimal risk or minor increase over minimal risk if specified conditions are met. The fact that only limited risk is tolerable regardless of the potential benefit, supports the view that the National Commission was not willing to go the full way to balance beneficence and respect for subjects…" (Veatch 1996, 52–3). Veatch is especially concerned with claims that the desire to secure social benefits as urged by the principle of beneficence may trump the obligation to respect persons. Veatch suggests that in very rare circumstances, justice claims might require or allow us to constrain autonomy, but beneficence-based claims may not constrain autonomy/respect for persons (53–4). Veatch recognizes that the National Commission might have understood the principles as *prima facie* duties that may be overridden at times and that some commissioners clearly favored balancing, but trading off one against the other is not allowed (2005, 195). He argues that even where it appears that the National Commission allows compromises on respect for persons (e.g., waivers of consent), subjects retain some rights that may not be offset by others' interests (196). A further limit on support for the balancing claim is the observation that the National commission did not provide a means for compromising on justice (196). They did not say, for example, that if one could secure great benefits for society, one should recruit institutionalized persons who could be enrolled and controlled easily in a study. He demonstrates that the National Commission and work that relies on National Commission documents implies that some balancing is permissible when principles conflict, but the limits on balancing "make no sense unless respect for persons and justice were independent necessary conditions for research to be ethical, conditions that could not be offset by balancing them against beneficence" (197). The Common Rule, Veatch argues, also indicates that the ethical conduct of research requires fulfillment of all three principles: "'In order to approve research covered by these regulations the IRB shall determine that *all* of the following requirements are satisfied' [45CFR46.111]. There follows a list of the famous seven criteria, including those that seem clearly grounded in the Belmont principles of respect for persons and justice" (196–7). These limits on trade-offs and the Common Rule's list of requirements together, Veatch argues, suggest that we should not approach the principles as three independent *prima facie* duties that may be balanced against one another.

The Belmont Report and the work that relies on it and its principles are ambivalent about the relationship among the principles. This ambiguity in *The Belmont Report* might have been strategic. In the face of conflict and given the need to achieve consensus and deliver a report to Congress, silence and ambiguity might have been important tools. Nevertheless, such ambiguity has serious implications today.

10.3 The Importance of the Relationship Among the Three Principles

Much is at stake. How we understand the relationship among the principles and the proper response to conflict among them has significant implications for evaluating research protocols. The possibility of conflict is real, and how one believes such conflict should be resolved may change what one sees as permissible research. For example, imagine a study to determine whether people with HIV have higher mortality rates in the ICU. Such a study may be very important in a setting in which ICU beds are scarce and it is assumed that HIV-positive individuals are not as likely to benefit from intensive care as others. If persons with HIV are more likely to die, then they may have lower priority than others in allocating ICU resources. One way to evaluate this is to test all persons admitted to the ICU over a period of time and compare mortality rates.[3]

If informed consent to test for HIV and record the results for research purposes is required, some persons may refuse to participate, decreasing the value and validity of the study. There may also be concerns that asking surrogates for permission, since many people admitted to the ICU might be unable to give informed consent, would put patients at risk. This study reflects a conflict between the principle of respect for persons and the principle of beneficence. Is it permissible to grant researchers a waiver of informed consent even for those persons who are capable of giving informed consent upon admission to the ICU so that they may test all persons and record their results for research and then use their medical records to compare mortality rates? If respect for persons is a side-constraint on research, we will conclude that at least those persons who are capable of giving informed consent have the right to be respected as free and autonomous persons and should be asked whether they are willing to be in the study.[4]

If we hold that the principles may be balanced when they are in conflict, then one might, depending on how one chooses to balance them, allow such a study to proceed without informed consent. If one adopts a balancing approach, numerous outcomes are possible because balancing requires that we assign value, significance, or weight to different interests, rights, and goals. When does the prospect of significant benefit to society or subjects or populations within a society justify violating the principle of respect for persons or the principle of justice? When is increased risk justified by a proportionate increase in benefit to others? When are we justified in decreasing risks to subjects even though there is a proportionate (or disproportionate) decrease in anticipated benefits? When should we insist on respecting individual autonomous agents even though doing so decreases the value or validity

[3] This example is based on a study reported in the *British Medical Journal* by Bhagwanjee et al. (1997).

[4] A separate issue concerns the treatment of individuals who are not free and responsible agents, such as children. That is not the focus here.

of a study? Balancing allows one to make these tradeoffs under some circumstances, but judging the circumstances under which tradeoffs are permitted requires additional moral analysis because it requires assigning value and significance to various rights, interests, and outcomes.

Given all that is at stake in determining the relationship among the principles for the ethical conduct of research and the possibility of conflict among them, and given the evidence of different accounts, we have been mistaken in not addressing this issue more explicitly in the research ethics literature. When Engelhardt proposed three principles for the ethical conduct of human research and the National Commission adopted a version of his first two principles, he articulated clearly the relationship among the principles – the first principle was essential and limited the range of research that could be pursued. Our research regulations have moved away from treating the first principle as a side constraint, allowing research without consent even on persons who have decision-making capacity under certain circumstances. We learn from Engelhardt that, in a morally pluralistic society, the only grounds for authority among moral strangers is permission. This is not to say one ought to *value* permission, he argues, but rather that this thin morality that binds moral strangers is a default position; it is all that is available to us in the face of deep, irresolvable moral differences (Engelhardt 1996). This is why he holds the first principle for the ethical conduct of research to be a necessary side constraint. There appears in the Common Rule and in its application to be a de facto decision to balance the principles and allow for violations of the principle of respect for persons in the name of other goals (Veatch 2005, p. 197). But there has not been an explicit justification of this balancing approach, only arguments as to why in some cases it is good overall to pursue such balancing. That justification presumes that pursuit of the overall good may trump individual rights that others see as side constraints on permissible interaction. Insofar as balancing respect for persons, which recognizes the authority of individuals over themselves, against other principles, which invoke conceptions of the good and of justice, requires that we have a particular conception of the good and assign values to different goods, rights, and interests, we face additional concerns. We live in a pluralistic society in which we find that people sometimes have radically different conceptions of the good and will assign different values to the same interests, rights, and goals. In light of these differences, we should encourage a more robust discussion of the relationship among the principles for the ethical conduct of research. The role of the first principle as a side constraint may be very important in a pluralistic society such as ours, given that balancing requires that we invoke particular conceptions of the good and the right and impose them on others. To have borrowed two of Engelhardt's three principles without recognizing the importance of his arguments regarding the relationship among the principles has left us with a significant gap in our understanding of the ethical conduct of human research.

References

Ackerman, T.F. 1996. Choosing between Nuremberg and the National Commission: The balancing of moral principles in clinical research. In *The ethics of research involving human subjects: Facing the 21st century*, ed. H.Y. Vanderpool, 83–104. Frederick: University Publishing Group.

Beauchamp, Tom. 1978. Distribute justice and morally relevant differences. In *National Commission for the Protection of Human Subjects of Biomedical and Behavioral Research, The Belmont Report,* Appendix 1, 6.1–6.20.

Bhagwangjee, Satish, David Muckart, Prakash Jeena, and Prushini Moodley. 1997. Does HIV status influence the outcome of patients admitted to a surgical intensive care unit? A prospective double blind study. *British Medical Journal* 314: 1077–1084.

Engelhardt Jr., H. Tristram. 1978. Basic ethical principles in the conduct of biomedical and behavioral research involving human subjects. In: *National Commission for the Protection of Human Subjects of Biomedical and Behavioral Research, The Belmont Report,* Appendix II, 8.1–8.45.

Engelhardt Jr., H. Tristram. 1996. *The foundations of bioethics*. New York: Oxford University Press.

Engelhardt Jr., H. Tristram. 2000. *The foundations of Christian bioethics*. Beverly: M&M Scrivener Press.

Jonsen, Albert. 1996. The weight and weighing of ethical principles. In *The ethics of research involving human subjects: Facing the 21st century*, ed. Harold Vanderpool, 59–82. Frederick: University Publishing Group.

Jonsen, Albert. 1998. *The birth of bioethics*. New York: Oxford University Press.

Jonsen, Albert. 2004. Interview with Duane Alexander, *Oral history of the Belmont Report and the National Commission for the Protection of Human Subjects of Biomedical and Behavioral Research.* www.hhs.gov/ohrp/archive/docs/InterviewAlexander.doc. Accessed 23 Apr 2013.

National Commission for the Protection of Human Subjects of Biomedical and Behavior Research. 1978. The Belmont Report.

Veatch, Robert M. 1996. From Nuremberg through the 1990s: The priority of autonomy. In *The ethics of research involving human subjects: Facing the 21st century*, ed. Harold Vanderpool, 45–58. Frederick: University Publishing Group.

Veatch, Robert M. 2005. Ranking, balancing, or simultaneity: Resolving conflicts among the *Belmont* principles. In *Belmont revisited*, ed. James F. Childress, Eric M. Meslin, and Harold T. Shapiro, 184–204. Washington, DC: Georgetown University Press.

Chapter 11
Non-certain Foundations: Clinical Ethics Consultation for the Rest of Us

Lisa M. Rasmussen

11.1 Introduction

> A woman arrives at the hospital emergency department with her unconscious brother, stating that because he is a member of Jehovah's Witnesses, he should not receive any blood, even if that is the only way to save his life. Should the emergency team respect her instructions?
>
> A homeless man with no known family or other decision makers is dying at the hospital, unable to state his wishes for end of life care. Should the hospital attempt to revive him in the case of cardiac arrest, or assume that he would not want such intrusive treatment?

These scenarios represent some of the cases posing ethical challenges in clinical medicine. In the context of the United States and other countries with laws regarding medical decision-making, individuals with decision-making capacity usually have the legal right to make decisions for themselves. Mainly for this reason, most patients direct their own care (albeit often with guidance from medical care providers and input from friends and family) and need no help from a clinical ethics consultant. But in many other cases, patients lack this ability, and either the appropriate surrogate decision maker is not clear or the clear surrogate is not sure how to direct care. Even in cases where patients possess capacity, sometimes they are not sure of their own values, or are not sure of the implications of the values they *do* have, or the medical team disagrees with the values they have, and they want some help thinking about these issues.

Clinical ethics consultation has arisen partly as an aid to decision-making in these contexts. What could be the problem with a field that seeks to clarify or advise and which also lacks power to enforce particular conclusions? A good deal is wrong

L.M. Rasmussen, Ph.D. (✉)
Department of Philosophy, University of North Carolina,
9201 University City Blvd., Charlotte, NC 28223, USA
e-mail: lrasmuss@uncc.edu

L.M. Rasmussen et al. (eds.), *At the Foundations of Bioethics and Biopolitics: Critical Essays on the Thought of H. Tristram Engelhardt, Jr.*,
Philosophy and Medicine 125, DOI 10.1007/978-3-319-18965-9_11

with it, according to Engelhardt, most centrally the philosophical incoherence of any claims to the possession of moral truth. The next section considers Engelhardt's arguments against the practice of clinical ethics consultation.

11.2 Engelhardt's Account of the Sociohistorical Context of the Rise of Bioethics

Five texts form the main body of Engelhardt's work critiquing clinical ethics consultation (CEC): *The Foundations of Bioethics*, 2nd ed. (1996) provides Engelhardt's arguments regarding the justification of normative claims in general; "The Ordination of Bioethicists as Secular Moral Experts," (2002), "Credentialing Strategically Ambiguous and Heterogeneous Social Skills: The Emperor Without Clothes," (2009), "Core Competencies for Health Care Ethics Consultants: In Search of Professional Status in a Post-Modern World," (2011), and "Why Clinical Bioethics So Rarely Gives Morally Normative Guidance" (2012) together offer a context for the rise of the field and a series of arguments regarding the practice.

This context is a rich, important, and unique feature of Engelhardt's scholarship on clinical ethics consultation, and we do well to consider it. His claim is that a sociohistorical account of how clinical ethics consultation (and bioethics in general) was born and grew so rapidly explains its character and shapes how we should view the practice. Different essays accent different strands of the story, but in the main Engelhardt argues that a constellation of the secularization of society, the deprofessionalization of medicine, the "social establishment of health care" (2002, 71), and what turned out to be futile Enlightenment hopes for the success of rational arguments to establish moral truths together created a vacuum into which bioethics and clinical ethics consultation stepped or were sucked. Each of these facets constitutes a failure of traditional authority that leaves individuals without familiar sources of guidance.

With the secularization of society, the diminishment of attendance at and reliance on religious institutions meant that traditional religious leaders were no longer the assumed source of moral authority for increasing numbers of Americans. The deprofessionalization of medicine (ushered along by legal cases which abolished it as a self-constituting and -monitoring profession[1]) meant that it could not maintain its internal ethics, and instead invited a consideration of ethics external to its traditional practice. The establishment of health care as a social institution, via health insurance plans and federal funding, both undermined the traditional doctor-patient relationship in which the two held each other accountable (they don't call it "third-party payer" without reason) and created a need for an externally justified set of values with which to guide medical practice. The background Enlightenment hope, that rational discourse and argument could reveal and justify universal moral norms,

[1] See (Engelhardt 2002, 72, particularly footnote 40), in which he discusses cases that "stripped medicine of many of its guildlike self-regulatory privileges….".

Engelhardt argues is false (1996).[2] The culmination of all of these forces, Engelhardt argues, is an aching need for moral guidance.

One reading of this historical sketch is that it clearly reveals the impoverished world in which secularists live; the attendant normative conclusion is that everyone ought to seek a noumenal authority (and be sure to find the right one). That is, it reveals the barren ground of a world without metaphysical commitments. I am sure this is Engelhardt's reading, being as he is a member of the Orthodox Christian faith. But I will suggest that a second reading of this history is possible, one that merely marks a change over time from the recognition of one set of authorities to others. That reading accepts that a general challenging of traditional authorities is a feature of twentieth-century American history in particular, and that medicine has radically changed in the same period of time, both in terms of its social place and in terms of the tools it possesses for establishing its value commitments. On this interpretation of Engelhardt's historical account, however, there is no obvious normative conclusion regarding the legitimacy of clinical ethics consultation. So, even to the extent that one agrees with Engelhardt's account of this history of the rise of bioethics, it is possible to come to a different conclusion than he does about the field. Below, I describe his account of the field resulting from this history and the conclusions he argues that it yields; in the following section I offer a different interpretation based on an assessment of and alternative to Engelhardt's metaethical assumptions.

11.3 Engelhardt's Critiques of Clinical Ethics Consultation

In *The Foundations of Bioethics*, Engelhardt's main argument is that, as he summarizes it elsewhere, "Secular sound rational argument cannot resolve the important moral controversies we face, nor is there a non-controversial understanding of when consensus and by whom would be normative and for what..." (2009, 294). The implication of this conclusion is that there is no possibility for a professional practice based on giving normative guidance based merely on conclusive rational arguments. To the extent that clinical ethics consultation seems to presuppose such a possibility, then, it is illegitimate.[3]

This foundation, with the addition of arguments presented in the four articles referenced above, yields the following (much-) condensed set of claims:

---

[2] However, see McCullough (Chaps. 1 and 14, this volume), for dissent regarding Engelhardt's interpretation of "the" Enlightenment.

[3] However, a clinical ethics consultation which exists within a robust moral community consisting of "a body of men and women bound together by common moral traditions and/or practices around a shared vision of the good life" – a clinical ethics consultation among moral friends, in other words – makes no such claim to rational proof (Engelhardt 1996, 7). Engelhardt explicitly notes this possibility: "In health care institutions with a strong religious affiliation or with very particular morally concrete mission statements, ethics consultants may very well be expected to play a frankly normative moral role by giving actual normative moral direction" (2009, 300).

> Clinical ethics consultants cannot justify normative claims regarding moral truth. Because they cannot justify these normative claims, if they encourage or fail to discourage the belief that they possess moral expertise, they are engaging in dishonest behavior. Clinical ethics consultants are strategically ambiguous about their skills and are seeking licensure because that affords them both economic and reputational benefits. Honesty requires that consultants disclose to patients and families the tendentious nature of moral assertions as a way of establishing an informed consent to consultation. Moreover, because the field is not a profession (as it lacks a conceptual center), consultants are employees and should behave accordingly, including performing as an employer wishes them to rather than on the basis of an internal, independent professional ethics.

The first claim is, as mentioned, an implication of the arguments in *The Foundations of Bioethics*. If Engelhardt is right that "secular sound rational argument cannot resolve the important moral controversies we face," then clinical ethics consultants who do not possess moral authority for those they advise cannot justify the advice they give. Advice in the form of "Action X in this circumstance is morally permissible," for example, cannot be justified as universally true in a pluralistic society, because there are varying understandings of what constitutes moral permissibility.

Engelhardt also claims that consultants know (or ought to know) about this failure of rational argument to resolve moral controversies, so the fact that they allow others to believe they do in fact possess such authority is dishonest. The solution to the dishonesty would be to seek an informed consent to the service:

> Candor, indeed informed consent about the nature of their ethics, would require ethics consultants at least to acknowledge the problem of specifying which ethics it is about which health care ethics consultants should have the competence to consult, as well as why many hold that the project of securing a foundation for that ethics has failed. Absent such a disclosure, ethics consultants may mislead as to what the ethics of ethics consultation is about (2011, 133).

However, he argues, this dishonesty serves practitioners well: "It would be against the financial and professional standing of many such ethicists to acknowledge this difficulty confronting applied ethics, for it would open up the recognition that secular clinical ethicists belong to different secular moral sects" (2009, 301). The dilemma Engelhardt poses for the field here is that consultants should either recognize that they are dishonest and rectify that fact, or be assumed to be intentionally engaging in dishonesty about the landscape of morality so as to benefit from others' relative unawareness of that fact.

There is no option of an internal professional morality to guide the practice either, Engelhardt argues:

> Ethicists do not have a professional integrity, history, and character that are independent of the institutions purchasing their services, as is the case with regard to the professional integrity of physicians and lawyers who can practice independently of hospitals or law firms (2011, 142).

What consultants provide, he argues, is instead a "protean cluster of services… bundled under the term 'clinical ethics'" (2009, 293), a "complex amalgam of disciplines and services" (2012, 152) rather than a profession with a conceptual coherence at its heart. Clinical ethics consultants are merely "jacks of enough trades to be of interest to employers such as hospitals" (2009, 299). This center will not hold; it cannot unify the provision of robust moral advice that purportedly grounds the practice. The consequence of this fact is that, having no professional internal morality, consultants should serve their contracting institutions – for instance, "[h]ealth care ethics consultants should have an appreciation of how to work as mediators…in order to achieve the goals of those who hire them…within the constraints established at law and in public policy," and "hospital ethics consultants should recognize that they are service providers whose particular roles, and therefore competencies, are most plausibly defined by the purchaser of the services who determines the mission and the context in which the services are to be provided" (2011, 142).[4]

What we are left with after considering Engelhardt's arguments is a picture of a practice that sidled into the clinical medical encounter and allowed others to falsely assume it offers a moral expertise so as to maintain and improve its financial and professional benefits within the space it seized. It seems possible that this account is true of some consultants, or at any rate, that it *could* be true given the lack of professional sanctions against any practitioners meeting this description.[5] But there is an alternative perspective on the field and its legitimacy that arises from a different metaethical assumption than Engelhardt makes. I consider such an alternative account in the following section.

11.4 An Alternative to Engelhardt's Metaethical Assumptions

Engelhardt argues that a historically occasioned vacuum sucked up into being the field of clinical ethics consultation, but a different account of history might lead to different conclusions. Historians may find reasons to challenge Engelhardt's account, but I find it persuasive on the whole. It is undeniably true that people do not rely on the authorities they used to; that medicine is far less self-regulating than it

[4] This labeling of consultation as "*arriviste*" opportunistically ignores the effect of time. For someone so historically situated and inspired, Engelhardt is suspiciously atemporal when he compares the professions of clinical ethics consultation and medicine, because these internal professional values had to be worked out in medicine and law in just the way they are being worked out in clinical ethics consultation. One can easily imagine an Engelhardt of old critiquing the early scribes of the Hippocratic Oath in similar fashion; after all, different versions of the Oath swear to a Christian God or the Greek gods or promise to teach the sons of fellow physicians for free or not (Reiser et al. 1977, 3–4).

[5] This is a significant part of the reason consultants say they seek professional status – because a profession can establish a self-understanding and exclude those who fail to comply with it, thus reassuring consumers of the service that they can rely on a consistent product. However, Engelhardt undercuts the force of this point by observing that "there is no hard evidence of serious harm to others from the current fluid character of this group of service providers," and argues that credentialing consultants with a brittle and static set of skills "runs the risk of curtailing the natural adaptation of this service enterprise to changing needs and conditions" (2009, 304).

used to be; that the delivery of health care has changed profoundly from the days of the country or private-practice doctor who settled bills with his patients face-to-face; and that the Enlightenment dreams of dispositive rational arguments regarding moral conclusions have been shown to be of the "pipe" variety.

So what? Engelhardt would have us look with horror on this bleak landscape of groundless moral beliefs and urgently seek authorities to grant us foundations. *Contra* Enlightenment thinkers, he concludes that "thick" or robust morality exists only in a community with received metaphysical foundations; it cannot be found via reason's discovery of free-floating moral Truths. A common theme in Engelhardt's work is the notion that this drives us – or ought to drive us – into deep, rich communities in which sources of authority and claims to truth are clear and recognized. Of course, this makes sense, for in the post-modern condition of a lack of accepted authorities, one solution is to find an authority to which to submit oneself. But, also of course, this is only a suitable solution if one assumes that such an authority is necessary. For those (e.g., atheists or agnostics) who instead assume such an authority is inaccessible, impossible, or anathema, another solution will have to be sought. In other words, why must we seek foundations only of the sort Engehardt identifies?[6]

An alternative approach is to accept the world we have. Viewing our world (heuristically if not actually) as a "fallen" (or "non-certain") world helps us to appreciate the degree of confidence we can expect in our moral conclusions. It is clear that reason alone – understood as the use of logical argument based on axiomatic yet universally true foundations – will not yield absolute confidence in any normative position. Confidence would require two things: (a) a premise that certainty is required to justify normative claims, and (b) a premise that the set of foundations [X] yields certainty.[7]

The latter is a claim of faith, and its truth conditions are not ones mutually agreeable in a pluralistic world. One person's assertion that she has identified the right set of foundations is usually met by another's assertion that *he* has identified a *different* right set. Depending on their theological obligations in defense of the faith, calamity may ensue. I will not engage these arguments, because my target is instead the first premise.

Premise (a) is a metaethical position that articulates how we may justify moral claims. On this premise, morality is never a half-measure, never merely the best we think we can do. One is certain, or one knows nothing. There cannot be degrees of confidence, because there is no ordinal axis upon which we could measure the degrees. As Engelhardt puts it, in the post-modern condition, we must recognize "the collapse of classical metaphysics as the study of being as it is in itself, and its replacement with the examination of being insofar as it is *for us*, that is, insofar as it offers itself within appearance, within language, within culture" (2012, 159; emphasis added). The result is that

[6] Laurence McCullough (Chaps. 1 and 14, this volume) also engages the question of the assumptions at the heart of Engelhardt's work. His challenge to Engelhardt's interpretation of the "Enlightenment Project" focuses on the role of certainty and reason in the search for common moral grounds.

[7] As Engelhardt puts it, "Justified action requires choosing one particular account of proper action" (2002, 81).

> being can only be for us as sociohistorically conditioned and therefore framed within diverse narratives. The modern moral philosophical project shatters into a plurality of accounts, narratives, moral visions, and bioethics, all set within a plurality of self-enclosed hermeneutic circles which in their diversity constitute post-modernity. Each offers an alterative experience of reality, morality, and bioethics.
>
> Absent possessing a nous, there is no escape. No definitive choice can be made among the alternative narratives, moralities, and bioethics on the basis of unbiased, discursively rational, principles (2012, 160).

If we try to make a definitive choice, we either beg the question, engage in circular argument, or depend on an infinite regress. Certainty is right out.

If one wants certainty, then one should find an authority that promises certainty. But if one rejects that requirement, then one must find an alternative to certainty. What constitutes a non-certain justification for moral claims? One response to this may be to claim that "non-certain justification" is an oxymoron: something is either justified and certain or it is uncertain and therefore not justified. But again, this assumes that justification is only complete, never partial. Engelhardt's own solution to life within a pluralist society, as articulated in the *Foundations of Bioethics*, is the "Principle of Permission" (1996, ch. 2).[8] According to Engelhardt, "secular morality is the authority of consent" (1996, 68), which "gives no value to permission. It simply recognizes that secular moral authority is the authority of permission;" it "is nothing more or less than the authority of those who agree to collaborate" (1996, 69). In this way, Engelhardt provides a "grammar" of secular morality with no content (which is vital if he is to avoid his own critiques that other accounts assume a content that cannot be defended without resulting in an infinite regress, etc.).

However, for clinical ethics consultation in particular, it is not clear how to operationalize the Principle of Permission without relying on such problematic content. Who should agree? Must it be the person who will be most affected by the choice? What does this imply for surrogate decision-making? For decision-making by minors?[9] For incapacitated patients? However these categories are fleshed out, they will require value assumptions in order to be used. For example, what counts as being "affected by the choice"? What approach should we adopt regarding children – are they merely the property of their parents? Or is society obligated to protect them as vulnerable? Do consequentialist considerations point to an on-balance superior approach with respect to children? There is no default answer to such questions, but the questions must be answered every day. As Khushf argues elsewhere in this volume, "[w]hile [a] specific decision involves contingency, the *need* for a decision is not contingent" (Khushf, Chap. 7 this volume, pp. 87–144; emphasis added).[10]

---

[8] Even here, he assumes that certainty is required for justification, and derives his solution from a transcendental argument. See Khushf (Chap. 7, this volume) for an extensive consideration of the mechanism and success of Engelhardt's transcendental argument.

[9] See Garrett (Chaps. 13 and 23, this volume) for an extended exploration of the implications of Engelhardt's work for pediatric bioethical decision-making.

[10] See Khushf's excellent essay for a thorough treatment of the mechanism of and problems with Engelhardt's transcendental argument for the principle of permission. Khushf argues that one of the problems with Engelhardt's argument is that "it assumes that demonstration of contingency

A possible Engelhardtian response to this challenge is to acknowledge that the principle is not dispositive in every circumstance, and that it only specifies the clear condition for mutual cooperation: when two adults of sound mind agree to cooperate, that agreement is what binds them to whatever they contract, and sanctions punishment if one violates the agreement.[11] If this is correct, then the approach is not very helpful for much of the domain of bioethical decisionmaking, which often involves patients without these paradigm markers of personhood or, if that is too contentious a term, paradigm markers for "permission-granters." Return to the two cases at the outset of this paper – what help is the Principle of Permission there if we are forbidden from invoking contingent content? Even when patients have much more capacity than these cases portray, there can still be uncertainty regarding when permission is given. As it turns out, either (a) contingent content is necessary to articulate when permission has been granted, or (b) the Principle of Permission does not (and cannot) offer criteria for when it is met.[12] In the first case, if it turns out that contingent content is necessary for the Principle of Permission, an account would need to be given for why *that* contingent content, but not the contingent content in other moral theories, is acceptable as a basis for secular morality. In the second case, if it turns out that the Principle of Permission is utterly unable to provide (or is uninterested in providing) the conditions for its own satisfaction, it is irrelevant to the practice of bioethics, and clinical ethics consultation in particular. In either case, the Principle of Permission is not helpful for moral issues that arise in clinical ethics consultation, for the driving characteristic of most of these cases is that *a decision must be made*. To decline to decide is merely to decide to let other forces make a determination. We cannot avoid invoking contingent content in order to make these decisions, and so we must try to find a justified means of doing so.[13] So a different metaethical assumption is possible: we can try to articulate a *modus vivendum* notion of justification in a "fallen" world, and abandon the quest for certainty. Like a Buddhist who must breathe and walk to live, and thus participate in the murder of living entities in the air and on the ground, we cannot live in a morally pure space; that is the essence of a fallen world. We *cannot function* without acting on some values that surely not all will share. And when two of us act on values that we do not share, we might try to go to the meta-level and at least agree on how we can navigate

associated with a standard disqualifies any ethical theory using such a standard from serving as an ethic for moral strangers" (Khushf, Chap. 7, this volume, 100). But because in some cases decisions must necessarily be made, contingency is omnipresent, and Khushf argues that "Engelhardt needs to provide criteria that enable us to distinguish which kinds of contingency are problematic" (Khushf, Chap. 7, this volume, 100). In this respect, I think my response to Engelhardt's work is very similar to Khushf's. I have certainly benefited from reading Khushf's work during the preparation of this essay.

[11] As Engelhardt puts this point, "Without such permission or consent there is no authority" (Engelhardt 1996, 122).

[12] Again, see Khushf, Chap. 7, this volume (particularly Sect. 7.4.1, "Meta-Controversies About Controversies") for an excellent discussion of the apparent reliance of Engelhardt's arguments on moral content.

[13] McCullough (Chaps. 1 and 14, this volume) calls this a "reliable bioethics."

our disagreement. But often, the problem is merely iterated at the meta-level, since we also may not share a concept of how our second-order conflicts should be resolved. What constitutes justification when certainty is impossible?

This is no easy question, and calls to mind a claim of Engelhardt's: "The more one lives within the secular pluralist embrace of a cosmopolitan society, the more the fabric of taken-for-granted morality will be a cento woven haphazardly out of pieces of diverse moral visions" (1996, 34). Engelhardt's mention of a cento ("patchwork") may provide some guidance. A cento is a poem composed entirely of lines from other poems. It is thus derivative. Yet, as anyone with even a passing knowledge of modern popular music can aver, the sampling of other work does not necessarily yield a work without meaning. Even a patchwork quilt can do the job of covering the body, after all; sampled music is its own art form, and can even elevate and change the meaning of the originals from which it derives its material. In a moral cento, then, perhaps a moral *modus vivendum* could be stitched together to provide some guidance. Perhaps it even looks something like the Principles of Bioethics (Beauchamp and Childress 2012). So long as it is not imposed, since the standards of justification for imposition on others by force are more demanding (though defining what "imposed" means is itself challenging), it nevertheless can serve as a practical tool through which individuals can work out their disparate values.

I do not have a systematic account of "non-certain" justification at the ready. The work of this paper has been mainly to challenge Engelhardt's metaethical foundation as itself contingent, thus creating the space within which to articulate an alternative to grounding bioethics on the Principle of Permission. However, the work so far does lead to some implications for the field of clinical ethics consultation, which I consider in the next section.

11.5 Implications of a "Non-certain" World for the Practice of Clinical Ethics Consultation

Where do these considerations leave us? In his 2011 paper assessing the 2nd edition of the Core Competencies for Healthcare Ethics Consultation (ASBH 2011), Engelhardt suggests that a clinical ethics consultant who acknowledges the limits of the practice should not claim or in other ways represent that he is trained in or able to identify a timeless, universal moral truth. Instead, consultants should tell their clients that the main source for their recommendations is law and public policy, to the extent that these requirements constrain the options. As he puts it, "In this context, hospital ethics consultants are those who give advice about how to negotiate controversial decisions regarding medicine and the biomedical sciences, within the constraints of the ethos that is established at law and in public policy" (2011, 141). What follows from this claim is a set of "Engelhardtian Core Competencies," including knowing local law and public policy; recognizing grey zones in law and policy and using them "to advance the goals of those who engage their services" (141)[14]; familiarity with mediation; and adherence to theological dictates in the case of religious hospitals, among other things.

Many of Engelhardt's recommendations seem correct. Certainly interpreted minimally, it's correct that consultants need to act within constraints of law and policy. But even interpreted more broadly, working within "the ethos" that results from such established law and policy actually provides a great deal of freedom within which consultants, caregivers, patients, and families can interact while also capturing many broadly shared values. One might also look at this list and conclude that consultants already meet most of these "Engelhardtian" competencies; some are already listed in the Core Competencies, 2nd ed. (ASBH 2011) (hereafter CC2). For example, Engelhardt's first item is matched in CC2 by Skill Competency A-2: "Access relevant ethics literature, policies, guidelines, and standards" (ASBH 2011, 25) and Knowledge Competencies 6, 8, and 9 (27). "Familiarity with mediation" in some sense appears throughout CC2, but they use the term "facilitation" rather than mediation. The CC2 has not taken up the question of how to behave as a consultant within a religious institution, and nor will I except to say that, first, it is a reasonable means of protecting the mission of religious institutions, but, second, much will depend on the funding of that hospital and the options open to patients using it.

One "Engelhardtian core competency" deserves separate treatment. In the course of pointing out that consultants should recognize grey zones in law and policy, Engelhardt argues that they should use those grey zones "to advance the goals of those who engage their services." The force of the point is that lacking an internal morality, consultation must be a "service for hire" and thus subject to employment conditions. However, this implies that employers alone set the terms of employment. Imagine an employer who wanted consultants aggressively to advise people to refuse or terminate life-support in order to save money, since consultations seem to achieve that goal (Schneiderman et al. 2003). Engelhardt's arguments suggest that consultants should simply comply. I think a historical account of the rise of clinical ethics consultation could go a long way towards establishing that whatever consultation is, respect for patients over against those with more power (ranging here from physicians to institutions) is one of its taproots. To exchange patient self-determination for institutional self-interest would radically undermine the field. Even if that history weren't compelling, however, it is still the case that consultants could band together and simply (a) decide to call themselves by a certain term, and (b) declare that this practice under this term would offer certain services and not others. Independence from an institution's whims is high on that list.[15] (Though of

[14] Engelhardt in fact proceeds very far in this direction: "…as long as health care ethics consultants act within the constraints of law and public policy, it is no more unethical to be a well-paid partisan ethicist for hire than for a lawyer to be the well-paid advocate of a client" (2011, 141); and "Ethicists do not have a professional integrity, history, and character that are independent of the institutions purchasing their services, as is the case with regard to the professional integrity of physicians and lawyers who can practice independently of hospitals or law firms" (142).

[15] See, for example, the proposed code of ethics for the field (ASBH 2014).

course, institutions could refuse to hire consultants who wouldn't agree to the institution's terms. "Banding together" would just make that potentially harder to do.) Engelhardt's arguments have not shown that consultants must accede to institutional pressures in this way.

The very uncertainty of secular moral judgments means that as much space as possible should be preserved in clinical ethics consultation for individuals to pursue their own certainties.[16] Fortunately, given legal protections for patient autonomy and surrogate decision-making, that is already a potent foundation for the field. However, the appreciation that those protections are also grounded in an acknowledgement of deep moral pluralism should affect consultants' approaches to other aspects of the practice – for example, by inducing a deep humility rather than arrogance regarding the certainty of the recommendations they give.[17] But, further, Engelhardt is absolutely correct to suggest that consultants should not pretend to be in possession of timeless, universal, certain moral truths. The amount of uncertainty inherent in such judgments should be communicated to those who use the service, together with the basis for whatever recommendation is being made. This also suggests that the use of terms such as "ethical" in documents like the CC2 must be fleshed out, even if only to acknowledge that there is deep disagreement over that valuation.

Another fundamental point this discussion establishes is that conversation and exploration of moral values is a vital part of the field. If recommendations made by a clinical ethics consultant are non-certain, and others must live with the consequences of these choices, no one can make such recommendations in a vacuum. There may be side-constraints on the recommendations, but even so, there may be a great deal of latitude within which to fashion a solution that is justified, even if not certain.

11.6 Conclusion

By offering a sociohistorical account of the field, Engelhardt serves clinical ethics consultation well. This account demonstrates some of the forces that gave rise to the field and shaped its current guise, illuminating for us some of the tensions in the field. However, I have argued that the accusation that the field is unjustified or illegitimate because it cannot provide certainty regarding moral recommendations

[16] This does not mean that we should behave as atomic individuals and leave each other to our own devices with minimal interaction. This "non-certainty" is compatible with a wide range of interactions between people, including strongly, considerately, lovingly, or harshly worded conversations regarding what one person feels about another's choices. It is also compatible with refusing to participate in a choice or situation that one finds reprehensible.

[17] I hasten to acknowledge the difficulty this may pose. Anecdotes are readily available of consultants who attempt the "on the one hand/on the other hand" approach and are deemed useless by clinicians. But the need to be helpful to healthcare providers does not erase the need to be honest and transparent, nor does it mean that a brief acknowledgment of the ways a variety of moral foundations could cut in a given situation would be out of place.

is problematic for two reasons. First, Engelhardt's own argument for a solution to the lack of secular moral certainty does not succeed, and second, clinical ethics consultation is shaped around decisions that must be made. What results is the need for decision-making mechanisms that are non-certain but strongly justified. Though I have not offered a theory of non-certain justification, I have suggested that we take up Engelhardt's critical term, "cento," and appropriate it by articulating the patchwork of values that must be respected in order to make a justified recommendation in clinical ethics consultation.[18] This approach cannot offer us certainty, but if we begin by acknowledging that we cannot have certainty in secular morality, that will fundamentally shape how we search for answers to moral questions. Obviously Engelhardt will contest what I have said here, and I welcome that exchange. It has been a pleasure, and is still instructive, to be steeped in and grapple with Engelhardt's work intensely again.

References

ASBH. 2011. *Core competencies for healthcare ethics consultation*. Glenview: American Society for Bioethics and Humanities.

ASBH. 2014. *Code of ethics and professional responsibilities for healthcare ethics consultants*. http://www.asbh.org/uploads/files/pubs/pdfs/asbh_code_of_ethics.pdf. Accessed online 30 Mar 2015.

Beauchamp, Tom, and James Childress. 2012. *The principles of biomedical ethics*, 7th ed. New York: Oxford University Press.

Engelhardt Jr., H.T. 1996. *The foundations of bioethics*, 2nd ed. New York: Oxford University Press.

Engelhardt Jr., H.T. 2002. The ordination of bioethicists as secular moral experts. *Social Philosophy and Policy* 19(2): 59–82.

Engelhardt Jr., H.T. 2009. Credentialing strategically ambiguous and heterogeneous social skills: The emperor without clothes. *HEC Forum* 21(3): 293–306.

Engelhardt Jr., H.T. 2011. Core competencies for health care ethics consultants: In search of professional status in a post-modern world. *HEC Forum* 23(3): 129–145.

Engelhardt Jr., H.T. 2012. Why clinical bioethics so rarely gives morally normative guidance. In *Bioethics critically reconsidered: Having second thoughts*, ed. H. Tristram Engelhardt Jr.. Dordrecht: Springer.

Rasmussen, Lisa M. 2011a. Clinical ethics consultation's dilemma, and a solution. *Journal of Clinical Ethics* 22(4): 380–392.

Rasmussen, Lisa M. 2011b. An ethics expertise for clinical ethics consultation. *Journal of Law, Medicine and Ethics* 39(4): 649–661.

---

[18] I have, however, begun to work in this direction in a variety of papers. See my (2011a) for an argument about proscriptive standards in clinical ethics consultation, and my (2011b) for arguments regarding the non-certain expertise which consultants can possess. In a separate paper in preparation, I am arguing for the inclusion of a checklist as a means of achieving rigor in a consultation, where the checklist serves as a repository for important features of cases that should not be overlooked.

Rasmussen, Lisa M. In preparation. *Achieving rigor in clinical ethics consultation may include following a checklist*. Presented in May 2015 at the 11th Annual international conference for clinical ethics consultation, New York, NY.

Reiser, Stanley J., Arthur J. Dyck, and William J. Curran. 1977. *Ethics in medicine: Historical perspectives and contemporary concerns*. Boston: MIT Press.

Schneiderman, L., Todd Gilmer, Holly D. Teetzel, Daniel O. Dugan, Jeffrey Blustein, Ronald Cranford, Kathleen B. Briggs, Glen I. Komatsu, Paula Goodman-Crews, Felicia Cohn, and Ernlé W.D. Young. 2003. Effects of ethics consultations on nonbeneficial life-sustaining treatments in the intensive care setting. A randomized controlled trial. *JAMA* 290(9): 1166–1172.

[illegible]

Chapter 12
Moral Strangers, Proceduralism, and Moral Consensus

Fabrice Jotterand

12.1 Introduction

I would like to start this essay in the honor of H. Tristram Engelhardt Jr. with a few autobiographical comments by way of introduction to what I address in the following. These comments are based on some observations I made during the 20 years I spent in the United States. I grew up in Switzerland but was raised between two cultures since my mother is Italian and my father is Swiss. I moved to Boston in 1995 for my education. My first encounter with American culture occurred in what is considered a bastion of liberal democracy by many, particularly the iconic "Cambridge, Massachusetts," the location where John Rawls develop his influential *Theory of Justice* (1971) at Harvard University. Not aware of the symbolism of "Cambridge, Massachusetts"[1] in my early years in the United States, I distinctly remember my wife (also from Switzerland) and I commenting many times on how "conservative" people in Massachusetts were compared to the homeland. Our background and our political culture rooted in European (aka Swiss) assumptions provided the lenses through which we made our evaluation. To our estimation,

---

[1] I made an interesting observation early on as one of Tris' assistants. We attended a conference at Notre Dame and, as traditionally happened, an evening cannot take place without a trip to a bar. I remember ordering a Samuel Adams beer (produced in Massachusetts), which created some convulsions on Tris' face and some remarks such as "Oh a Yankee beer!"...indicative of his somewhat annoyance with my choice. Puzzled I carried on and enjoyed my beer. It took me few years to understand that I made a major "faux-pas" and the symbolism of my choice. A Shiner Bock (produced in Texas) would have been a more judicious choice!

F. Jotterand, Ph.D., M.A (✉)
Department of Health Care Ethics, Rueckert-Hartman College of Health Professions, Regis University, 3333 Regis Boulevard, G-5, Denver, CO 80221, USA

Institute for Biomedical Ethics, University of Basel, Basel, Switzerland
e-mail: fjotterand@regis.edu

L.M. Rasmussen et al. (eds.), *At the Foundations of Bioethics and Biopolitics: Critical Essays on the Thought of H. Tristram Engelhardt, Jr.*, Philosophy and Medicine 125, DOI 10.1007/978-3-319-18965-9_12

Massachusetts was still very much "conservative" and religious compared to old Europe. We grew up in a secular society immersed in a long tradition of Enlightenment ideology, shaped by socio-democratic ideas, and characterized by multiculturalism and pluralism (ideas, I might add, that do not necessarily represent my political views).

These particular philosophical and political assumptions determined, to a certain extent, how I approached my education in the United States and the set of issues I wanted to investigate in my scholarly endeavors. More importantly, however, these assumptions did not motivate me to investigate the possibility of securing a common morality. Since the Enlightenment, European history has provided multiple historical examples of the failure to produce a common morality, which translated into ideological and religious wars and political conflicts between European states.

While in Europe, I already established, albeit in an unsophisticated way as I was early in my academic training, that a robust common morality outside particular traditions (theological, philosophical, cultural, or political) was intellectually untenable. The intellectual agenda of modernity, characterized by rationalism and a resolute effort to break with the Judeo-Christian heritage of Western culture, did not produce the grand narrative it hoped for (Lyotard 1984). To the contrary, many scholars have questioned the agenda of modernity and recognized the fragmented nature of contemporary culture and the irreconcilable ideologies shaping political, philosophical, and moral debates (Hunter 1994; MacIntyre 1984; Engelhardt 1996).

Recognizing the potential for moral nihilism, my intellectual journey led me to further reflect on issues related to the nature of the good, the just, and human flourishing. In this quest for answers, I came across Alasdair MacIntyre's *After Virtue* in the later part of the 1990s while in Boston. This work deeply influenced my thinking and provided a coherent (and convincing) analysis of our current predicament as well as a way out of moral nihilism through what MacIntyre calls "the construction of local forms of community within which civility and the intellectual and moral life can be sustained…" (MacIntyre 1984, 263; see also 1989). MacIntyre laid for me the philosophical foundation to further examine the postmodern condition, especially as I got interested in medical ethics and bioethics.

My encounter as a graduate student of H. Tristram Engelhardt, Jr. at Rice University and my reading of the second edition of his magnum opus, *The Foundations of Bioethics* (1996), put the final nail in the coffin of modernity, at least in my own thinking. In Chap. 2, "The Intellectual Bases of Bioethics," Engelhardt dissects the contemporary context of the moral fiber of our pluralistic and postmodern society and examines the necessary conditions to achieve moral consensus. He concludes that the hope of the Enlightenment project to provide a content-full morality through reason alone has failed and any attempt to assert the moral authority of public policy grounded on sound rational arguments is untenable. Engelhardt's deconstruction of the philosophical agenda of modernity deeply resonated with my own views. My puzzle, however, was and still remains a work in progress, how to rebuild and provide moral guidance to social practices such as medicine, science, and technology despite moral pluralism and the incommensurability of current bioethical debates.

These autobiographical comments are meant to honor H. Tristram Engelhardt, Jr. and acknowledge the influence of his work in my wrestling with the philosophical, moral, and political predicament of Western culture. My years as his assistant, or rather as his slave, as he liked to remind me, have been a privilege and a unique opportunity to learn and be challenged by a great mind of our time. In this essay, I would like to recognize my debt to Tris in my intellectual development and in my scholarship, particularly in framing a proceduralist approach to ethics. That said, I would like also to outline some points of divergence. While I am sympathetic with his diagnosis of the predicament of Western culture and its implications for bioethics, I would like to raise some critical points concerning the notion of moral strangers and his approach to procedural ethics. In this chapter, first I outline Engelhardt's diagnosis of the nature of secular morality in Western culture, which by default is procedural, and examine the concept of moral strangers. Second, I critically assess Engelhardt's proceduralism and argue that his framework does not take into account the possibility of overlapping frameworks between various moral communities. Hence, third, I argue for a weak form of proceduralism based on the work of Wildes (2000), which allows the establishment of moral discourse through a web of partial understandings of moral issues, in spite of moral disagreements. I conclude my essay by recognizing the significance of Engelhardt's criticism of mainstream bioethics and underscore the importance of his legacy for the future of the field.

12.2 Procedural Ethics and Moral Strangers

Bioethics emerged in the United States in the cultural context of the 1960s characterized by a questioning of traditional moral authority driven by "progressive" political ideas. As Engelhardt writes, bioethics was a "post-traditional moral and political movement, grounded in and motivated by theological dissent and liberal-democratic political aspirations" (Engelhardt 2013, 56). The roots of this dissenting movement took place within Roman Catholicism as many clerics attempted to fill the moral vacuum left by the social disruption that occurred in America (and in most of the rest of Western culture).[2] Engelhardt, himself a Roman Catholic at that time,

[2] See W.T. Reich (2013) for an outline of the origins of bioethics in the culture of the 1960s. He points out that "[t]he rejection of the moral authority of the previous generation and the calling into question of the moral authority (or the absoluteness of the authority) of major social institutions including church and state were major characteristics of the 1960s counterculture. ...In the 1960s America...was experiencing an enormous force of alienation, especially alienation from society, its values, and the authorities of its traditional institutions. ...At the very least, an examination of the socio-cultural origins of bioethics in the 1960s should make us realize that bioethics did not arise simply as a response to a set of biomedical moral problems, but as a result of an enormous moral upheaval in our society, an awareness of the power of medical technocracy over our lives, and a healthy skepticism as to whether the power-oriented religious and civil authorities could

participated in the development and construction of bioethics but soon realized that he did not align completely with its moral, theological, and political causes. His were intellectual interests that focused on the history and philosophy of medicine, and phenomenology. One particular issue that caught his attention was the issue of the foundations for a common morality in an increasingly pluralistic and secular society. As Engelhardt writes,

> [m]y background was…in the history and philosophy of medicine, as well as in work regarding Kant, Hegel, Husserl, and more generally phenomenology. As a consequence, I tended to approach bioethics with fewer politically-driven conclusions in mind. My engagements in bioethics were motivated primarily by an interest in solving conceptual puzzles. One such puzzle was my amazement that I could not secure by sound rational argument the lineaments of traditional, Western Christian morality regarding such issues as abortion, infanticide, and euthanasia. I found no foundation for a canonical, content-full, secular morality. Moreover, it became clear that moral pluralism was real and intractable to secular sound rational argument. *De facto*, there was no common morality (Engelhardt 2013, 65).

Recognizing the limits of secular morality, Engelhardt concludes that by default the only way to secure moral agreement between moral strangers is through "a purely procedural morality grounded in permission" (Engelhardt 2013, 66). By no means does Engelhardt advocate moral nihilism. He maintains that despite the impossibility to determine, in secular terms, the source of moral authority, moral truth exists as his criticism appeals to "a moral epistemological skepticism, not a metaphysical skepticism" (Engelhardt 1996, 45). In short, secular morality, based on rational arguments, does not offer an authoritative content-full moral framework to provide any guidance to secular society since debates over many issues still remain unsettled.

12.2.1 Moral Strangers

In the *Foundations of Bioethics*, Engelhardt articulates the conditions for the resolution of moral issues within our pluralistic context characterized by competing understandings of the good, the just, and moral flourishing. He examines particularly the context of bioethics, which is, in his view, a late-Enlightenment attempt to provide a common morality in the field of medicine. Since a common ground (tradition, philosophy, religion, etc.) for establishing moral authority is not possible in contemporary bioethical reflections, Engelhardt argues the only justifiable alternative is the principle of permission. It provides the basis for a morality of mutual respect, which regards individuals as the source of "general secular moral authority"

---

solve those problems without contributions from all available intellectual and moral resources" (Reich 2013, 89, 91). For a full analysis of the birth of bioethics see Jonsen (1998).

(Engelhardt 1996, 119). In addition, he points out that the traditional moral theories that influenced the West have failed for three main reasons. First, all these theories are mostly conceptual in form and therefore they cannot generate any content or any guidance. Secondly, these theories are descriptive rather than prescriptive; they do not have the strength to convey any moral authority. Finally, all of them have particular assumptions or specific rankings of the good that are sometimes in opposition between theories (Engelhardt 1996, 42).

Because of how moral accounts differ in substance, people who do not share a common morality are by default "moral strangers." That is, moral strangers are persons who do not share enough common moral ground to agree on moral issues. They are "persons who do not share sufficient moral premises or rules of evidence and inference to resolve moral controversies by sound rational argument, or who do not have a common commitment to individuals or institutions in authority to resolve moral controversies" (Engelhardt 1996, 7). Moral strangers are intrinsic to the moral fabric of secular morality, which attempted to locate agreement through a common grammar able to sustain a common moral language recognizable and recognized by all. However, the Enlightenment project did not produce a common account of morality established by human reason, but rather resulted in a polytheism of moral understandings, each of them assuming particular ideas of morality and the good (Engelhardt 1996, 37). All moral theories assume what each tries to justify, i.e., a particular set of assumptions and a specific content.

On the basis of the recognition of the fragmented character of contemporary morality and in order to find peaceable resolutions to moral issues in a pluralistic society without endorsing particular moral, religious, or metaphysical assumptions, an appeal to permission as the source of authority is the de facto position.[3] Consent constitutes the basis for the resolution of moral litigations and represents a "necessary and sufficient condition (sufficient when combined with the decision to collaborate) for a general secular ethics…[which] is nothing more or less than the authority of those who agree to collaborate" (Engelhardt 1996, 69). Conversely, while Engelhardt argues that people have the right to decide about the character of the resolution of moral issues, he also makes the point that moral agents have the right to be left alone and therefore are entitled not to participate in any particular community (Engelhardt 1996, 70).

It follows that Engelhardt is not concerned to set the conditions for a peaceable society. His goal is to provide the condition for the resolution of moral issues between moral strangers (Engelhardt 1996, 70). Persons are the source of authority (the good, i.e., beneficence, is ultimately determined by what one agrees to) in morality, and permission, as a transcendental condition for morality, is the necessary condition (i.e., procedure) to solve moral issues between moral strangers.

---

[3] It must be emphasized that for Engelhardt consensus has a moral force in so far as there is univocal consensus. When consensus is the result of "a balance of political power" in which people of different moral traditions and assumptions are forced to collaborate, the moral legitimacy of consensus is difficult to establish (Engelhardt 1996, 63).

12.3 Assessing Engelhardt's Proceduralism

The force of Engelhardt's analysis and arguments is that it removes the impression that there is one unifying account of morality from which certain principles can be derived as in the work of Beauchamp and Childress in *The Principles of Biomedical Ethics* (2008). The principles of justice, autonomy, beneficence, and non-maleficence are subject to interpretation and therefore, unless one shares a common source of moral authority, these principles have different meanings and meet various expectations bestowed by individuals. Even the principle of beneficence is almost empty by itself if we don't consider how the process of moral reasoning takes place, that is, how moral assumptions about the good, the right, and the just will depend on the moral agent. For instance, the diagnosis of a defective fetus constitutes a clinical judgment concerning a particular medical condition. However, although the diagnosis relies on empirical data, the physician's assessment of this clinical case takes place within a cluster of moral values that inform the prognosis and how he or she communicates to the mother the bad news. Whether the physician will suggest, as the best course of action, the abortion or the nurturing of the fetus will depend on his or her particular moral, political, and religious commitments.

Furthermore, Engelhardt's procedural ethics takes differences in moral reasoning seriously and set boundaries for the protection of these differences so as to provide the conditions for moral communities to coexist. He is willing to take the risk that certain moral communities might be involved in morally reprehensible acts to preserve specific moral commitments considered "unacceptable" according to secular standards. For instance, most secular people would abort a fetus with Down syndrome or or a severe handicap because, among other things, of the cost to society to care for such individuals and the poor quality of life the child would endure. Engelhardt, however, argues that if a community is committed to the principle of the sanctity of human life on a religious ground, and therefore is opposed to the abortion of deformed fetuses, the state or any institutions have no right to impose any particular moral visions on that community and should respect, if not protect, these moral commitments. The danger that Engelhardt rightly perceives is the imposition of moral values and norms on communities that do not share, for instance, secular socio-democratic values.

That being said, the concept of moral strangers articulated by Engelhardt is not without problems. While it is certainly the case that people do not share a common morality, there are social practices (e.g., medicine, science, and technology) that require common reflections and a willingness to collaborate at the moral level for their implementations in society. The concept of moral strangers, if understood as the impossibility to collaborate at the moral level (i.e., foundational level) but as the possibility to collaborate at the political level (i.e., procedural level), can lead to a very narrow understanding of morality. That is, some moral issues can be reflected upon exclusively within particular communities (whether ideological, political, or religious), hence indirectly promoting a moral *laissez faire* in which people are disengaged from moral reflections in the "market of ideas" while engaged in procedural reflections.

Does this mean that people cannot solve moral problems due to the lack of a common and content-full morality? Of course, Engelhardt will answer negatively since he articulates a "proceduralist morality" grounded on the individual as the source of authority and permission as the safeguard for the respect of such individual authority. But morality, Engelhardt stresses, cannot be understood as content-full but rather as procedural. He describes morality, in our particular secular context, as a peaceful resolution of moral controversies, which requires the agreement or consent of those engaged in situations requiring moral deliberations.

To sum up, a proceduralist ethic provides an attractive alternative to moral theories such as principlism since it recognizes the plurality of moral accounts and provides a peaceful approach to moral quandaries. Contrary to other theories that attempt a priori to ascribe a particular content to principles without any reference to a specific moral tradition or community, Engelhardt articulates an ethic in which the principles of permission and beneficence do not require any content (of the good, one of the main problems in contemporary moral theory). Content is given only by what one agrees. The main criticism, however, is that Engelhardt does not take into account that indeed there can be overlaps between various moral communities that can offer the basis (limited, I recognize) for a common reflection on various moral issues in medicine, science, and technology. An alternative to Engelhardt's strong procedural is Kevin Wm. Wildes' version of procedural ethics (moral acquaintances), since it offers, in my view, a richer understanding of morality that takes into account these "limited overlaps."

12.4 Weak Proceduralism

12.4.1 Moral Acquaintances

Wildes articulates an approach to bioethics that captures the pluralistic condition of our society while at the same time acknowledging that indeed different moral communities (philosophical, religious, or scientific) do overlap to a certain extent. The main thrust of his argument is captured in the following quote:

> In spite of strong differences, there is no reason, a priori, to hold that communities cannot overlap. A consideration of methodology in bioethics needs to address how to account for this overlap. Health care, in secular societies, is a collaborative enterprise and moral problems are not contained only within the boundaries of particular communities. Nor do most men and women live, strictly, within the boundaries of a particular moral community. The categories of moral ecumenism and moral acquaintanceship provide a way to understand and map the different intersections that can and do take place (Wildes 2000, 141).

Wildes does not argue for a single communitarian bioethics or morality. He points out that there are various "bioethical communities" competing in the market of ideas marked by a lack of general moral agreement. However, he emphatically makes the point that there may be some overlap between rival moral accounts and moral traditions.

In order to account for the absence of moral unity, the fact that people do not live in isolation, and that certain practices (medicine for instance) contain a social dimension, Wildes articulates an alternative to both principlism and Engelhardt's strong proceduralism. As Wildes puts it "there is a common morality that is less robust than many assume [Beauchamp and Childress] but more vibrant than Engelhardt concludes, and procedures are a fruitful way to identify the common ground" (Wildes 2000, 56). Wildes follows Engelhardt in his procedural ethics but differentiates himself by adopting a middle ground position – what I would call a weak form of proceduralism. He develops an alternative to the concepts of "moral strangers" (Engelhardt) and "serious moral agents" (Beauchamp and Childress) by proposing the concept of "moral acquaintances."

Moral acquaintances are people of different moral traditions who can come together to resolve moral questions not in terms of a naïve moral ecumenism but rather by including philosophical differences in moral deliberations. Moral acquaintanceship occurs when "there is managed agreement, which is reached when people seem to arrive at the same judgments without sharing the same moral framework and ranking of values. This agreement is characteristic of moral acquaintances" (Wildes 2000, 152–153; see also Guttman and Thompson's conditional proceduralism, 2000).

This type of agreement is not based on a common morality in the traditional sense, i.e. the sharing of common values and norms. Rather agreement is reached through a series of procedures not grounded on substantial claims about moral norms and values (content) but through deliberations (procedures) that create spaces for moral disagreements and recognize points of overlap (Wildes 2000, 172–173). The degree of overlap between moral perspectives depends on the extent to which moral agents share others' moral commitments and the level at which moral judgments occur.

Wildes distinguishes two types of acquaintanceship: type A1 describes people who may understand the moral perspectives of others but do not share them; while in type A2, people understand each other's morality and *partially* share some aspect of it (Wildes 2000, 139). In his view, many people in bioethics assume a type A2 of acquaintanceship that distinguishes people from moral strangers, but at the same time does not separate or isolate people in their moral commitments, and presupposes some overlap between moral communities.

12.4.2 Moral Acquaintances: Middle Ground Position

A middle ground position constitutes a possible approach to moral reasoning because it considers the social dimension in which social practices take place. Furthermore, it avoids some of the problems intrinsic to foundationalism, which limits the ability of various moral communities to reflect on some crucial ethical issues related to the development of these social practices.

Of course, we live among competing moral systems that do not necessarily share a sufficient moral language able to sustain a content-full common morality. However, rather than simply adopting a status quo position, or a type of moral regress (that is, the idea that any moral commitment must be justified and grounded on various assumptions), it is crucial to provide the conditions for the creation of moral spaces in which people can affirm their moral convictions while at the same time reach a limited or managed consensus or comprehension of other moral perspectives – acquaintanceship (A2).

Wildes recognizes the challenge of moral pluralism but rightly refutes the two extremes in which differences are either simply ignored or so emphasized that a dialogue between moral communities becomes impossible. In the former case, Wildes points out that sooner or later these differences will resurface anyway so this position is not very helpful as it does not allow robust moral reasoning. The latter case, what he calls a "communal relativism," is not helpful either because the social dimension of certain practices is indisputable since they are not "private or local enterprises" (Wildes 2000, 162).[4]

Some could argue that discursive reasoning and the abandonment of particular irreconcilable moral commitments for the sake of political consensus could constitute the basis for ethical principles and moral actions. John Rawls, for instance, exemplifies this political move. He argues that in a modern democracy the distinction between what he calls "a pluralism of comprehensive religious, philosophical, and moral doctrines" and "a pluralism of incompatible yet reasonable comprehensive doctrines" is necessary in order to insure a neutral framework in which political consensus can take place (Rawls 1993, xviii). The doctrines of the former kind are the source of disagreement and cannot constitute a basis for social collaboration. The latter kind represents the necessary conditions for social consensus and consequently establishes a morality in itself in modern democracies (Rawls 1993, 1997).

Conversely, one could confine moral reflections exclusively in relation to the teaching of a particular tradition and mores (intrinsic to a specific community – "communal relativism") independently of what professional values and obligations require. The content of moral discourse and moral actions is restrained by individuals' (e.g., physicians', nurses', scientists') socio-political or religious background belonging to that particular community. In other words, such an understanding of morality holds that the outcome of the decision making process within a professional setting is almost exclusively the result of the practitioner's own moral commitments, based on the moral tradition of his/her community independently of professional obligations and values.

These two positions are problematic. In the first approach, the quest for political consensus raises the issue of the danger of emptying morality of its content and depriving it of rigorous moral analysis. This "political move" transforms morality into a set of procedures designed to provide a justification for what is socially

[4] Not only does Wildes argues that this position is not helpful, but he also stresses that it can be dangerous because it can make assumptions concerning particular communities that might not be true for some others (Wildes 2000, 126).

suitable and acceptable for the sake of a particular social order without acknowledging the deep moral disagreements characteristic of competing moral theories. More importantly, if we accept the second approach ("communal relativism"), it implies that one's understanding of medicine, science, and technology, for instance, and certain ethical issues related to them are understood only within particular communities independently of what these practices requires for professionals. For instance, we could image a physician belonging to a community that encourages, on a moral ground, the killing of people considered as burdensome for society (people in a vegetative state, for instance). It does not follow, however, that that particular individual can justify the killing based on his personal convictions while acting as a professional in a clinical setting. As a doctor, this individual is obliged to act according to some particular professional standards.

One might object that a professional association may impose on a minority of physicians the professional obligations to practice what would be considered as morally wrong actions (i.e., abortion) for them. However, a distinction here is necessary. It is important to distinguish between refraining from partaking in unethical actions (which has no consequences for one's moral integrity) and imposing on others, through specific actions, one's moral views (i.e., the moral obligation to kill burdensome people for society). In the latter case, moral wrong is acted upon the patient and the family (by imposition) while in the second case one is free to refrain from participating in a specific action, thus leaving the decision to others and creating a moral space in which one can act as a professional and as a moral agent. Furthermore, even if a professional association would impose particular obligations contrary to one's convictions, there is always the possibility to resign or simply not be a member of the association. In the United States, for instance, there is not an obligation to be member of the American Medical Association (AMA) in order to practice medicine.[5]

What is important to keep in mind is that practices such as medicine or biomedical research are practiced by a variety of people of different socio-cultural backgrounds who are required to respect fundamental professional principles and a set of ethical norms regulating their practice. Undoubtedly, our social context reveals various communities with different competing, and sometimes incompatible, moral understandings. Nevertheless, despite these differences, it does not follow that some overlap between communities and moral traditions cannot occur. Health care (and by extension science and technology) is a collaborative enterprise that does not limit moral problems to particular communities. Moral discourse in bioethics, from a collaborative perspective, can take the form of acquaintanceship. In this type of moral relationship, people do not necessarily share moral views but rank values (e.g., freedom, justice, etc.) differently and understand the differences that separate them from others. The result is that a moral discourse can be established between moral acquaintances through a web of partial understandings of moral issues, in spite of moral disagreements.

[5] Interestingly not all physicians in the United States are members of the American Medical Association. Statistics show that membership rose from 51 % in 1912 to 73 % in 1963. In 1990, membership was less than 50 % (Krause 1996, 45).

12.5 Consensus and Proceduralism

Engelhardt and Wildes agree that a procedural ethic represents the only option for moral discourse in a pluralistic and secular society between people of different moralities. They both contend that in order to engage in human relationships that require moral reflections, the principles of permission and consent are essential in moral deliberation. However, there is a crucial difference between the two perspectives. Engelhardt's foundationalism is very restrictive as to the possibility of moral discourse in society. Unless people share a content-thick morality they are not able to solve moral and social issues because of the lack of "sufficient moral premises" to sustain a moral discourse. Hence the alternative: proceduralism and the principles of permission and consent that do not require a content-thick morality since content is determined by what moral agents agree.

The division between a content-thick morality (moral friends) and a content-thin morality (moral strangers) is too simple because people of various moral traditions have reached agreements despite their different moral assumptions. Furthermore, a proceduralist ethic is based on a certain type of agreement that necessarily requires "an overlapping moral consensus" (Wildes 2000). This means that proceduralism is not a methodology in ethics that is void of moral assumptions. Let's consider person X and person Y. They mutually agree to enter into a business relationship that involves the use of X's resources to take care at home of his (X's) mother who is in a terminally ill condition (cancer) and incompetent. X and Y disagree on what type of care ought to be provided to the mother since Y thinks it would be better to let her die on the ground that it is a waste of resources and that keeping her alive in this state is only a prolongation of her misery. Now, the reason why X hires Y is because Y is a friend (a long time family friend but a moral stranger), needs money, and has the medical expertise to take care of X's mother at home.

In this case, two people enter into a business relationship which involves respecting an agreement, i.e. X will pay Y to take good care of X's mother. (For the sake of the argument, I will not take into consideration that X could choose a person of the same moral sentiments as his; this would solve the problem between moral strangers. However, people always interact and make decision with moral strangers, reaching a certain consensus despite different moral sentiments.) These two persons are moral strangers: they do not share what is morally required in relation to the mother. How can we explain that these two people, despite being moral strangers, can agree on what has to be done? After all, Y could indeed use certain medical procedures that could harm the mother if not kill her but decides to respect X's wishes.

This case illustrates that although people are moral strangers, they can still enter into a mutual agreement that requires *moral* consensus. However, this agreement is not purely procedural as such. It assumes certain moral premises that justify the type of agreement made. In the case above, the agreement between the two parties assumes the consent of X to have Y taking care of the mother and the toleration of conflicting moral visions. It requires trust from X that Y will not kill his mother and honesty from Y in the sense that he will perform his duty as a medical professional according to the wishes of X.

The key point here is that any type of agreement requires certain moral assumptions (trust, keep promises, respect of others' liberty, etc.) which secure and justify consent. In order to avoid assertion of power by one of the parties, there must be a certain overlap or consensus that allows the resolution of moral questions among people of different moral sentiments. Without such overlap, or some commitment to moral content, agreements would not be possible. It follows that procedures are not abstractions that bind people independently of some moral commitments. To the contrary, as Wildes explains:

> ...*[P]rocedures* [are] *moral practices*...[that] embody certain moral commitments. The use of procedural ethics, which has been so important to the contemporary practice of medicine and health care, challenges the field of bioethics to examine the moral assumptions underlying common procedural resolutions (Wildes 2000, 163–164).

Thus, while the language of a thin common morality is problematic without qualification and while it is not clear how to delineate clearly the locus of moral agreement, even within the context of a weak form of procedural ethics, the possibility of moral discourse is possible between moral acquaintances. Procedures imply moral assumptions concerning the nature of agreements in general. These moral assumptions reflect the nature of the fabric of our pluralistic and secular society, which include (1) a commitment to liberty (a necessary condition for the principle of consent); (2) the respect of the rule of law, which secures that procedures are fair and impartial; (3) recognition of the limits of the moral authority of the state – individuals are the source of moral authority and not the state; (4) toleration, that is, a recognition that there are competing moral accounts and that no one can impose a moral perspective without the approval of a moral agent – by the same token toleration also secures the moral integrity of individuals since they are respected in their moral commitments (Wildes 2000, 167–173).

One could assume from the moral commitments listed above that Wildes' procedural ethic is intrinsically individualistic, that is, they constitute the necessary conditions for the protection of individuals' liberties independently of social considerations. Wildes, however, is critical of this form of strict individualistic approach. He points out that an individualistic approach is not suitable because "it is inappropriate, indeed incomplete, to consider individual human beings apart from the community. The communitarian approach supports the importance of autonomy but emphasizes that no autonomous person lives in isolation" (Wildes 2000, 125). What he calls the "communitarian turn," in his view, attempts to recapture some important elements (in particular the meaning of being human) in moral reasoning that have been omitted since the Enlightenment.

12.6 Engelhardt's Legacy

Despite the apparent success of bioethics, questions remain as to the intellectual foundations of the field, particularly in its ability to provide the necessary intellectual substance to deliver moral guidance to medicine, science, and technology in a postmodern and pluralistic society. If the birth of bioethics has been an attempt to fill the moral vacuum left by post-Enlightenment thought, the development and growth of bioethics appears as an erosion of the philosophical basis of ethical reflections.

The field of bioethics would not be the same without Engelhardt's scholarly contributions, diagnosis of our predicament, and "coups de gueule." His willingness to go against the grain, to challenge the bioethics establishment, can only testify to his enormous intellectual contribution. At the onset, he recognized the challenge arising from the collapse of traditional values in American culture and the distinctive political identity of bioethics rooted in liberal democratic ideology. He rightly diagnosed that a proceduralist approach is the only option available to secure peaceable agreement in our pluralistic and secular culture. However, as I have outlined in this chapter, Engelhardt's purely proceduralist ethic fails to recognize that the disagreement between moral strangers is not as strong as he assumes but less obvious than what Beauchamp and Childress contend. Hence, a middle ground version of procedural ethics (Wildes' concept of *moral acquaintances*) is attractive because it provides a richer understanding of morality that acknowledges our pluralistic moral condition but also affirms that indeed sometimes moral commitments overlap between people of different communities and traditions.

References

Beauchamp, T. L., and J. F. Childress. 2008. *Principles of biomedical ethics*, 6th ed. New York: Oxford University Press.

Engelhardt Jr., H.T. 1996. *The foundations of bioethics*. Oxford: Oxford University Press.

Engelhardt Jr., H.T. 2013. Bioethics as a Liberal Roman Catholic Heresy: Critical reflections on the founding of bioethics. In *The development of bioethics in the United States*, ed. J.R. Garrett, F. Jotterand, and D.C. Ralston. Dordrecht: Springer.

Gutmann, A., and D. Thompson. 2000. Why deliberative democracy is different. In *Democracy*, ed. E.F. Paul, E.F. Miller, and J. Paul. Cambridge: Cambridge University Press.

Hunter, J.D. 1994. *Before the shooting begins. Searching for democracy in America's culture war*. New York: The Free Press.

Jonsen, A. R. 1998. *The birth of bioethics*. New York: Oxford University Press.

Krause, E.A. 1996. *Death of the Guilds*. New Haven: Yale University Press.

Lyotard, J.-F. 1984. *The Postmodern Condition: A Report on Knowledge*. Minneapolis: University of Minnesota Press.

MacIntyre, A. 1984. *After virtue*, 2nd ed. Notre Dame: University of Notre Dame Press.

MacIntyre, A. 1989. *Whose justice? Which rationality?* Notre Dame: University of Notre Dame Press.

Rawls, J. 1971. *A theory of justice*. Cambridge, MA: Harvard University Press.

Rawls, J. 1993. *Political liberalism*. New York: Columbia University Press.

Rawls, J. 1997. The idea of public reason revisited. *The University of Chicago Law Review* 64: 765–807.

Reich, W.T. 2013. A corrective for bioethical malaise: Revisiting the cultural influences that shaped the identity of bioethics. In *The development of bioethics in the United States*, ed. J.R. Garrett, F. Jotterand, and D.C. Ralston. Dordrecht: Springer.

Wildes, K. Wm. 2000. *Moral acquaintances*. Notre Dame: University of Notre Dame Press.

Chapter 13
Cracks in the Foundations of Engelhardt's Bioethics

Jeremy R. Garrett

13.1 Introduction

There are two basic ways to critically evaluate the "foundations of bioethics": (1) critique various fundamental axioms that, whether explicitly proposed or implicitly assumed, structure the field of bioethics (a *critical* project), or (2) develop and defend stronger premises upon which the field might rest (a *constructive* project). Throughout his distinguished career, H.T. Engelhardt, Jr. has been continuously engaged in *both* projects. Indeed, it is fair to argue that, in the three decades following the publication of the first edition of his seminal work, *The Foundations of Bioethics*, no other philosopher or bioethicist has engaged these foundational tasks so deeply or extensively. On the one hand, Engelhardt's relentless critiques of the prevailing methods and values of (secular) bioethical discourse have convinced many to reject all "thick" content-full bioethical claims as unjustified. On the other hand, his positive "thin" account of bioethics as consisting entirely in what "moral strangers" can agree upon via a principle of permission has resonated with many (even if they simultaneously regret that reasoned argument cannot deliver more). Together, these two tasks decisively shape nearly every work in Engelhardt's extensive scholarly corpus on (secular) bioethics.

However compelling these two projects may appear individually, though, a clear tension manifests when undertaken together. The sweeping nature of Engelhardt's critical project raises serious worries about whether any constructive project, including

J.R. Garrett, Ph.D. (✉)
Children's Mercy Bioethics Center, Children's Mercy Hospital,
2401 Gillham Road, Kansas City, MO 64108, USA

Department of Pediatrics & Department of Philosophy, University of Missouri,
Kansas City, MO USA
e-mail: jgarrett@cmh.edu

L.M. Rasmussen et al. (eds.), *At the Foundations of Bioethics and Biopolitics: Critical Essays on the Thought of H. Tristram Engelhardt, Jr.*,
Philosophy and Medicine 125, DOI 10.1007/978-3-319-18965-9_13

his own minimalistic permission-based bioethics, can survive its all-consuming reach. This tension reveals a number of notable "cracks" in the foundations of Engelhardt's bioethics, each tracing back to fundamental underlying assumptions embedded in his view. In each case, these assumptions are too "thick" for a purely procedural bioethics; they cannot adequately be defended by claims of default justification but only through content-full arguments, the kind that Engelhardt's critical project rules out as illegitimate in principle.

My strategy for making this argument will begin with Engelhardt's handling of paradigm cases (i.e., interactions between fully mature, independent, and autonomous individuals) before moving to his treatment of non-paradigm cases (with a particular focus on interactions involving children). First, I critically evaluate basic assumptions about the structure of the moral universe that motivate Engelhardt's procedural morality and argue that there are at least three sets of undefended premises that, by the standards of his own critical project, undermine its success even in idealized paradigm cases. I then examine how these general features of Engelhardt's theory, in particular his ultra-thin conception of consent underlying the principle of permission, yield particularly objectionable implications when applied in non-paradigm cases, including in pediatric bioethics. In other words, I will argue that core assumptions that already look unsupported and/or questionable in paradigm cases look even more dubious when taken to their logical endpoints in non-paradigm cases where they fail to take seriously the dependency and vulnerability of children and other non-autonomous beings.

The upshot of my argument, then, is that Engelhardt's twin projects generate a dilemma. If the critical project is justified, then the result is not a default procedural morality, but rather a nihilistic rejection of all morality, including the presumed authority of permission-giving. However, if the constructive project (with all of its morally thick elements) can survive the critique, then arguably much else can, too –including principles and values that Engelhardt repeatedly has rejected throughout his scholarly career. Either way, addressing the cracks in the foundations will require fundamental revisions to the basic blueprints for Engelhardt's bioethics.

13.2 A Brief Overview of Engelhardt's Secular Bioethics

A clear statement of Engelhardt's overall project can be found in the Preface to the Second Edition of his *Foundations*. Responding to various misperceptions of the First Edition, Engelhardt emphasizes that his libertarian bioethics does not derive from any substantive premises about the value of liberty itself, but instead falls out simply as the *default* conclusion of taking moral diversity and pluralism seriously:

> The account offered in *Foundations* does not provide a content-full ethics by which men and women can live their concrete moral lives. Rather, it justifies a moral framework by which individuals who belong to diverse moral communities, who do not share a content-full moral vision, can still regard themselves bound by a common moral fabric and can appeal to a common bioethics. It offers a moral perspective that can reach across the diversity of

> moral visions and provide a moral lingua franca. By not endorsing a particular moral vision, it seeks to avoid the difficulties that beset secular accounts that do: they beg the question, arbitrarily affirm a particular point of departure, or invoke an infinite regress. (Engelhardt 1996, ix)

According to Engelhardt, then, all content-full moral theories, including principlism, utilitarianism, Kantianism, communitarianism, and even libertarianism when derived from substantive values or principles, are doomed to fail when applied to a large and diverse society with many competing visions of the good and the right. Thus, if the project of deriving a content-less moral framework that can bind moral strangers also fails, then moral nihilism is the inevitable result.

Having taken his readers to the precipice of nihilism, though, Engelhardt reassures us that we need not go over the edge. Rather, a way forward can be found through a framework of "bare" permission-giving. According to Engelhardt, "… when individuals attempt to resolve controversies and do not hear God (or do not hear him clearly) and cannot find sound rational arguments to resolve their moral controversies, they are left with the device of peaceably agreeing how and how far they will collaborate" (1996, x). Again, though, it is critical to understand that while "the morality that binds moral strangers has by default an unavoidably libertarian character…, this is not out of any value attributed to freedom or individual choice," but instead because "the only morally authorized social structures under such circumstances are those established with the permission of the individuals involved" (1996, x). Thus, secular bioethics is made possible, and *only* made possible, by what Engelhardt calls "the principle of permission," which holds that "secular moral authority is derived from the permission of those involved in a common undertaking"(1996, xi).

The principle of permission plays an equally central role when Engelhardt's bioethical framework is applied in the pediatric context, where he grounds his conclusions regarding extensive parental decision-making authority in a libertarian social-constructivist account of the family. As with his more general bioethics, Engelhardt here makes another "default" argument, claiming that his libertarian pediatric bioethics "presumes no particular normative view of the value of liberty, but only a recognition of the authorizing force of permission" (Engelhardt 2010, 512). However, permission functions slightly differently in the context of the family than it does among adult moral strangers. According to Engelhardt, "family members, in remaining members of a family, convey legitimacy to the family's authority over its members through the act of remaining within the family. By remaining with his family and by not seeking exit from the family…, a child agrees to place himself under the authority of the family" (Engelhardt 2010, 512). Importantly, this act or process of tacit consent is unconstrained by antecedent views of children's liberty, equality, rights, welfare, or any other substantive value. Libertarian pediatric bioethics, according to Engelhardt, "accepts actual choice as legitimating whatever choices are made," including choices "not to be free and not to be equal" (Engelhardt 2010, 512). Thus, on Engelhardt's view, children "are at liberty to submit to family authority in whatever areas, including pediatric decision-making, they choose" and effectively consent to a given decision whenever and so long as they do not exit the family.

13.3 Three Key Assumptions in Engelhardt's Bioethics

What should be clear from my brief summary of Engelhardt's bioethics is that he places enormous weight on the necessity and sufficiency of a procedural morality rooted in a "presumption" of the authorizing force of permission. Underlying this general presumption, I argue, are three sets of undefended assumptions that need to be critically analyzed and evaluated, including assumptions regarding:

1. the background circumstances of morality;
2. the necessity and sufficiency of procedural morality;
3. the legitimacy of certain actions when procedural morality breaks down.

In some cases, I simply call attention to the fact that important assumptions are being made that bear considerable moral weight in Engelhardt's framework; in other cases, I critique the assumptions and argue that there are in fact good reasons to reject them.

13.3.1 Assumptions About the Background Circumstances of Morality

Engelhardt makes a number of important assumptions about the background circumstances of morality that then inform his constructive project of grounding the foundations of bioethics in the principle of permission. For present purposes, I'll focus on three such assumptions and analyze them in turn.

13.3.1.1 Assumption #1 – Moral Disagreement Is Ubiquitous, Fundamental, and Intractable

According to Engelhardt, even the most cursory examination of human social life will bring one into contact with the fact of moral diversity. Moreover, Engelhardt assumes that this moral diversity is neither rare nor shallow nor resolvable; to the contrary, moral diversity is ubiquitous, fundamental, and intractable. Because so much of Engelhardt's argument, like that of many cultural relativists, depends on this initial picture of moral diversity, it is worth analyzing these three purported features more carefully.

First, there is the assumption that moral diversity is *ubiquitous*. On Engelhardt's view, moral diversity is extensive and widespread, the norm rather than the exception. People and groups disagree far more often than they agree and, hence, regularly find themselves locked in conflicts with others regarding matters of ethics. This empirical claim is significant to Engelhardt's argument, as his positive project for conflict resolution via permission-giving looks less compelling to the extent that there is less disagreement from the start.

Second, there is the assumption that moral disagreement is often *fundamental* in nature. Engelhardt claims that different people and cultures not only disagree about moral conclusions, but they often disagree *because* they are committed to *fundamentally different basic norms* and not simply because they operate with different factual beliefs or metaphysical or conceptual frameworks. The disagreement is fundamental, then, because it "would persist even if both parties agreed about all of the relevant non-moral facts about the disputed issue" (Timmons 2002, 48). To illustrate this distinction between fundamental and nonfundamental moral disagreement, Timmons offers the example of a society where one's parents are, as a matter of course, put to death at a certain age. This looks like a case of drastic and significant moral diversity when compared with practices in Western societies. However, the disagreement actually may not be fundamental in nature. Suppose that this society in question has a deeply ingrained, widespread belief that awaiting their parents is "an afterlife full of activities like hunting and playing and that one has the body in the afterlife that one has had in this life" (Timmons 2002, 49). In such a case, the society does not hold fundamentally different basic norms about respecting one's elders or parents; indeed, on their view, they are actually treating their elders or parents with the highest respect. The disagreement, then, is *nonfundamental*, deriving from differing metaphysical beliefs about death and the afterlife. This distinction between fundamental and nonfundamental moral disagreement is important to note here because nonfundamental moral disagreement is often, at least in principle, rationally resolvable through appeals to evidence and observation. Thus, if much of the moral diversity in the world turns out actually to be nonfundamental in nature, then there may be many ways of resolving moral conflict aside from deferring exclusively and entirely to the principle of permission.

Finally, there is the assumption that moral diversity is often *intractable* in nature. Here the idea is that moral diversity, perhaps because it is assumed often to be fundamental, cannot be resolved in itself. It is a fact of life with which we simply are stuck and thus forced to work around via individual acts of permission-giving. The importance of this assumption of intractability, as with the other assumptions above, can hardly be overstated when evaluating Engelhardt's argument. If much of the moral diversity in the world turns out to be tractable, whether because it is nonfundamental or because even fundamental moral disagreements can resolve over time and under the right circumstances, then the case for Engelhardt's principle of permission will shrink accordingly.

13.3.1.2 Assumption #2 – The Categorical Distinction Between Moral Friends and Moral Strangers Constitutes an Exhaustive Difference of Kind

The distinction between moral friends and moral strangers is fundamental to both Engelhardt's critical and constructive projects. According to Engelhardt, for any two given people in the world, there exists one of two moral relationships: moral friend or moral stranger. Moral friends are "those who share enough of a content-full morality so that they can resolve moral controversies by sound moral argument or by an appeal to a jointly recognized moral authority whose jurisdiction they acknowledge as derived from a source other than common agreement" (Engelhardt 1996, 7). Moral strangers, on the other hand, are those "who do not share sufficient moral premises or rules of evidence and inference to resolve moral controversies by sound rational argument, or who do not have a common commitment to individuals or institutions in authority to resolve moral controversies" (Engelhardt 1996, 7). Lacking the thicker moral resources for resolving (ubiquitous) moral conflicts that are available to moral friends, then, moral strangers are left with nothing more than the principle of permission if their conflicts are not to degenerate into exchanges of aggression and force.

Implied in these definitions is the idea that these two categories constitute a difference of kind, divided by a threshold point where the quantity and quality of moral resources shared between two people are either sufficient or insufficient for resolving moral controversies. Importantly, Engelhardt does not qualify his claim here, so we are left wondering whether moral friendships are formed on a case-by-case basis or if they require that the two parties be able to resolve *all* (possible? actual?) moral controversies via their shared content-full morality. The implications of this lack of qualification could hardly be more profound. If moral friendships require that two parties be able to resolve all moral controversies via their shared content-full morality, including all controversies about how to rank goods and values and interpret rules and principles, then it is doubtful that there are any moral friendships that obtain in our world. It would, frankly, be shocking if any two people shared the exact same ranking of all goods and values and precisely the same interpretations of all rules and principles. And Engelhardt has claimed elsewhere in his work that "moralities need only order key human goods and right-making conditions in different fashions to be different" (2011, 250) and, similarly, that "different sets of guiding judgments as to how one ought to rank values and moral principles will constitute different moralities" (2007, 123). However, if moral friendships form on a case-by-case basis for any particular moral conflict (such that the same person could be a moral friend to me in some situations and a moral stranger to me in others), then the resulting picture of moral relationships will be messy and complex while also undermining Engelhardt's claims about the comparative value of maintaining attachments to thick moral communities.

This point raises a larger question about Engelhardt's conceptualization of the moral universe. If the basis for moral diversity is not just a matter of divergent moral *conclusions*, but rather diverse moral *rationalities*, then one must worry about the usefulness of Engelhardt's categories. It is doubtful that any two people share the exact same moral rationality, even if they, by some chance, held most of the same moral conclusions. There will certainly be some differences in their webs of moral belief, whether in terms of different foundational premises, rules of inference, value rankings, factual beliefs, and so on. Even when two people are comfortable "just" appealing to God or the same thick religious or cultural tradition, this does not entail the existence of identical moral rationalities. Moreover, any appeal to the authority of a particular human (or even supernatural) authority or agency will be balanced by background beliefs and values. For example, two people might declare, *in the abstract*, "let the Bishop decide," but there are surely some decisions that the Bishop could make that would be rejected when they severely conflict with fundamental axioms at the core of someone's web of belief. All of this raises a number of fundamental questions for Engelhardt's account. Is there really such a thing as a moral community where all unanimously agree? If not, then why not just view the state as being a larger version of a non-unanimous community? And, if unanimity is insisted upon in all decisions of any consequence, then won't communities be exceedingly small (far smaller than the sects and denominations that figure so prominently in Engelhardt's discussion of communities of moral friends)? Does Engelhardt drastically exaggerate both the dearth of resources shared by moral strangers and the abundance of resources shared by moral friends? These and other related questions reveal the significant role that Engelhardt's assumptions about moral relational categories play in his bioethics.

13.3.1.3 Assumption #3 – Obligations Not to Harm Are Stronger Than Obligations to Benefit

In his analysis of the principles of bioethics, Engelhardt makes a third important assumption, though one that initially appears to have an argument made on its behalf. Reflecting on the priority and relative weight of the principles of beneficence and non-malevolence (or perhaps more accurately, non-maleficence), Engelhardt makes the following claim:

> …the obligation to be non-malevolent is stronger than the obligation to be beneficent. In the case of the failure to be beneficent, one does not live up to the core goal of morality, achieving the good. But in the case of malevolence, one acts against this goal. (Engelhardt 1996, 110)

He goes on to claim that "this much seems plausible even without an appeal to any content-full understanding of good or evil" (Engelhardt 1996, 110).

However, it is unclear on what basis such a claim can be defended. For one thing, Engelhardt here assumes that the "core goal of morality" is "achieving the good," which is by no means a content-less premise shared by all moral theories. For another, Engelhardt's claim rests on the undefended assumption that there is a stronger obligation to refrain from acts contrary to the goal of morality than to act in support of that goal. Again, it is not clear on what basis this assumption can be defended absent a content-full argument about the moral nature and quality of acts and omissions. Such assumptions, however elementary, foundational, or "plausible" they may seem, do not fall out of a value-neutral system of logic in the same way as, say, the law of excluded middle. Rather, they are value-laden assumptions about how to decide and how to act.

Appreciating this fact is important in its own right, since it highlights just how impoverished a purely procedural morality will be when it cannot smuggle in these kinds of assumptions. However, it's also important because, as I shall argue momentarily, Engelhardt uses these assumptions to lend support to a number of important claims about the legitimacy of certain actions when procedural morality breaks down.

13.3.2 Assumptions About the Necessity and Sufficiency of Procedural Morality

Consent plays a central rôle in Engelhardt's treatment of paradigm bioethical cases involving interactions between adults. Given the "limits of secular moral reasoning," he concludes, consent becomes the only "means (within certain constraints) of giving moral authority to common undertakings without establishing the moral worth or moral desirability of any particular choices" (Engelhardt 1996, vii–viii). According to Engelhardt, then, the act of giving permission legitimizes actions between moral strangers that otherwise would be illicit, and it carries moral authority without embodying any particular moral content.

Engelhardt's views on consent and permission rest on a number of important assumptions about the necessity and sufficiency of voluntaristic procedural morality, assumptions that cannot be adequately defended through claims of default justification but only through content-full arguments. For example, consider several such assumptions made in the following important passage in *Foundations*:

> This account of the morality that can bind moral strangers, the one element of the Enlightenment hope that survives, is still sufficient to justify a wide range of human practices and health care policies. Indeed, *it can justify all practices that draw their authority from bare consent* or from the necessary forbearance from using individuals without their consent, which lies at the foundation of the very possibility of a general secular morality. Thus, a justificatory account can be given of such practices as free and informed consent, the market, and limited democracies. In the acquisition of permission, consent, or agreement, all that will be foreclosed is coercive force on the part of the one seeking agreement. *One may take market advantage of those coerced by nature or third parties*, insofar as those entering the agreement are not responsible for such coercion. It will not be improper to employ inducements or to engage in peaceable manipulation to garner permission, agreement, or consent. Persuasion, inducements, and market forces are means of making it worthwhile for individuals to agree to join in particular undertakings. Such manipulations form part of the fabric of a secular society of persons acting with common authority, as long as such manipulations are peaceable, *as long as they do not involve any threats of force or unconsented-to-interventions* that are undertaken against the possibility of free agreement... (Engelhardt 1996, 71, emphasis added)

In this passage, one finds at least three significant assumptions about procedural morality. First, there is the assumption that permission-giving is *sufficient* to justify any action or practice so long as even "bare" consent is obtained. The reference to "bare" here seems to imply that consent or permission does not need to be informed or satisfy any other qualitative measure. Simply uttering the relevant words, and perhaps even forming the intention to do so, is viewed as sufficient justification for whatever follows. Second, there is the related assumption that the validity of permission is not compromised by the coercion of nature or third parties. Finally, there is the assumption that "force" is essentially physical in nature and, hence, that any psychological or financial aggression is morally irrelevant.

The important point to recognize here is that *not one of these three assumptions can be derived from procedural morality itself or from a morally neutral system of logic or reasoning*. They each rest on content-full judgments about the necessity and/or sufficiency of ("bare") consent and permission or about the nature of freedom and coercion. One can hardly avoid being struck by how Engelhardt here appears to be engaged in the same type of content smuggling of which he charges Kant. In a lengthy footnote in *Foundations*, Engelhardt charges Kant with incorporating "into his notion of rational action numerous positive understandings of acting rationally" (1996, 132, note 6). But he appears to do something structurally identical with his construction of the principle of permission, incorporating into his account of its moral necessity and sufficiency numerous positive understandings of acting freely or forcefully.

13.3.3 Assumptions About the Breakdown of Procedural Morality

In addition to making assumptions about the necessity and sufficiency of permission-giving, Engelhardt's framework also incorporates content-full normative assumptions about the breakdown of procedural morality. So, for example, when someone violates the principle of permission and initiates non-consensual force against another, Engelhardt makes substantive ethical claims about the legitimacy of self-defense and punishment that cannot be supported by internal appeal to procedural morality. In such cases, I will argue, one absolutely must have a thick theory of the good and right in order to take a position on these matters, the kind of theory which Engelhardt's broad-based moral skepticism rules out in matters between moral strangers. Simply put, ethical propositions regarding the legitimacy of self-defense or punishment cannot be derived by "default" through value-free inferences.

It will be helpful here to identify and analyze some prototypical examples of such assumptions in Engelhardt's claims about *self-defense* (i.e., the use of force against a present or imminent violation of the principle of permission) and *punishment* (i.e., the use of force as retribution for a past violation of the principle of permission). In such cases, he claims that "…anyone who acts malevolently and without permission has no grounds to protest when visited with…defensive, punitive, or retaliative force" (Engelhardt 1996, 110–111). Why? According to Engelhardt, violators of the principle of permission have no grounds to protest retaliatory force because their action places them "outside the peaceable community"; they "cannot consistently appeal to a principle they have rejected in order to condemn the users of force" (Engelhardt 1996, 109).

However, there are at least two ways in which this argument fails. First of all, the argument assumes that the *only* principle that the violator could appeal to in condemning the retaliating party is the principle of permission, but this clearly is not the case. There are at least three principles that the violator could appeal to in condemning the retaliation that do not simultaneously undermine his or her own initiation of force. One principle would be rooted in ethical egoism and would see right actions as those which promote the interests of the actor and wrong actions as those which set back those interests. A second principle would be one which permitted the initiation of force but condemned the use of retaliatory force, perhaps through an esoteric interpretation of Biblical ethics (e.g., a principle that combined various Old Testament dictates to attack and plunder other tribes with Jesus' admonition to turn the other cheek when someone strikes you first). A third principle, specific perhaps to the case of punishment, would be one that endorsed a restorative or rehabilitative response to past violations. This principle would enable the violator consistently to protest retributive acts, so long as he or she was willing to restore or rehabilitate when violated against. However problematic these views might be in their own right, they are *potential* principles that are available to the violator of the principle of permission. One cannot rule them out by default, but only by appealing to content-full moral arguments.

An additional, and likely more powerful, objection to Engelhardt's argument about self-defense and retributive punishment concerns the failure to account for the dimension of *proportionality* in the use of force. Engelhardt seems to assume that no type or degree of retaliatory force can be protested (consistently, at least) once a violation of the principle of permission is enacted or even threatened. Even the most minor violations could "legitimately" receive deadly force in return. For example, imagine a case where someone violates the principle of permission by pinching a non-consenting person who is not wearing green on St. Patrick's Day. Even if it is plausible to conclude that the violator has no consistent grounds to protest being pinched in return if they lack green attire, it seems quite implausible to conclude that this violator would have "no grounds to protest" being tortured and shot dead in retaliation for such an "offense." Yet unless Engelhardt is willing to allow for content-full views of ethically sound proportionality to operate within his framework (something that cannot be derived from procedural morality alone), it appears that he is vulnerable to such counterexamples.

Thus, for reasons of overlooking additional principles as well as neglecting the issue of proportionality, Engelhardt's conclusions about the legitimacy of self-defense and punishment either must be rejected or he must allow for some content-full moral argument to be introduced (and, of course, critically evaluated) in their support.

13.4 Evaluating an Example of Engelhardt's Treatment of Non-paradigm Cases: Permission, Property, and Parental Authority in Pediatric Bioethics

There is a regrettable pattern among leading moral and political philosophers to focus exclusively on paradigm cases at the level of ideal theory. When their theories arrive in the vicinity of children and other "marginal cases," they note that these are "special cases" that will need to be addressed at another time, a time which rarely ever comes. And when these theories do actually deal with these "special cases," they often do so in a tortured way that introduces ad hoc qualifications that water down or contort their more general arguments so that they don't produce the counterintuitive results that would seem naturally to flow from them.

It is a testament, then, to Engelhardt's intellectual courage and willingness to occupy the margins of bioethics that he both addresses such cases head-on and that he does so with relatively little theoretical wiggling and backtracking. Rather than retreating from the natural logic of his more general premises about permission and proceduralism, he instead extends these premises relatively uncompromised into the context of pediatrics and other non-paradigm cases.

However intellectually admirable, though, I will argue that it is in this extension of his more general framework to address non-paradigm cases that one can truly appreciate the salience of the key undefended assumptions discussed above. These core assumptions, which already look unsupported and/or questionable in paradigm cases, look even more dubious when taken to their logical endpoints in non-paradigm cases where they fail to take seriously the dependency and vulnerability of children and other non-autonomous beings. In what follows, I will defend this claim with a special focus on Engelhardt's pediatric bioethics, though much of what I have to say would also generalize to his treatment of other non-paradigm cases.

As a general rule, accounts of pediatric bioethics are fundamentally shaped by views about parental authority: the wider the scope of parental authority, the more that pediatric bioethics shifts from substantive to procedural morality, from engagement with independent concepts of children's rights and best interests to concern with parents' permission to examine and treat their children. According to Engelhardt, the only view of parental authority that is compatible with moral pluralism and the limits of secular moral reasoning is one where the scope of parental decision-making is extremely wide and largely unconstrained. So long as children "agree" to remain within and not exit the family, then this bare consent legitimates "whatever choices are made" for them by their parents (Engelhardt 2010, 512). On Engelhardt's view, then, parents have robust authority to make decisions for their children until those children remove themselves from that authority at a sufficient point of maturity and independence. Hence, for Engelhardt, pediatric bioethics is primarily a matter of parents exercising their vast range of authority via giving or withholding permission for others to examine and treat their children. Parental permission-giving grants to clinicians the authority to do things to the child that otherwise would be illicit, and there is no other means for conferring this authority. This does not entail that medical clinicians are obligated to provide whatever services that a parent might demand, but it does entail that parents have the ultimate authority regarding anything that is actually provided to their children.

In analyzing his account more carefully, there are at least five important components of Engelhardt's pediatric bioethics, each of which stems from more general commitments in his treatment of paradigm cases. Together, these five components constitute what he calls a "libertarian social-constructivist view of the family" (Engelhardt 2010, 511–513).[1] First, the view is fundamentally shaped by the demands

[1] Importantly, Engelhardt's own thick content-full view of the family probably falls within what he calls the "categorical view," embracing as it does "traditional heterosexual and gender-essentialist understandings of marriage" consistent with the traditional teachings of Orthodox Christianity (Engelhardt 2010, 512). However, given his views of moral pluralism and its implications for

of moral pluralism; it allows nearly all content-full views of the family to obtain when chosen by individuals, but rejects any such view when it is imposed for all members of a political community. Second, it has no prior, independent commitments to values like liberty and equality (or, we might assume, welfare and utility); it recognizes the default authorizing force of permission, but does not embrace this on the basis of the value of freedom or autonomy. Third, the libertarian social-constructivist view of the family is "grounded in the actual, nonrationally reconstructed consent of the family members"; hence, it accepts the barest form of tacit consent as valid with no qualitative criteria (i.e., being informed, being capable of understanding, and so on) needing to be satisfied. Fourth, it accepts such actual consent as "legitimating *whatever* choices are made" (emphasis added). Finally, members of a family are "at liberty to submit to family authority in whatever areas, including pediatric decision making, they choose." Taken together, these five components deliver a view of parental authority with a wide and virtually unconstrained scope of decision-making.

Indeed, one might see the libertarian social-constructivist view as committed to the same basic account of the family as the one proposed by another widely-known libertarian theorist, Murray Rothbard. In Chapter 14 of his work, *The Ethics of Liberty*, Rothbard unblushingly draws out what he takes to be the logical endpoint of the libertarian view of parental authority:

> …a parent does not have the right to aggress against his children, *but*…should not have a *legal obligation* to feed, clothe, or educate his children, since such obligations would entail positive acts coerced upon the parent and depriving the parent of his rights. The parent therefore may not murder or mutilate his child, and the law properly outlaws a parent from doing so. But the parent should have the legal right *not* to feed the child, i.e., to allow it to die. The law, therefore, may not properly compel the parent to feed a child or to keep it alive.

Such a view, if adopted, makes short work of most (secular) pediatric bioethics. If the legitimate scope of parental decision-making authority permits decisions not to feed, clothe, or educate children, then surely it also permits decisions to withhold any medication or treatment from them as well. And, since this view rejects any positive claims to be provided with a good or service by another person absent that person's agreement, it necessarily permits clinicians to withhold their goods and services from parents who request them. These two implications together would immediately address (arguably) 95 % or more of the ethical controversies in the average pediatric facility.

Regardless of the exact percentage, however, it certainly appears likely that, on this view of decision-making authority, (secular) pediatric bioethics will be almost entirely a *procedural* matter of determining what choices the relevant decision-making authorities want to make. These authorities include both the parents, who have full and unlimited negative-style rights of non-interference regarding anything that is done to their children, and the clinical team, who have full and unlimited

social and political interaction, the only view of the family that he could support at the level of political arrangements and public policy is the libertarian social-constructivist view.

rights regarding what services, devices, and treatments they actually provide upon request. In practice, this entails the following implications:

1. parents can withhold permission for any diagnostic or therapeutic intervention for their children;
2. clinicians and other potential decision-makers, including the state, cannot override parental dissent;
3. parents can permit and request (but not demand) the provision of any intervention, perhaps excluding those which would kill or severely disable/injure their children;
4. clinicians can deny any parental request for any reason, including reasons of individual conscience, professional integrity, or mere preference or prejudice.

Needless to say, these are stark and momentous implications that diverge significantly from standard views of pediatric bioethics, which in each case would attach qualifications and provisos to the relevant claim. As such, it is important to critically evaluate their validity by examining the basic core assumptions about ownership, parental authority, and procedural morality that give rise to them.

One core assumption underlying Engelhardt's pediatric bioethics is that parents "own" their offspring, first by expending "labor" and "extending themselves into" their children, and then later (following the point that children become "self-conscious") by their submission as "indentured servants" to "parental authority in exchange for parental support" (Engelhardt 1996, 157). This Lockean and Hegelian-inspired theory of property rights inherits numerous general problems, many of which were famously cited (without effective rejoinder) by Robert Nozick in his own Lockean account in Chapter 7 of *Anarchy, State, and Utopia* (1974). Among the general problems that apply unquestionably to Engelhardt's bioethics is the general question of the scope, force, acquisition/transfer, duration, and enforcement of property rights. One cannot derive conclusions regarding, for example, the right of simple use versus the right to exclude or the right of individual ownership versus collective ownership by neutral default inferences; property rights of all kinds might obtain in different situations and contexts and one can only determine specific property claims by evaluating *substantive* values and principles.

Additionally, Engelhardt's theory faces additional problems when applied in the special case of children, which have been developed and elaborated upon by Moskop (1997), Nelson (1997), and Hanson (2009, 26–35). For my purposes here, I shall focus only on one objection to the kind of parental "ownership" of children endorsed by Engelhardt: namely, that such a view cannot be defended by default, but only by presupposing "a certain contentful claim or theory of the good" (Hanson 2009, 26). In any case involving a moral conflict about pediatric decision-making, one cannot rule out the potential ownership or authority (temporary or permanent) of the child by a clinician or other decision-maker (e.g., the state) without making substantive assumptions about competing values or principles (e.g., assumptions that these agents are performing less labor, or less important or relevant labor, in caring for the child than are the parents).

Moreover, regardless of whether the relevant ownership-conferring labor is genetic/biological or nurturing/caring/rearing, there is also the potential for disagreements to arise between two parents as "co-owners" of a child (e.g., a mother supports one decision while a father supports another). In such cases, which are not rare in pediatric bioethics or other non-paradigm contexts, there is absolutely no way to determine who is in moral authority to make decisions in a default content-less manner. Procedural morality will run its course without resolving the case and any further action or inaction will have to be guided by substantive arguments.

Thus, Engelhardt's pediatric bioethics faces a dilemma. On the one hand, if Engelhardt's view of parental ownership is valid, then it both (1) limits his ability to provide resolution to an important range of cases that arise in this context (i.e., conflicts between those in co-authority to make decisions) and (2) undermines his larger claims about the illegitimacy of content-full arguments in secular bioethics (since his views of parental ownership can only be established by making substantive assumptions or arguments). On the other hand, if Engelhardt's view of parental ownership is invalid because it is limited and content-full, then he is left without a neutral "default" ground for his libertarian pediatric bioethics.

In addition to these problems with his underlying views of ownership, Engelhardt's account of bare (tacit) consent in pediatric bioethics is also objectionable. At the very least, his account depends upon assumptions about the moral sufficiency of such "consent" that are content-full. One cannot derive such an important moral claim (i.e., that simply residing within a family authorizes any decisions that parents might make for children) by default simply from the fact that a child remains in a family.

What is more, the specific content of these content-full assumptions is itself quite dubious, neglecting as it does any consideration for the vulnerability and dependency of young children through the age of potential independence. Typically, views regarding the moral sufficiency of consent presume a morally validating level of equality between parties (e.g., that there be some genuine possibility of not agreeing) and that both/all parties have sufficient capacity. Absent these basic conditions, one cannot presume anything substantive from even expressed, let alone tacit, agreement. Obviously, these basic conditions are often not present in pediatrics where dependency and vulnerability limit capacity and create inequality. Such features of children plausibly give rise to additional decision-making constraints that do not pertain to decision-making between competent adults. Children's lack of equal standing and lack of awareness that exit is even a live possibility undermine the claim that, by remaining in a family, they are "consenting" in any morally relevant and sufficient way to the decision-making authority of the parents/guardians.

This would be true, strictly speaking, in any case regardless of its momentousness; however, it is particularly salient in the kinds of life-or-death cases that rise to the level of ethical controversies in pediatric bioethics. As Hanson (2009, 37) has noted, our typical views of children's "indenture" to their parents may plausibly extend to things like "curfews, rules on dating and sexual interaction, and rules on the use of various mood-altering substances" but not to issues like removing or refusing life-saving treatment (captured in Hanson's imagined parental speech: "As long as you are under my roof, you will refuse life-saving treatment, and perhaps die, if I think that is the best thing for you, even if you do not.").

Even after the age of potential independence, Engelhardt's views about tacit consent are questionable. At the very least, if non-exit from the family is taken to be morally sufficient to support any parental decision regarding their children's health, then it is plausible to conclude that this choice should be made explicit to children and adolescents at the borderline of potential independence. Thus, if Engelhardt is serious that, "members of a family are at liberty to submit to family authority in whatever areas, including pediatric decision making, they choose," then clinicians (or perhaps agents of the state) would seem to have moral authority to inquire and seek to understand whether children/adolescents really want to submit to family authority when their own wishes diverge from those of their parents. This would, of course, require the insertion of third-parties into family decision-making in a way and to a degree that seems deeply at odds with Engelhardt's views regarding the autonomy of the family. However, it is difficult to see how the "choice" of an adolescent to remain within their family carries the kind of life-or-death moral weight that Engelhardt thinks it does if such choices are not made explicit. And, of course, even if such an explicit choice is made, one cannot presume anything regarding the moral sufficiency of the act without drawing on content-full premises that would be ruled out by Engelhardt's larger bioethical methodology.

13.5 Concluding Remarks

In conclusion, I have argued that Engelhardt's overall account generates a dilemma. On the one hand, his critical project, if successful, implies a principled rejection of all morality, including his own proposed framework, since no morally significant conclusions can be inferred by default without leveraging content-full premises (explicitly or implicitly) at some point. On the other hand, since Engelhardt's constructive project relies on morally thick assumptions at its core, its success would undermine his principled rejection of content-full bioethics and suggests that conclusions must be supported on the basis of substantive rather than procedural morality. And taking either horn of this dilemma would dramatically reshape Engelhardt's account of bioethics.

In the end, then, Engelhardt's most lasting contribution to the foundations of bioethics may be neither his critical nor his constructive projects as such, but rather the powerful way in which he has called our attention to the important and inescapable need to examine such matters. If, as I have argued here, there really are no default propositions that can be arrived at through neutral content-less argumentation, then Engelhardt's foundational challenge is even more pressing and significant. Responding adequately to this challenge might require radical rethinking of key bioethical assumptions and approaches, perhaps developing novel integrative approaches to value conflict (Garrett 2014) or alternative moral epistemologies to the foundationalist approach presumed by Engelhardt's critique. What is clear, though, is that bioethicists cannot simply ignore the philosophical and ethical foundations of their field, no matter how difficult and perplexing such work might be.

References

Engelhardt Jr., H.T. 1996. *The foundations of bioethics*, 2nd ed. Oxford: Oxford University Press.

Engelhardt Jr., H.T. 2007. Bioethics as politics. In *The ethics of bioethics: Mapping the moral landscape*, ed. L.A. Eckenwiler and F.G. Cohn, 118–133. Baltimore: The Johns Hopkins University Press.

Engelhardt Jr., H.T. 2010. Beyond the best interests of children: Four views of the family and of foundational disagreements regarding pediatric decision making. *Journal of Medicine and Philosophy* 35: 499–517.

Engelhardt Jr., H.T. 2011. Confronting moral pluralism in posttraditional Western societies: Bioethics critically reassessed. *Journal of Medicine and Philosophy* 36: 243–260.

Garrett, Jeremy R. 2014. Two agendas for bioethics: Critique and integration. *Bioethics* 1–8. doi:10.1111/bioe.12116.

Hanson, Stephen S. 2009. *Moral acquaintances and moral decisions: Resolving moral conflicts in medical ethics*. Dordrecht: Springer.

Moskop, J.C. 1997. Persons, property, or both? Engelhardt on the moral status of young children. In *Reading Engelhardt: Essays on the thought of H. Tristram Engelhardt, Jr*, ed. B.P. Minogue, G. Palmer-Fernandez, and J.E. Regan, 163–174. Boston: Kluwer Academic Publishers.

Nelson, J.L. 1997. Everything includes itself in power. In *Reading Engelhardt: Essays on the thought of H. Tristram Engelhardt, Jr*, ed. B.P. Minogue, G. Palmer-Fernandez, and J.E. Regan, 15–29. Boston: Kluwer Academic Publishers.

Nozick, Robert. 1974. *Anarchy, state, and utopia*. New York: Basic Books.

Rothbard, Murray N. 2003. *The ethics of liberty*. New York: NYU Press. p 100

Timmons, Mark. 2002. *Moral theory: An introduction*. Lanham: Rowman & Littlefield Publishers, Inc.

Part II
Personal Reflections

Chapter 14
Two-Score Years Ago

Laurence B. McCullough

Meeting and working with H. Tristram Engelhardt, Jr., 40 years ago changed my life in ways for which I am ever grateful. I have had the privilege of elsewhere publishing a personal portrait of my friend and colleague, Herr Professor Dr. Med. Dr. Phil. Engelhardt (McCullough 1997). Here I want to touch on two aspects of my history with Tris: my first days with Tris in Galveston, Texas; and as the beneficiary of the most impressive and effective academic placement service I have ever encountered and ever expect to encounter.

14.1 First Days in Galveston, Texas

In May of 1974, I received a phone call that started me on my way to Galveston, Texas, and work with Tris. Tris' Doktorvater, Chet Lieb (1925–1992) (Professor of Philosophy at the University of Texas at Austin and, at that time, Vice President and Dean of Graduate Studies at The University), called me and said that Tris had a new position for a research assistant at the Institute for the Medical Humanities in the

Tris will be very, very upset with even such an oblique and mathematically inaccurate literary allusion to the Gettysburg Address and its author, whom I revere. When I lived in the Washington, DC area during the 1980s, one of my favorite things to do was to visit Mr. Lincoln at night, just like President Nixon (whom my father once looked in the eye in the 1956 campaign and concluded was evil incarnate but, being a loyal Republican, voted for him. Loyalty counts; ask Tris). Tris, it seems, has another view of the Savior of our Union. On this, Stephen Wear and I agree, Tris is wrong, but we still love him. Tris will, however, be pleased that this little essay has four footnotes.

L.B. McCullough (✉)
Center for Medical Ethics and Health Policy, Baylor College of Medicine,
One Baylor Plaza MS 420, Houston, TX 77030-3411, USA
e-mail: Laurence.McCullough@bcm.edu

L.M. Rasmussen et al. (eds.), *At the Foundations of Bioethics and Biopolitics: Critical Essays on the Thought of H. Tristram Engelhardt, Jr.*, Philosophy and Medicine 125, DOI 10.1007/978-3-319-18965-9_14

University of Texas Medical Branch at Galveston, or UTMB as all culturally informed folk know it. I had met Tris just once before, at a dissertation defense in the Department of Philosophy at The University, from which Tris received his PhD, interrupting his medical studies at Tulane to do so. I took the position for several reasons, among which was that I needed the money. Another was that my father had taught me and my siblings to grab good opportunities that might come our way and that only "damn fools" turned down such opportunities. Little did my father know how right he was. To find out why, dear reader, read on.

The official reason for my being hired was to assist Tris with one of the first summer seminars in bioethics and medical humanities sponsored by the National Endowment for the Humanities. These seminars played a crucial – and, I fear, long forgotten – role in creating programs in bioethics and medical humanities in American medical and other health professions schools or, to be highfalutin, academic health centers. The participants, members of medical and nursing faculties, were to come to Galveston for a multi-week, intense seminar, led by Tris, and then return to their academic institutions and promote the teaching of bioethics and medical humanities. This resulting seeding of the field has paid enormous dividends, a fine example of our tax dollars actually doing not only some good, but lasting good.[1]

Not one to waste resources, Tris put me to work on other projects as well. When it comes to his research assistants (to identify many of them, see page numbers in the index to the second edition of *Foundations*, under 'slave,' courtesy of Mark Cherry, now of St. Edward's University in Austin), Tris has little or no commitment to the 13th Amendment to the Constitution of the United States, in part because, as

[1] Readers should know that, among Tris' many extrinsic denominations (look it up!), is the Golden Fleece Award in 1976 by Senator Proxmire of Minnesota (which is in the United States of America and therefore means that the extrinsic denomination is not much of an extrinsic denomination) to the National Endowment for the Humanities (NEH) "For spending at least $750,000 this year on grants to doctors and others to attend vacation-like, month-long seminars." (See http://www.taxpayer.net/user_uploads/file/Awards/GoldenFleece/fleeclst.pdf, accessed May 31, 2013.) Like most statements made by Americans, this is false. The participants in the 1974 Summer Seminar at UTMB stayed in the Flagship Hotel, on a pier over The Gulf. They were from the United States and so they panicked when they saw what they took to be large roaches – Galvestonians do not pay attention to tokens of this type less than two feet in length – "all over" their rooms. The Hotel (now demolished and replaced by some kind of pleasure palace), sensitive to the needs of visitors to Texas from international destinations, had had an exterminator come in and treat the rooms the day before our participants arrived. On the first morning of the seminar, Tris explained to the Seminar participants that they should be grateful for this respect for them as persons but his argument, complete with quotations from Latin and German (lots of Hegel) sources, fell on deaf ears. Then he made them work very hard for the month, with mountains of reading and hours of intense seminars in a windowless room in the Blocker Collection in the UTMB Library. It's actually a spectacularly beautiful room, lined with bookcases holding fabulous rare books. Notwithstanding, all Texans know that, if you want to vacation on The Gulf, you go to Matagorda Bay or Laguna Madre and remember the heroes of Coleto Creek and La Bahia. As Leibniz would conclude, this extrinsic denomination is *non bene fundadatum. Sic transit gloria metaphysica et Americana Proxmirae.* (All of these italicized words appear in your Lewis and Short on your bookshelf; ask Tris if you do not know what I am referencing. '*Proxmira*' is one of those rare first-declension, masculine nouns and appears several times in little-known footnotes in the works of Hegel and therefore in *Foundations*.)

a Texan with generations-deep roots in our country, Tris holds no truck with the constitutions of foreign powers. Tris will, for example, gladly instruct you that Independence Day comes, as every culturally informed person knows, each March, not in July. Who needs Bunker Hill (really Breed's Hill) when we have San Jacinto?[2] In part, too, Tris believes in hard work. He may even be the Platonic Form of hard work, a matter to be settled by an entity far, far above my pay grade.

Some of my other work for Tris included reading papers from a conference on medicine and philosophy at UTMB that Tris had convened earlier that year. These papers become Volume I of the now more than 120-volume Philosophy and Medicine book series (published by Reidel, then Kluwer, and now Springer, as the post-World War II publishing industry has consolidated internationally, like so many other manufacturing industries), which series is now increased by the present volume. One of the papers was Tris' own, which he asked me to read, by the next morning. I took it back to my room in one of UTMB's medical fraternities – there were such institutions in ancient times – presided over by a distinguished mutt named "Buddy," who could dive gracefully from the low board and outswim most of the fraternity brothers, because he, unlike them, was not disabled by altered mental status secondary to ETOH ingestion. Some of those fraternity brothers may now be your physicians, dear reader.

Tris will, if you ask and maybe even if you do not ask, tell you the story about what happened the next morning. First, you need to know that Tris' secretary in the Institute, the wonderful Ms. Marjorie Huffman, thought that I had attended a military academy, so good were my manners those first few days. Little did Marjorie know that I was a fish altogether out of water, in full struggling-to-survive mode in the new and completely unfamiliar setting of a medical school and medical center, not to mention Tris' mild, soft-spoken demeanor and undemanding ways. (To this day, I go into tachycardia when someone says to me, "Just one more thing." My cardiology colleagues come on the run to my aid.) The setting was also imposing: UTMB is built in hurricane-architecture style, resulting in massive, solid buildings that need to, and do, remain standing when storms come in from The Gulf. My parents had taught me to be on my best behavior in such circumstances. So I was.

Until the morning in question. I went into Tris' office, not closing the door behind me. Marjorie's desk was in the adjoining space, a few feet from the open office door.

[2] Tris will tell you that the Texian army was victoriously commanded at swampy San Jacinto, a few miles to the east of where we live (I in Houston and Tris in oh-so-tony West University Place on a street named for Lafayette College, which, in turn and in shame from which Tris has yet to escape, is named for a frog, Marie-Joseph Paul Yves Roch Gilbert du Motier, Marquis de La Fayette, who, as a God-help-us-all American general, helped win the War of Independence and American freedom from the tyrant George III and his ever-ready minions of evil in Parliament in London) by Big Drunk, aka The Raven, who saw two eagles soaring over the camp of the Napoleon of the West and his troops, many of which were about to be mercilessly cut down in revenge for Goliad, not the Alamo (whose defenders, including some Irishmen, refused quarter, met their fate, some while trying to escape on the final day of the siege, and passed into legend). Yorktown in Virginia plays second fiddle, although it is one of my favorite history destinations in the United States of America; please don't tell Tris I said that.

Those who know me know that I sometimes do not speak in a quiet voice, a survival-positive adaptive trait when growing up in a large Irish-American household. In what I still, correctly, regard as a normal voice – but which those outside the Tribe do not – I responded to Tris asking what I thought about his paper.

This is a scholarly report of what happened next. For a truthful account, with many – and constantly rewritten – grace notes, ask Tris. (Do not ask Tris to sing grace or any other notes; your auditory system will take weeks to recover.) I responded to Tris' query by explaining, apparently with alarm, that someone had been putting out bovine extirpation under Tris' name. Well, that phrase was not my precise word choice. I used a word that begins with a 'b' and ends with a 't' and that Texans can stretch out to at least fifteen syllables but Americans can enunciate only in two. Tris was unmoved, as I expected. Marjorie – out of sight around the corner but not out of earshot – *was* moved, Tris told me later. Needless to say, I was no longer the polite graduate of a military school.

More importantly, I learned that summer, and the summer of 1975 for a second offering of the NEH seminar, about the exhilaration of intense intellectual work and teaching in the health professions. Tris taught me that the lives and health of patients serve as an unyielding corrective to all that we do in medical ethics research and teaching. Little did I suspect that Tris was turning me into a medical educator. I am very, very glad that he did.

14.2 The Platonic Form of Placement Services

Success in one's academic life requires mentors who work to get one placed in one's first and subsequent academic positions and who prepare one to become a successful academician, in teaching, research, and getting published. In the two summers I worked for Tris, he undertook an intense program of preparing me to find an academic position. He taught me how to write multiple drafts of ethics papers for the medical literature, in an era in which 'multiple drafts' meant retyping on a typewriter (look it up!) from beginning to end. For Tris, 'multiple drafts' means 1 short of the nondenumerable infinite.

In the summer of 1975, Tris geared up to full fellowship-application and job-placement mode, a sight to behold. He showed me how to organize a curriculum vitae and how to present it as a visually pleasing, informative document, a crucial set of skills that I use to this day. He instructed me to have my CV copied onto bond paper (look that up, too!), which I did. One medical school to which I applied sent back my CV to Tris, with a note from Tris' history of medicine colleague who was running the job search, stating that he did not want to keep the original of my CV, had made a copy, and was returning the original. I came to know later that Tris spent a great deal of time working on individual letters of support for me. He wrote original letters, not photocopies, and on bond paper that he then sent to each institution to which I applied. He also made phone calls on my behalf, even if that meant placing international calls to the US of A. To call his efforts tireless would mark me down for egregious understatement.

Tris' efforts paid magnificent dividends, in what was then an already increasingly tight academic market. I was offered and accepted my first academic appointment at Texas A&M University, in its then-new medical school and Department of Philosophy. I was then very fortunate to be awarded a post-doctoral fellowship, funded by the NEH, at the Hastings Center, then in teeniny Hastings-on-Hudson in a mansion that was dominated by a magnificent beech tree and enjoyed a spectacular view of Henry Hudson's river. I recall that winter observing the tidal effects (Marc Lappè (1943–2005) had to explain this to me) of the Atlantic Ocean on Captain Hudson's river, in the form of ice floes going forcefully upstream. Tris assured me that, were I to ask him with better manners than I displayed that morning during my first days in Galveston, James A. Knight, MD (1918–1998), A&M College of Medicine's founding dean, would hold my position for a year. Jim said of course and off to the Northeast I went, returning Southwest, toward home,[3] to start at A&M in the fall of 1976.

Tris went back to work on my behalf after, in the fall of 1978, John McDermott, then Head of the Department of Philosophy at A&M, showed me an ad for a position in bioethics in the medical school and Kennedy Institute of Ethics at Georgetown and said that I should apply. I did. I was fortunate to be hired into that position by Warren Reich, who, although an American, is originally from Alabama, which is not far from Texas, is also on The Gulf, and sent its native son to command the garrison at The Alamo in San Antone and write the immortal words of defiance dear to Texans who refuse, like Tris, to live only in the present, "Liberty or Death." Tris will be glad to provide a detailed scholarly commentary, not too much of it made up, on this last missive from the doomed heroes of 1836 from within the walls of what is now known as The Shrine. Tris can explain that too; allow several hours.

I became the first of many of Tris' research assistants, many of them his *Doktorkinder*, to benefit from the Engelhardt placement machine. I am confident that I speak for that very long line of colleagues who had the very good fortune to come after me and therefore whom Tris similarly prepared and supported in becoming professors when I say to Tris, *vielen dank*![4]

---

[3] This allusion is to the great Willie Morris (1934–1999) (Morris 1967), a native Mississippian (Yazoo City, to be precise, which has a branch office of the Texas Farm Bureau, of which Tris is a member, and which ought to be the name of historical ruins deep in the swamp known as The Big Thicket and therefore is), who became one of the most famous naturalized Texans in our country's history. Mr. Morris edited the *Texas Observer*, a liberal rag that remains must-reading to learn what is really happening in Texas government. Then he bumped his head, went into a fugue state, moved to New York City in the United States of America, to edit *Harper's* magazine, and become very famous. The *Observer*'s overt left-leaning political commitments mean that, to this very day, Tris eagerly awaits each new issue, dropping everything, even Hegel, to read each fresh *Observer* from cover to cover when it is delivered to his home by (gasp, intubate, maximum settings!) the United States Postal Service.

[4] This is not a footnote.

References

McCullough, L.B. 1997. A personal and professional portrait of H. Tristram Engelhardt, Jr. In *Reading Engelhardt*, ed. J. Reagan, B. Minogue, and G. Palmer-Fernandez, xi–xix. Dordrecht: Kluwer Academic Publishers.

Morris, W. 1967. *North toward home*. New York: Houghton Mifflin.

Chapter 15
A Recollection with Appreciation

George J. Agich

I first met Tris Engelhardt in the early 1970s at a dissertation defense of a mutual friend at the University of Texas at Austin Philosophy Department. I was working on my PhD there. Tris had completed his PhD there in record time some years earlier and had recently moved from Tulane Medical School to the University of Texas Medical Branch (UTMB) as an Assistant Professor. At a celebratory party for the successful candidate later that evening, Tris and I got into a rather lively discussion about embodiment. He laid out and defended Hegel's categorial treatment. I argued for Merleau-Ponty's account that I thought positively augmented Husserl's. Sometime after that, we began corresponding on the same topic—of course, never really coming to agreement. This correspondence importantly provided me with a link that I used later.

The Institute for the Medical Humanities at UTMB Galveston had received a National Endowment for the Humanities development grant and Chester Burns, project director, sent announcements to Austin to inform graduate students in philosophy that they were offering Fellowships. This was a time when there was no bioethics, medical ethics, or even applied philosophy in American graduate programs. So it is not surprising that my own graduate studies mainly mirrored my disparate interests in analytic philosophy, the history of philosophy, and phenomenology. I was well into my dissertation on the systematic significance of Kant's *Critique of Aesthetic Judgment* when I heard the news. A good friend of mine, who was working in ethics, encouraged me to apply—he was committed to staying in Austin since his wife was gainfully employed and would not countenance his leaving—because he knew how unenthused I was about accepting a position teaching mainly introductory philosophy courses, the only prospect that I had at the time.

G.J. Agich, Ph.D. (✉)
Philosophy Department, Bowling Green State University,
6805 Via Correto Dr, Austin, TX 748749, USA
e-mail: agichg@bgsu.edu

L.M. Rasmussen et al. (eds.), *At the Foundations of Bioethics and Biopolitics: Critical Essays on the Thought of H. Tristram Engelhardt, Jr.*, Philosophy and Medicine 125, DOI 10.1007/978-3-319-18965-9_15

I thought about this more and more and decided to write to Tris. I told him about my long-standing interest in phenomenological psychiatry and epistemology of science and wondered whether I might be able to use some of the time as a fellow in gaining clinical experience of psychiatric syndromes. Tris responded telling me, in brief, that the name of the game was medical ethics and that in writing an application letter to Chester Burns I should indicate my interest in that. He also extracted a promise that if I came to Galveston as a fellow, we would spend time returning to the embodiment dispute that we carried on earlier – I think Tris was intent on correcting what he regarded as my errant ways. Following this advice, I was accepted and went to Galveston. I was offered a fellowship and became the first fellow in the Institute for the Medical Humanities and participated in the new medical ethics teaching program. Soon after arriving and settling into my rather loosely defined duties as a Fellow, Tris linked me with Robert White, a training analyst and director of consultation-liaison psychiatry service. Robert took me under his wing and taught me about functioning effectively in a clinical setting among many other things.

While at the Institute, I had many informal interactions and meetings with Tris, and these experiences and interactions were the launching ground for my career in bioethics. One of my favorite memories involves our lunchtime discussions of Hegel's *Phenomenology of Spirit,* which we read together. I don't recall what I ate, but I do recall that he frequently had peanut butter, jelly, and jalapeño sandwiches. During one of these discussions, my comment on the text was met with chagrin. He insisted that I had taken a similar and, in his view, indefensible position in our correspondence on embodiment a few years earlier. I, of course, denied this and tried to steer the conversation back to Hegel's text, but that was not easy. A few weeks later, in going through my dissertation files, I came across carbons of my letters to him and his responses. Sure enough, I found the point I made, embarrassingly similar to one I had recently made, and his response then, which mirrored his current view. No doubt Tris has a remarkable memory and I learned that one should not challenge it lightly.

One of my functions was to review and comment on the many manuscripts in draft that he prodigiously produced. This function was not part of my duties as fellow and I would sometimes object to the intrusion into my own time, but I was always generously rewarded with a lesson on German wines and many fine bottles of Spätlese and an occasional Auslese. He would often bring the typescripts to me fresh from his secretary—no word processors then. He encouraged me to make corrections and suggestions on the freshly typed pages. After I had reviewed five or so sheets, he'd come into my office or insist that I come to his, and we would discuss the points I'd made. He'd tell me whether he thought the points to be sound or not, but even when he rejected my corrections or suggestions, the original typescript was ruined with my markings. This, of course, meant that his secretary spent her time typing and retyping the same manuscript. Sometimes, more than half the paper would sit with corrections while she finished transcribing the final pages of the draft including references. Since there were major changes early in the paper, the final pages were worthless since the changes altered the pagination sequence. It seemed each paper underwent multiple retypings. I must say I always thought his secretary was a candidate for sainthood. When I'd tell Tris that, he'd lecture me about the role of a secretary and chide me for not appreciating her function as someone who produced typed pages for a living.

If I learned nothing from Tris, which of course is absolutely far from the truth, I came to develop a healthy respect for and commitment to writing during the year and a half that I spent as a Fellow. Tris was working on building a publication record for tenure at that time. It really seemed as if each week or so a new offprint would be added to the stack of publications that sat in the corner of his office. His productivity was absolutely phenomenal and his ability to work on multiple topics and projects simply amazing. This was an education in itself that stood me well later in my career. As a role model in academic productivity and scholarly commitment, Tris is simply the best.

Among my observational activities as a Fellow, I was allowed to sit in on some administrative meetings about the ethics curriculum and the Institute's activities. During one of these meetings, Tris kept leaning over to me asking about a topic or subject either he or I were working on or simply had an interest in. I found these sideways comments enormously distracting since I was diligently trying to follow the course of the meeting, and I often whispered so to him. At the same time that he made comments about various matters to me, he appeared to be making notes on another of his draft manuscripts. At one point, an influential member of the committee stated serious reservations about the ethics curriculum and the time devoted to it. I recall Chester Burns cringing at this challenge. Tris simply turned from his manuscript, focused on the man, and recounted in considerable detail much of the foregoing and the substance of the report that Chester had presented to the committee. He then asked, with a seemingly sincere expression of innocent surprise and curiosity, how anyone could draw such a conclusion since the evidence that had been presented by Chester Burns clearly pointed in the opposite direction! The committee eventually agreed. Tris whispered to me that *non sequiturs* should never go unchallenged, especially when they lack foundation. The report that the Institute delivered was finally accepted by the committee with congratulations for a job well done.

Since these interactions with Tris at the Institute, we had a long and fruitful, though episodic, relationship over the years. He has been a colleague, mentor, and friend. One last recollection involves the periodic phone calls that I received from him in my early days as a faculty member at Southern Illinois University Medical School. I was fortunate as a young faculty member to have secretarial support. The secretaries answered all incoming calls. Tris, on hearing a new voice answering the phone, would comport himself as some Texas official or deputy sheriff. He would inquire of the secretary in a roundabout way whether she knew if any of my activities in Springfield might violate my "parole" or cause them to have concern about me. He knew I would come to Galveston periodically with my wife to visit her family, so another of his lines was to have his secretary confirm that I was coming to keep my appointment with a parole or investigating officer. Our department had three secretaries and each of them was treated at least once to these "inquiries." This caused no small amount of consternation. I had to assure each of them, and my department chair who seriously lacked a sense of humor, that the call was from a friend with a perverse sense of humor.

These recollections are part of a deep and longstanding friendship. He has always responded to any request or need and his help was significant at crucial stages of my career. It has been a great joy as a friend to know him and an honor to have him as a long-time colleague.

Chapter 16
H. Tristram Engelhardt, Jr.—A Personal Reflection

John C. Moskop

It is an honor and a privilege to contribute to this Festschrift celebrating the contributions of H. Tristram Engelhardt, Jr., to bioethics over the past four decades. Tris' work has left an indelible mark on the modern field of bioethics that he helped to found. His prolific, highly original, and carefully argued scholarly writings continue to engage and challenge scholars around the globe. His tireless editorship of bioethics journals and books has enabled many hundreds of scholars to publish their own work and has greatly enriched the bioethics literature. His teaching and mentorship of graduate students and junior colleagues have brought scores of talented and dedicated professionals into the field. For all of these contributions, I am very grateful.

I am proud to be one of the graduate students who was inspired to enter the field of bioethics through working with Tris. One spring day in 1976, Larry McCullough, my friend and classmate in the philosophy graduate program at the University of Texas at Austin, approached me and asked if I would be interested in a summer job as a teaching assistant for a Dr. Tris Engelhardt at the University of Texas Medical Branch in Galveston. Larry had been working for Dr. Engelhardt, but he had accepted a faculty position, and so wanted to help Tris find a new assistant. Larry assured me that I would enjoy the job, and I took it, not realizing that it would redirect my career and life path. I found Tris to be a highly charismatic teacher with an engaging style and an infectious enthusiasm for the emerging field of "medical ethics." Working with Tris on his scholarly papers gave me a firsthand introduction

J.C. Moskop, Ph.D. (✉)
Department of Internal Medicine, Wake Forest School of Medicine, Winston-Salem, NC, USA

Clinical Ethics Committee, Wake Forest Baptist Medical Center, Medical Center Boulevard, Winston-Salem, NC 27157, USA
e-mail: jmoskop@wakehealth.edu

L.M. Rasmussen et al. (eds.), *At the Foundations of Bioethics and Biopolitics: Critical Essays on the Thought of H. Tristram Engelhardt, Jr.*, Philosophy and Medicine 125, DOI 10.1007/978-3-319-18965-9_16

to new and challenging ethical problems in health care. Helping Tris to organize and host national conferences, first in Galveston and later at the Kennedy Institute of Ethics at Georgetown University, gave me an opportunity to meet and to learn from leading bioethics scholars. Tris encouraged my growing interest in bioethics, helped me to identify open positions in the new field, and wrote strong letters of recommendation on my behalf. I was a successful job candidate, first at the University of Calgary, and then at the new East Carolina University School of Medicine, and my bioethics career was launched, thanks to the inspiration, generous mentorship, and faithful friendship Tris provided.

Working for Tris in the 1970s, and on projects with him thereafter, was never dull. I rented a garage apartment from him behind his home in Galveston, and so sometimes agreed, with my future wife Ruth, to babysit for Tris and Susan's young daughters. They encouraged us to speak German with Lisa and Christina, and even with baby Dorothea! I also occasionally received a late-night knock on my door—Tris was working late, as usual, and wanted to deliver the next day's assignment without delay!

As his many friends can attest, Tris has a disarming sense of humor, and he loves to engage friends, colleagues, assistants, students, and even casual acquaintances, in unconventional ways that keep them on their toes! One example stands out in my mind. Tris had invited me to participate in an international workshop on the concepts of human dignity and the sanctity of life at the University of Bielefeld, Germany, in October 1992. After a 24-hour-long, difficult plane and auto trip to Bielefeld on the day before the workshop, I finally arrived at the local hotel, feeling exhausted and unwell. Tris' first words to me, when I stopped by his room, were: "Did you bring your gun?!" I was, of course, well aware of the fact that, as a Texan, Tris is a strong proponent of the right to bear arms, but I couldn't imagine what he had in mind! He proceeded to explain that, when local disability rights activists heard that Professor Helga Kuhse (who had defended the practice of infanticide with her Australian colleague Peter Singer) would be one of the faculty participants at the workshop, they petitioned the Rector of the University to cancel the workshop and threatened to disrupt it if he did not. Tris was undeterred and favored a strategy of overt resistance to these opponents of the free expression of ideas, much to the horror of our German colleagues, who wanted to avoid a confrontation with protesters and the adverse publicity it would evoke. The protesters refused our invitation to an open discussion with them about the workshop topics. Instead, they threw paint on Kurt Bayertz, the German co-director of the workshop, and they physically restrained several of us from entering the University building where the workshop was to take place. The workshop start was delayed, but we were guided by a University staffer into the building through a private back entrance, and so the workshop did eventually take place, despite threats of continuing protests (Bayertz 1996). (This experience took a highly ironic turn for several of us who traveled directly from Bielefeld to attend the inaugural congress of the International Association of Bioethics in Amsterdam a few days later. At that congress, Peter Singer was lionized as an international leader in the field and was easily elected President of the new professional organization. What a difference a few hundred kilometers can make!) This is the only time in my career that the clash of ideas ever came to physical violence, and it offers just one illustration of my claim that working with Tris is never dull! Many thanks, Tris, for these life-enriching experiences; I look forward to future adventures together!

Reference

Bayertz, K. (ed.). 1996. *Sanctity of life and human dignity*. Dordrecht: Kluwer Academic Publishers.

Chapter 17
Post-apocalyptic Tris

Eric Juengst

Popular culture is full of post-apocalyptic visions these days, especially the fiction and movies aimed at young adults. The plot usually features a variegated band of survivors crossing a wasteland in search of a place of peaceful plenty that may or may not exist. Ruined cityscapes poke out of the sand. The survivors have nothing in common and don't even get along well with each other, at least at first. The communities they encounter along the way tend to be militarized dictatorships, patriarchal cults, or cannibals. There is a mysterious, lanky man in scuffed boots who claims to know the way, on the basis of arcane knowledge from Before.

Every time I encounter one of these stories, I think of Georgetown in the 1970s, when I was a young adult, working as a research assistant for Tris Engelhardt. To hear Tris tell it, we were the survivors of both the end of the pre-modern religious worldview and the post-modern collapse of the Enlightenment project, wandering the twentieth century's pluralistic wasteland, trying to negotiate with moral strangers of possibly cannibalistic temperament, in search of a community of moral friends who could tolerate our own mixed bag of values and commitments.

We graduate students were a microcosm of the larger pluralism: Midwestern Dominican priests, city-wise Sisters of Charity, veteran military officers, long-haired Appalachian hedonists, nurses and surgeons. Bioethics at that time was not attracting many straight-from-college graduate students, although I was one: a biology major, lured into a philosophy program by visions of cryogenics, genetic engineering, and "test tube babies." Even as we struggled to make ourselves intelligible to each other, Engelhardt set about poking our most cherished ideas –the few compasses, canteens, and theoretical lenses we thought might be useful in the

E. Juengst, Ph.D. (✉)
Center for Bioethics, School of Medicine, University of North Carolina,
333 MacNider Hall, Chapel Hill, NC, USA
e-mail: ejuengst@med.unc.edu

L.M. Rasmussen et al. (eds.), *At the Foundations of Bioethics and Biopolitics: Critical Essays on the Thought of H. Tristram Engelhardt, Jr.*, Philosophy and Medicine 125, DOI 10.1007/978-3-319-18965-9_17

wasteland– with his "analytic finger" and showing how they shattered when they fell to the hard ground. What was left? Mutual tolerance and transparent negotiation worked pretty well amongst us: the cohort just self-segregated amicably into the "Square Circle" of the religious and the secular regulars at the Garrett's Railroad Bar Seminar on Thursday nights, and real friendships developed across those lines. But it was not clear that the other communities of moral strangers on the horizon would be restrained enough by Nozick's precepts of procedural justice to keep their teeth out of us.

Back then, I assumed that Tris's destination for us would be a kind of apolitical countercultural Texas, where hippies held black powder shooting matches, vending machines dispensed recreational drugs, and folks circulated through a thousand odd churches and faiths as their whims took them. (In fact, looking back, I realize I thought we were going to today's Austin!). But along the path I fell in with another guide who thought he'd found a particular way out of the moral wasteland for biomedical ethics (Al Jonsen, excavating the pre-apocalyptic history of casuistry), and so I turned west toward San Francisco instead.

Imagine my surprise, then, to see Tris end up in Byzantium. If we had passed him in our graduate school wanderings calling to us from the fortified walls of some latter day Constantinople, even the Square Circle amongst us would have given him wide berth. To watch him reach that place through a career of self-reflection and argument, however, has been really interesting. In (the few) post-apocalyptic stories with happy endings, the destination never turns out to be a return to the Before times. Communities that try to carry on as before, with underground brass bands and painted-on smiles, are always portrayed as mistakes or worse. The ideal communities for the future of civilization are always something new: a polyglot mixture of pre-collapse peoples and cultures, inventing new traditions and institutions adapted to the changed world around them. They can't go back to older systems, because the premise is that they have all been thoroughly destroyed.

In fact, as Tris might point out, the new moral communities in those stories are very much like the utopian community experiments that Enlightenment thinking inspired in the nineteenth and twentieth centuries. But these never work, because – as everyone who has tried to "live in community" knows well – human communities are not benign environments. For me, this was the ultimate lesson of Tris' secular moral skepticism. For all our contemporary bioethical interest in "relational autonomy," "community engagement," and "communitarian ethics," each community is, like our band of graduate students, a microcosm of the larger phenomenon of pluralism. Even when all community members profess the same faith, some will see injustice or ignorance in the ways others profess it and try to get them to change, on grounds just as incommensurable as the arguments of moral strangers. Moreover, unlike our graduate student cohort, many substantive moral communities are not bound by minimal principles of mutual respect and peaceable negotiation in those attempts. Instead, they condemn the heretical to eternal hellfire and look forward to Judgment Day when, as Tris says, they can learn whether mesquite, live oak, or trash cedar fuels the furnace. So even if the walled compounds of warlords, gurus, and preachers dotting our cultural wasteland agreed to a minimal policy of non-interference and economic exchange, this does not bode well for the individuals and families inside those compounds – even the utopian ones.

Tris's point about the oppressive nature of coerced choices was formative for me, and I never saw why it should not be true within as well as between substantive moral communities, since it is inside those contentful communities – such as health care institutions, public health programs, the health professions, or the biomedical research enterprise – that most bioethical questions arise. As a result, it came as a surprise that our lanky libertarian guide from the 1970s would ultimately find such a hierarchical and, well, *un-Texian*, community and promote its teachings as his own contentful bioethical views.

Of course, in the face of our inability to argue rationally to a moral consensus, Engelhardt's turn to religion as a foundation for bioethics is understandable and, within contemporary academia, courageous. Greek Orthodox Christianity is a perfect religion for him as well, with its deeply braided cultural history, its disciplined spirituality, its strong moral views, and its rich tradition of ritual and ceremony. Good for him, I say. Having struggled to employ and teach the ostensibly secular morality of standard bioethics for 30 years, I can imagine the liberation – and terror – of embracing something so old, rich, and personally compelling as early Christianity.

But still. The movies have a point. If we have really been through an apocalypse, nothing from Before should have been spared the whirlwind. The only way to get to Orthodoxy within Engelhardt's philosophical cosmology is to argue that Orthodox Christianity was *not* destroyed by the cataclysms that demolished all our other foundations, and has come through the dark ages of modernity and post-modernity intact. But if the Orthodox Church survives as a source of orientation and value, why not other sources from amongst the world's many moral and religious traditions? Perhaps the world is not as ruined as we thought. In fact, the lesson of Tris' story may be that the world's moral cultures are too stable to be dislodged by philosophy's analytic finger, too resilient to shatter even when toppled, and not so opaque as to be mutually unintelligible after all, particularly in practical settings like health care. As Tris's own later work in Christian bioethics shows, for those who still feel like contemporary health care calls for some new moral traditions, there may be much more than mere foundations to build upon.

Meanwhile, for those of us still wandering the wilderness, there is always the itinerant craft of bioethics. For that, we salvage tools from the ruins around us, cobble together arguments to meet the needs of the day, and discuss the results over cheap beer at our annual Gatherings of the Clans. We don't claim to be doing either rocket science or high art; just fashioning sturdy and useful solutions to the questions at hand, always in the expectation that they will be improved upon next time by our apprentices and colleagues. It is an unsettled and often unsettling life, but as long as you keep a weather eye out for Orthodox Mounted Posses, academic slave traders, and other post-apocalyptic hazards, it can be a rewarding one. Those of us who have Tris to thank for cracking our canteens will always be grateful to him for setting us on this path.

Chapter 18
Ode to Tris

Mary Ann Gardell Cutter

In seventy-nine
At a place in D.C.
Tris Engelhardt taught
philosophy.

What a master teacher
And committed scholar
Author of *Foundations*
And oodles (and oodles) of others.

At the Kennedy Institute
He ran a large crew
Jane and Elena, and
Mary Ann plus three.

Back in those days
We read Wulff and Fleck
Rawls and Nozick
On the 37th and O deck.

We argued about facts
And explanations in medicine
We worried that the postmodern
Condition would win.

Soon it came time
To leave D.C.

M.A.G. Cutter, Ph.D. (✉)
Department of Philosophy, University of Colorado, 1420 Austin Bluffs Parkway,
Colorado Springs, CO 80918, USA
e-mail: mcutter@uccs.edu

L.M. Rasmussen et al. (eds.), *At the Foundations of Bioethics and Biopolitics: Critical Essays on the Thought of H. Tristram Engelhardt, Jr.*,
Philosophy and Medicine 125, DOI 10.1007/978-3-319-18965-9_18

Off to Baylor
Tris would be.

I followed suit
And made my way
To the Center for Ethics
To be Tris's "sla" (ve).

Tris taught me "Texan"
To drive in gears
To eat jalapenos
My, those tears!

He taught me to teach
To love what I do
To believe in the possible
To pursue the new.

As cliché as it may sound
He taught me to see
That faith, family, and friends
Is what one nee(ds).

When one thinks of Tris
One thinks of autonomy
But I would say
One should think of generonomy*

(*the rule of giving)

His generosity's astounding
His humor a treat
His commitment to students
Hard for anyone to beat.

Indeed we can say
How lucky we've been
To have Tris as our teacher
And to be part of his kin.

And how lucky we've been
To know Susan as well
She has stood by our side
Come heaven or hell.

And so in honor of Tris
This ode that is penned
(surely not by a poet!**)
Has come to an end.

(**Tris taught me philosophy, not poetry.)

Tris, we are blessed by your presence
By your endless good will
By your commitment to students
By all the shoes you fill.

Chapter 19
The Engelhardt Experience

Ana Smith Iltis

> And the Lord God said, 'It is not good for man to be alone. I will make him a helper comparable to him.' Also, God formed out of the ground all the wild animals of the field and all the birds of heaven, and brought them to Adam to see what he would call them. Thus whatever Adam called each living creature, that was its name. So Adam gave names to all the cattle, to all the birds of heaven, and to all the wild animals of the field. But for Adam there was not found a helper comparable to him. Thus God brought a trance upon Adam, and he slept; and He took one of his ribs, and filled up the flesh in its place. Then the Lord God built the rib He took from Adam into a woman, and brought her to him. So Adam said, 'This is now bone of my bones, and flesh of my flesh. She shall be Woman, because she was taken out of Man.' For this reason, a man shall leave his father and mother and be joined to his wife; the two shall become one flesh (Genesis 2: 18–24, The Orthodox Study Bible, 2008).

There is no way to understand H. Tristram Engelhardt, Jr. and all he has done without an appreciation of his tremendous wife, Susan. Susan is extremely bright; works virtually non-stop in multiple capacities; knows the rules of grammar inside and out (and is not afraid to correct others); she is kind, loving, insightful, and knows how to say what people need to hear even when they don't want to hear it; she is a shining example of what it means to be a Christian; and, as anyone who knows Tris can imagine, Susan is unbelievably patient. Susan has made it possible for Tris to do so much of what he has done, not only in terms of teaching, publishing, and speaking, but of being a true mentor and friend. Of his many, many accomplishments, I want to share a bit about his great love for others as evidenced through his work as a teacher and mentor. Tris has taught and mentored many students and countless others who have found themselves under his tutelage, often by first finding

A.S. Iltis, Ph.D. (✉)
Department of Philosophy, Center for Bioethics, Health and Society, Wake Forest University, P.O. Box 7332, Winston-Salem, NC 27109, USA
e-mail: Iltisas@wfu.edu

L.M. Rasmussen et al. (eds.), *At the Foundations of Bioethics and Biopolitics: Critical Essays on the Thought of H. Tristram Engelhardt, Jr.*, Philosophy and Medicine 125, DOI 10.1007/978-3-319-18965-9_19

themselves in the line of fire at a conference where Tris responded to their papers or asked a question that sounded like it was meant to disembowel them. I suspect that everyone who has been fortunate enough to experience Tris' love for and dedication to students and scholars would agree that we truly would not be the same, as people, as teachers, and as scholars, without the benefit of "The Engelhardt Experience."

Studying with Tris truly was an Experience. Never before had I been on a plane and had someone ask whether I should be stowed beneath the seat or in an overhead compartment. This was asked while my dear friend Lisa Rasmussen and I were seated in first class with Tris after he had managed to secure upgrades for both of us using a single companion upgrade. He treats his students well. Nor had I found myself facing bewildered people who thought Tris had uttered "the f word." He loves to use his Latin, and often says things like: "It is a joy to see you *facies ad faciem*" ("face to face") or "*Qui facit per alium facit per se*" ("He who acts through another does the act himself"). To the uninitiated, these phrases sometimes sound questionable. Lisa and I were asked: "Did he just say the f-word to me?" We could not help but laugh.

I imagine the trash drop was a source of consternation for some neighbors. The trash drop in the garage was a central feature of life as a student. A small trash can sat in the garage, and we would pick up and drop off important papers at all hours using "the drop." I always wondered what the neighbors thought of us. Exactly what kind of operation was this? They would have been terribly disappointed to find submissions to *The Journal of Medicine and Philosophy* and other philosophy papers, including dissertation drafts!

By the time I became a student, a wise former student had accumulated pages and pages worth of Tris' favorite trivia. From the dates of various wars, to the dates of various Roman popes, to all manner of obscure pieces of information, we had it all documented. We were used to being asked to answer questions on the spot. We also learned that we could be mildly subversive by answering incorrectly. When as graduate students we sat in on an undergraduate class, we learned that sometimes providing an incorrect answer could bring us great pleasure. The look of bewilderment was priceless.

The great flood of 2001 in Houston, the result of Tropical Storm Allison, was the source of so many events that help capture "The Engelhardt Experience." The night before the city was truly under water, with I-59 a river of floating cars and tractor trailers, homes under water, and the Texas Medical Center flooded and mostly without power, it was raining hard. It had been raining all day and the roads were flooding. Tris had taken several of us to dinner, and Lisa, Tris and I had driven to Brennan's Restaurant in Houston together. We were in my car, and the drive back to the Engelhardts' home, where Lisa's car was parked, was harrowing. We could feel the water slamming up against the car and I could see stalled out vehicles everywhere. I desperately wanted to stop on a median and wait for the rain to stop. I was terrified to drive and I was terrified to ruin my car. Tris would hear none of it. He told me to get behind a taxi and just drive – never let my foot off the gas and keep the car in gear. I kept following his instructions and shouting back that I was scared and would be driving his old Volvo if anything happened to my car. We made it back safely. Had we stopped, we would have been among the casualties on the freeway, for the rain never did stop that night. By morning, the whole road was a river. He had high expectations of his students in all things, but I never could have imagined that driving small cars through flooded streets was among them! The flood led to more adventures. The building in which his office and *The Journal of Medicine and Philosophy* office were located flooded. We desperately needed to rescue our files and computers. Lisa and I spent several days moving the offices, hauling filing cabinets in a dark, hot, and humid stairwell (four floors' worth!), for the building had no power. We had bruises to prove our hard work. Even in that mess, we found humor.

These anecdotes give a little glimpse into what sometimes seemed like a crazy universe. Many people who caught only snippets of what it was to study with Tris could not understand why anyone would do it. I suspect my husband asked this question on a daily basis. But for many of us who had this great joy, we cannot imagine not having had "The Engelhardt Experience." Our time as students was not just a string of humor, chaos, hard work, crazy hours, and sometimes overwhelming requests and demands. It was Olympic training for an academic career and it was the start of lifelong friendships. Tris made sometimes seemingly impossible demands of us – to produce material quickly, to summarize the literature on a particular question, to revise our papers thoroughly and quickly and so on. He seemed always to want better and more work, but he was willing to take the time to show us what that meant and how to achieve it. I remember sitting for hours with Tris and other students (and the ever-patient Susan) in the Engelhardts' living room, reviewing work, being pressed to provide a better argument or explanation, and fighting exhaustion. Little did I know at the time what this kind of training would do for me. Many of us who have studied with Tris have substantive disagreements with him on many matters, but he remains steadfast in his commitment to our lifelong learning and in his friendship. He is, in this regard and in so many others, the definition of teacher, mentor, and friend.

Chapter 20
Studying with H. Tristram Engelhardt, Jr.: Lessons in an Engelhardtian *Weltanschauung*

Lisa M. Rasmussen

Undertaking graduate study with Dr. Engelhardt provided a window into a rich tradition of true disciplinary mentorship.[1] This was not the world of periodic email communication, neatly scheduled deadlines, measured progress and a final congratulatory handshake. For all his Germanic tendencies, Dr. Engelhardt inhabits (and therefore by association, we students inhabited) a chaotic, fertile, humorous, rigorous, unpredictable *Weltanschauung*. We received calls outside the normal boundaries of the work day, found our dearest assumptions poked at repeatedly, were humbled by Tris's stamina for work and late nights, were thrilled and intimidated by opportunities he sent our way, and learned enormous amounts of philosophy, culture, history, religion, oenology and Texarcana. It was an apprenticeship with a master craftsman who took responsibility for our enculturation and education in a personal way. It was never boring.

Along the way, one was regularly regaled with stories: for example how, in the era of 1970s gas rationing, while living in Georgetown, Tris arranged to be "discovered" in Jim Childress's driveway siphoning gas from Jim's car. Upon this discovery, Tris explained that he was merely shifting resources from those who had more (*viz.*, Jim) to those who had less (*viz.*, Tris), a performative counterargument to redistributive taxation. Or how, in a spirited moment, he discovered the principle of *bunnahabhain* and argued that it deserved to join the pantheon of bioethical principles (Engelhardt 2000, 44, fn. 4). Such stories encouraged our own pranksterism; I attached a World Wildlife Federation sticker to the back of his old Volvo rear window,

[1] My time as a student of Tris's overlaps almost entirely with Ana Smith Iltis's, so I am happy to corroborate the many wonderful stories she remembers (see Iltis, Chap. 19, this volume, pp. 255–257).

L.M. Rasmussen, Ph.D. (✉)
Department of Philosophy, University of North Carolina,
9201 University City Blvd., Charlotte, NC 28223, USA
e-mail: lrasmuss@uncc.edu

L.M. Rasmussen et al. (eds.), *At the Foundations of Bioethics and Biopolitics: Critical Essays on the Thought of H. Tristram Engelhardt, Jr.*,
Philosophy and Medicine 125, DOI 10.1007/978-3-319-18965-9_20

hysterical with the knowledge that this avowed carnivorous, gun-rights-proponent Texan might unwittingly be viewed as supporting such a cause.

A fundamental point about studying with Tris is that he treated us authentically and directly. There was no misunderstanding when Tris disagreed with you; he said it forthrightly and paid the high compliment of offering his most withering critique (which is distinct from offering a critique witheringly). His was not the even-handed attempt (perish the thought) to offer both sides of a debate. Instead, though he certainly would have considered it a mark of our sensibility had we agreed with him, his opposition and willingness to engage in frank disagreements allowed us to develop our best arguments. I cannot speak for others, but I felt that I made, and still make, much more progress by being disagreed with profoundly – because I learned the dimensions of my own argument – than I ever could have were someone merely to point out *possible* or *hypothetical* counterarguments. Tris would say "I don't care what you think, I care what you can argue for," and in that philosophical space, made it clear both that one needn't agree with him and that he would demand rigor no matter what one believed.

When I considered going to graduate school in philosophy to study bioethics, I didn't have an advisor, so I developed a *sui generis* methodology for figuring out which graduate schools to investigate: I looked up the call number for bioethics in the library, and went to examine the shelves. At that time, on those shelves, it seemed that nearly half of the books had been written by Rice University authors (Baruch Brody and Tris) and half had been written by Georgetown authors (Tom Beauchamp, Jim Childress, and Bob Veatch). So I applied to the philosophy departments at both institutions. When I visited Rice, I sat down in Tris's office. He began pulling books by his students off of the shelves and stacking them to the sides of the desk, so that I peered through an alley of publication at him. I felt myself incapable of living up to such standards, and left his office overwhelmed. Despite that initial vertigo, I went on to work under Tris's mentorship and made my own modest contributions to that pile, due to his encouragement and tutelage.

During that conversation, he said something which made ever more sense over the years I studied with him: that it was clear what physicians, nurses, lawyers, etc. brought to the field of bioethics, but that if one trained only as a "bioethicist," the nature of one's expertise was less clear. On the other hand, it was clearer what rigorous philosophical training provided, and how it could be mined in the consideration of issues in bioethics. Tris's work – in bioethics, the philosophy of medicine, public policy, political philosophy, etc. – is one long, impressive demonstration of what philosophical reflection can contribute.

The term "*Doktorkinder*" refers to those students who are one's doctoral "children," and in the sense that he took responsibility for us, we really are the *Doktorkinder* of H. Tristram Engelhardt, Jr. We still call home from time to time, get frustrated with him, and rejoice in sharing accomplishments with him. He was the first person after my husband and parents whom I called with the news that I had received tenure, and in a surprise to me, I welled up when letting him know. I'll spend the rest of my professional life trying to emulate his mentorship.

Reference

Engelhardt Jr., H.T. 2000. Autonomy: The cardinal principle of contemporary bioethics. In *Bioethics and biolaw*, ed. Peter Kemp, 35–46. Copenhagen: Rhodos International.

Chapter 21
H. Tristram Engelhardt, Jr.: The Man Behind the Scholar

Fabrice Jotterand

Throughout the years I had the privilege to work for Tris as his assistant, I had the opportunity to learn about Tris the man as opposed to Tris the scholar and educator. One thing that I always wanted to know is the reason he kept his old blue Volvo, which in many ways was falling apart, but like an old friend he could not let it go. I don't very often mention it but I am a car enthusiast, so when I see people who have a special relationship with their car it always intrigues me. Tris and his blue Volvo, a model of the late 1960s or early 1970s if I am not mistaken, was always a subject of conversation and laughs – the smell and interior of the car had their own characteristics!

I distinctly remember one day Tris calling me at the office, requesting that I pick him up in the Village near Rice University due to a "minor" incident on his way from the house to the university to teach a class. I don't recall the exact nature of the clash between him and the other driver, but to cut a long story short, Tris' old companion on wheels ended up being undriveable. But interestingly, he could not let his Volvo go, like someone at the bedside of a dying old friend after an accident. Tris attempted to find spare parts to repair his Volvo, which in my view was a futile exercise! It is often said that the car one owns reflects one's personality – looking back, I think this old blue Volvo reminds me of Tris' personality: sturdy, dependable, from another age…but with so much character!

I also experienced Tris the person during various trips. Stories include the rather animated debate at the bar of an airport between him and a female lawyer on questions pertaining to feminism, or the reading of Orthodox liturgy in the car on our way to a conference at the University of Notre Dame. However, a trip to Ireland

F. Jotterand, Ph.D., M.A (✉)
Department of Health Care Ethics, Rueckert-Hartman College of Health Professions, Regis University, 3333 Regis Boulevard, G-5, Denver, CO 80221, USA

Institute for Biomedical Ethics, University of Basel, Basel, Switzerland
e-mail: fjotterand@regis.edu

L.M. Rasmussen et al. (eds.), *At the Foundations of Bioethics and Biopolitics: Critical Essays on the Thought of H. Tristram Engelhardt, Jr.*, Philosophy and Medicine 125, DOI 10.1007/978-3-319-18965-9_21

stands out. After crossing the Atlantic, Tris and I arrived at the airport in Dublin ready to embark in the last stretch of our trip in a rented car. We were supposed to drive to a small town south of Dublin for a colloquium. Tris, in his inimitable style, directed the operation, like Sam Houston getting ready to secure the independence of Texas from Mexico. He requested that I drive the car, and he took charge of reading the map (no smart phone then!) to reach our destination south of Dublin… at least this is what I thought. I could sense the nervousness of my passenger considering the challenge of driving on the left side of the road in a busy city but decided not to pay much attention, focusing on the task at hand. After a certain period of time driving, I realized that I could still see a body of water on our right and started questioning whether indeed we were heading the right direction. If my sense of direction was correct, the Dublin Bay had to be on our left on our way south but reality showed otherwise. After some intense deliberations and a closer look at the map, we turned around and reached our final destination south of Dublin. In many ways, this trip reflects what it means to work for Tris. It is like going on a trip in unknown territory with many twists and turns, constantly expecting the unexpected but always reaching our destination. Tris has been a generous and an exceptional mentor even if the road took sometimes unforeseeable turns. For someone Tris judged as "one enslaved to the falsehoods of heresy," I can only say: Thank you Tris! Heresy has been the cost of my freedom!

Chapter 22
Apprenticing with H. Tristram Engelhardt, Jr.: A Tribute to the Founder of Our Thing

Jeremy R. Garrett

Serving as a graduate assistant to H. Tristram Engelhardt, Jr. was truly an academic apprenticeship unlike any other. In part, this was because the apprenticeship extended far beyond strictly academic matters like philosophy and bioethics; and, in part, this was because Tris is a force of nature, an endlessly fascinating man and loyally committed mentor who seemed to me never to require sleep or downtime. He wants his students to come out of the apprenticeship with experience publishing papers and books, teaching undergraduate and medical students, managing professional journals, organizing and participating in academic conferences, and so on, but he would count it as a disappointment if they did not also know the three necessary conditions for something's being a single-malt Scotch (and have sufficient experience in appreciating the resulting product of those conditions from a variety of distilleries); the major dates, events, and colorful characters (especially one "Big Drunk") in the history of the Republic of Texas (T.R. Fehrenbach's magisterial history, *Lone Star: A History of Texas and the Texans*, was esteemed on a level approaching the works of Kant, Hegel, and Husserl); and a host of other items regarding philosophy, history, medicine, religion, politics, economics, science, gastronomy, culture, world travel, personal finance, and, of course, Texana (to name just a few areas of major interest). One never knew exactly what to expect on any given day (aside from exhaustion at trying to keep up with Tris!), but it was always interesting and enlivening.

A central feature of indenture to Tris is one's initiation into the large circle of philosophers and bioethicists who've served before. Though notably diverse in

J.R. Garrett, Ph.D. (✉)
Children's Mercy Bioethics Center, Children's Mercy Hospital,
2401 Gillham Road, Kansas City, MO 64108, USA

Department of Pediatrics & Department of Philosophy, University of Missouri,
Kansas City, MO USA
e-mail: jgarrett@cmh.edu

L.M. Rasmussen et al. (eds.), *At the Foundations of Bioethics and Biopolitics: Critical Essays on the Thought of H. Tristram Engelhardt, Jr.*,
Philosophy and Medicine 125, DOI 10.1007/978-3-319-18965-9_22

terms of philosophical, religious, political, and cultural orientations, Tris' former students feel a sense of loyalty, common purpose, and solidarity with others who've shared (and survived) the apprenticeship. In part, this is because there was something bordering on hazing that one had to endure to join its ranks. First, there was the often demanding workday itself. Few moments in life inspired the same dread as receiving a call from Tris' home phone, often just minutes (sometimes seconds) following a previous call, as this usually entailed that two to three new items were to be added to one's to-do list for every item that one had managed to cross off. It certainly was not uncommon to find oneself plugging away at that list in Tris' office until well after the Rice undergraduate parties had died out across the street. However, the rite of passage did not end with onerous work conditions, as indenture also entailed being called a mere thing when being introduced to famous academics, being vigorously encouraged to grow some distinctive form of facial hair (as clean shaven men were not to be trusted), and having any "peculiar" moral commitments or perspectives (e.g., atheism, vegetarianism, and so on) constantly scrutinized and even made the object of playful jokes and sincere bafflement.

Yet, in addition to being toughened up some, indenture certainly had its fair share of advantages (Tris never hesitates to point out that we all entered the arrangement quite voluntarily). We unfailingly were provided with the proper means to restore our strength and health (Tris would stop everything upon learning that one of his students was low on Scotch or wine and immediately rectify the situation by drawing something out of his own personal collection and engaging in a Nozickian transfer of holdings). We were given opportunities to travel to exotic places around the world, such as Hong Kong, China, Europe, and South Bend, Indiana, and meet many interesting and brilliant people. While in those exotic locales, we were treated to a host of cultural and recreational diversions. One of my favorite such memories occurred during a trip to London to attend a medical ethics conference sponsored by Notre Dame. Not only did this trip mark my first visit to No. 4 St. James Street in the City of Westminster (where the former embassy building for the Republic of Texas still stands), but it also offered me a unique insight into the far-ranging perks of working for Tris after watching him negotiate a "family discount" for a boat ride on the Thames River (for 20–30 people distinctly unrelated by biology!) with a perplexed and slightly fearful woman running the ticket booth.

Of course, more than anything, there were the benefits that were especially valuable to a career in academic philosophy and bioethics. We were given unique and valuable opportunities to learn, publish, present, teach, attend and organize conferences, and manage multiple academic journals. We were introduced to an extensive network of philosophers, bioethicists, doctors, and others with whom various kinds of partnerships, collaborations, and even friendships have developed over the years. We received immediate, thorough, and usually penetrating feedback on our own scholarship, especially if Tris was headed out of Houston on an American Airlines flight somewhere (which he usually was) and could work without distraction. And, not to be overlooked, we had the most loyal and devoted advocate on the job market that one could possibly imagine. Imagine someone who is one part charismatic politician, one part driven lobbyist, and one part used car salesman (three categories that Tris would normally want little to do with, admittedly!) and that is Tris when one of his students is trying to make the leap to a career elsewhere (even when "elsewhere" regrettably takes them into Yankee territory).

One of Tris' most endearing, if sometimes frustrating, traits is his insistence on letting no small conversational detail go unremarked upon, even in casual small talk. Of the many standard examples with which all fellow students will be familiar is his inevitable reply to the ubiquitous, "Good to see you!" – namely, a robust and strangely sincere, "It's good to be seen by you!" I'm sure I speak for all of those who've indentured to Tris in saying that, while it is certainly good to be seen by him whenever the opportunity arises, it is even better to see Tris and to feel that stirring of intellectual energy and, indeed, familial loyalty in the presence of those cowboy boots, the oversized belt buckle, the long goatee, and other defining elements of his larger-than-life persona. May we all never forget how fortunate we are to have worked with and learned from a truly original thinker, mentor, advocate, and friend!

Chapter 23
Dr. H. Tristram Engelhardt, Jr.: Scholar, Gentleman, Friend

D. Christopher Ralston

I first met H. Tristram Engelhardt, Jr.—"Tris" to those who know him personally—on a campus visit to Rice University after I had been admitted to Rice but prior to accepting the offer of admission. Two philosophy graduate students accompanied me to a class that Tris was teaching. I had seen his picture on the Philosophy Department's website, and had of course read his *Foundations of Bioethics,* but was unsure what to expect of him in person. I liked him immediately. In particular, I was struck by his humor—he had me in stitches in no time—and energy. Repeatedly, he would ask his students the question, "do I have your permission to use you in an experiment?" which, if answered in the affirmative, would usually be followed by a humorous thought experiment in which the student in question featured prominently. The phrasing of that question is significant: consistent with the "principle of consent," which plays such a prominent role in Tris' account of "general secular morality," it was imperative to obtain a student's "consent" to be "used" in the classroom setting, albeit for pedagogical purposes. Later that day, Tris made a point of welcoming me personally, spending time with me and making himself available to answer any questions I might have.

So many things about that day typify the relationship I had with Tris in subsequent years. In all, I spent seven fruitful, productive years in Houston, plus an additional

---

The affiliations listed above are for identification purposes only. The views expressed in this essay are the author's and do not necessarily express the formal opinions of The Center for Bioethics & Human Dignity, Trinity International University, or Joni and Friends International Disability Center.

D.C. Ralston, Ph.D. (✉)
Christian Institute on Disability, Joni and Friends International
Disability Center, 30009 Ladyface Cr., Agoura Hills, CA, 91301, USA

The Center for Bioethics and Human Dignity, Trinity International University,
Deerfield, IL, USA
e-mail: cralston@joniandfriends.org

L.M. Rasmussen et al. (eds.), *At the Foundations of Bioethics and Biopolitics: Critical Essays on the Thought of H. Tristram Engelhardt, Jr.*,
Philosophy and Medicine 125, DOI 10.1007/978-3-319-18965-9_23

year in California, working on the PhD—and, at every stage along the way, Tris was in various ways crucial to my eventual success.

Intellectually, Tris was a fertile source of ideas, patiently helping me shape my inchoate thoughts into a finished manuscript that could be called a "dissertation." In this regard, his philosophy of medicine, with its emphasis (among other things) on the "social performative" role of medical language, proved particularly helpful to me as I approached my dissertation topic, a philosophical analysis of the concept of disability. Given Tris' expertise in the philosophy of medicine, he was an ideal mentor for this project.

On a personal level, three qualities, in particular, stand out when thinking about Tris Engelahrdt: his humor, his optimism, and his generosity of spirit. I have already mentioned his humor. As for his optimism, suffice it to say that the long road that is doctoral studies provided plenty of opportunities to become discouraged. Even now, I still find myself repeating the encouraging words that Tris offered me time and time again: "*I'm afraid it's all going to work out.*" In so many ways, that phrase encapsulates his outlook.

As several of the other contributors to this volume have noted in their essays, Tris' relationships with his students are invariably characterized by an unparalleled generosity of spirit. My experience in this regard was no different. For example, while I was preparing for my qualifying exam, and then later as I was working on my dissertation proposal and the dissertation itself, Tris was incredibly generous with his time, opening up his home for long conversations about philosophy and other matters, helping me anticipate and prepare for potential questions or objections, and above all, reassuring this nervous philosopher-in-training. Speaking from the perspective of many years of experience, he could confidently say, "*Chris, I'm afraid it's all going to work out.*"

There is much more that could be said here. It is well-known, for example, that Tris is a Texan through and through—complete with a standard outfit of cowboy boots, suit or sports jacket, and tie. What may be less well known is that he is also a gentleman, albeit one with a *definite* set of opinions about things—of which he will be sure to inform his listeners! True to his libertarian philosophy, though, Tris is committed to allowing his students the freedom to develop their own ideas and positions. To be sure, he will advocate for his views—vigorously so. But at the end of the day, he is genuinely committed to a spirit of open enquiry that is honest about real differences, while nevertheless being generous of spirit toward his interlocutor. Contemporary academia would do well to take note of this seemingly fading art.

Tris is also fiercely loyal and committed to his students, providing them with a rich array of opportunities for professional development. He is demanding but fair, trusts his students, and gives them a wide range of responsibility. As a result, his students learn much from hands-on experience. Under Tris' direction, the PhD program at Rice proved to be an invaluable professional training ground for me.

There are a number of things I particularly appreciate about Tris. First, I appreciate his commitment to facilitating his students' professional development—whether that be in the form of enabling them to attend professional conferences, edit academic books and journals, or even co-author books and articles with him. Instead of hoarding such opportunities for himself, Tris goes out of his way to include his students in the various enterprises in which he finds himself engaged professionally.

More importantly, I especially appreciate Tris' unfaltering willingness to defend his beliefs—theological, philosophical, political, and otherwise—even when they might be deemed unpopular or "politically incorrect." As other contributors to this volume no doubt will have noted, Tris is "countercultural" in the truest sense of the word: being willing to take a stand against those cultural, intellectual, and other tides which one believes to be fundamentally wrongheaded, even when such a stance is costly. Intentionally provocative yet intensely earnest in his arguments, he is consistently charitable (even if strong) and always maintains a twinkle in the eye, even in the most spirited of exchanges.

Tris' ideas have shaped mine in several key ways. First, and most obviously, I find myself using language, patterns of speech, and so forth, acquired from years of working closely with him. To this day, in response to someone telling me "It's good to see you," I can't help but think of Tris' most common rejoinder: "Well, it's good to be *seen* by you!" (Those who know Tris' vocal patterns will be able to *hear* that last sentence.) Substantively, Tris has shaped my philosophical and theological thinking in important ways. For example, though not as pessimistic as he regarding the possibility of finding "common ground" among the different voices in a pluralistic society such as ours, I am considerably more aware, after having studied under Tris, of the difficulties involved in such an endeavor. More generally, it seems to me that perhaps the chief virtue of Tris' philosophical approach is his diagnosis of the contemporary, "postmodern," post-Enlightenment sociocultural scene in which we find ourselves. Many will disagree with his proposed prescriptions, but his diagnosis of what ails us in this present moment—and how we got to this point—are, in my judgment, spot-on.

This is the first opportunity I have had to publicly thank Tris for his positive contributions to my life. I am honored to have been abzzle to be one of his students, and will be forever grateful for the positive impact he's had on me. I am blessed to be able to call Tris not only my mentor, but my friend as well. I join with the other contributors to this volume in saying, "Thank you, Tris!"

Chapter 24
H. Tristram Engelhardt, Jr.: Beloved Mentor and Teacher

Jennifer A. Bulcock

Most can agree that H. Tristram Engelhardt, Jr. has a brilliant mind and a big personality: he can rehearse a large portion of history in the form of a question; he is often the source of much controversy; and as a professor and scholar he has had a profound impact on the fields of bioethics, philosophy of medicine, and Christian ethics. While Tris' genius is evident in the written and oral presentation of his work, and his zeal for sustained academic debate and controversy is obvious at any conference he attends, Tris' deep devotion, loyalty, and love for his friends and students is often underappreciated by those who have not had the distinct honor of occupying one of these roles. It is this elemental facet of Tris' personality that I would like to honor in this essay.

For more than forty years Tris has been a professor and mentor of students and scholars in the fields of medicine and philosophy. As one of his current students I can attest to the fact that his exuberance and passion for teaching and mentoring has not waned over the decades—although his previous students tell me he has become much gentler and more reasonable with age.

In the undergraduate classroom Tris is a unique hybrid of philosopher, historian, entertainer, and die-hard Texan. He starts each of his classes with erudite trivia meant to help students "appear smart at cocktail parties" and to ensure that no student leaves his class without being well-versed in Texas history, including being able to recite the first five presidents of "our country." Tris' vast knowledge and experience within philosophy and medicine coupled with amusing tales from the "sins of his youth" and the electricity with which he delivers lectures, makes him an exceptional professor. Additionally, despite the fact that he is frequently lecturing across the globe in between class meetings, Tris remains extremely accessible to his

J.A. Bulcock (✉)
Department of Philosophy, Rice University, MS-14, PO Box 1892, Houston, TX 77025, USA
e-mail: jb7@rice.edu

L.M. Rasmussen et al. (eds.), *At the Foundations of Bioethics and Biopolitics: Critical Essays on the Thought of H. Tristram Engelhardt, Jr.*, Philosophy and Medicine 125, DOI 10.1007/978-3-319-18965-9_24

students and is always willing to answer questions, offer help, or just talk philosophy. Few professors are as dedicated, entertaining, and engaged with their students as Tris is and his students love him for it.

While Tris is generous with his undergraduates and colleagues, I believe it is his graduate students that benefit most from his generosity and enthusiasm. I know I have. As an advisor, Tris is dedicated, generous, and supportive. For many of his graduate students he takes on the role of a concerned, slightly eccentric grandfather, always mindful of your wellbeing and looking for ways to help you succeed in accomplishing your goals. He considers himself personally responsible for the success of his graduate students and does not consider his duties discharged until each of them has been granted tenure. In light of this commitment, he makes himself available at all hours regardless of where in the world he is, he provides whatever assistance and guidance is necessary to ensure his students' success, and he works arduously to ensure their intellectual and professional development both while they are in graduate school and beyond.

Tris is a constant advocate and motivator. He is always willing to read and discuss your work and frequently does so while traveling the world. On many occasions I have provided Tris with a dissertation chapter or conference paper the night before he has left on a trip and later received a phone call upon his arrival at the airport in Hong Kong or Germany asking me to turn to page six to discuss the claim I've made or to suggest a footnote. He works tirelessly to ensure his graduate students receive the attention they require to flourish and expects their energy and enthusiasm to match his own. Having published hundreds of articles and books of his own he is constantly prompting his students to write, think, and publish as much as they can. His encyclopedic knowledge of philosophy and history makes it possible for him to enthusiastically engage his students in any topic they find interesting, helping them to chase down footnotes and discover new avenues for thought. No topic is uninteresting and no argument without merit. Tris possesses a command and love for knowledge that is infectious.

As a mentor, Tris works hard to ensure his students present, publish, network, and develop a professional acumen. As with most things, Tris goes above and beyond to provide such opportunities to his students. He frequently invites students to attend conferences with him and encourages them to develop and submit papers on a regular basis. He is in the front row of any conference presentation his graduate students give. He does not neglect his graduate students so that he may reminisce with old friends or see the sights of the city in his spare time. He approaches conferences as a time to show off and engage his graduate students with other professionals in the field. Tris knows exactly when to expect his students at a conference and always makes sure they arrive safely and that they are well fed upon their arrival. Every conference meal is spent with Tris discussing a breadth of philosophical issues and meeting other scholars.

One of my most favorite memories with Tris is debating morality and the existence of God over many hours and good scotch at a bioethics conference. The following day a number of graduate students who had been observing our conversation expressed to me how much they wished their advisors would take the time to have such conversations with them and to listen and engage with them as honestly and openly as Tris had with me. Having such interactions with Tris has given me confidence in my abilities as well as a professional poise that many graduate students are not given enough opportunities to develop. So much of my education and professional development has occurred informally in a conference setting with Tris.

Furthermore, a logical extension of Tris' perceived duties to his students is cultivating a community of scholars within which more senior scholars offer encouragement, support, and guidance to younger scholars, providing them with a safe environment to discuss ideas and unparalleled mentorship—a rarity in academe. As one of Tris' current students, I have deeply benefited from this community and consequently owe a great debt to many of the contributors to this volume. I have received guidance from these exceptional scholars on everything from the topic of my dissertation to how to talk to a dean during a campus interview. Tris has not only assisted us all in becoming successful scholars, he has provided us with an exemplar of what good mentorship should look like. Consequently, Tris' legacy is evident not only in the academic scholarship he and his students produce, but perhaps more importantly through a distinct teaching and mentorship ethos, an ethos that deeply benefits young philosophers at a time when mentorship and teaching are not the foremost valued activities in higher education.

Tris has given many gifts to academic philosophy, but I believe the most exceptional has been the kindness and loyalty he has shown his students and the great successes he has helped them achieve both as individuals and professors.

Tris Engelhardt is without a doubt the kindest, most generous, most loyal, and most supportive person I have ever had the honor to know and work with. The great personal and professional debts I owe him I will most likely never be able to to repay. I am indebted to Tris for his unwavering confidence in me, for the hundreds of hours he has spent reading drafts of my work, for forcing me to undertake opportunities I didn't believe I was ready for, for many hours of good conversation, for many glasses of good scotch, for providing me with an education well beyond the philosophy contained in books, for creating a space in which scholars support and care for each other rather than competing, for providing an unsurpassed education in all things Texan, for respecting my ideas when they were diametrically opposed to his own, for providing me with colorful stories to tell of my graduate school experience, for his wisdom and understanding during difficult times, and for being an advisor I could respect, honor, and learn from.

Thank you, Tris.

Index

L.M. Rasmussen et al. (eds.), *At the Foundations of Bioethics and Biopolitics: Critical Essays on the Thought of H. Tristram Engelhardt, Jr.*, Philosophy and Medicine 125, DOI 10.1007/978-3-319-18965-9

MIX
Papier aus verantwortungsvollen Quellen
Paper from responsible sources
FSC® C105338

FSC
www.fsc.org

If you have any concerns about our products,
you can contact us on
ProductSafety@springernature.com

In case Publisher is established outside the EU,
the EU authorized representative is:
Springer Nature Customer Service Center GmbH
Europaplatz 3, 69115 Heidelberg, Germany

Printed by Libri Plureos GmbH
in Hamburg, Germany